Introduction to Organic Chemistry

Introduction to Organic Chemistry

SECOND EDITION

Charles H. DePuy
University of Colorado

Kenneth L. Rinehart, Jr.
University of Illinois

John Wiley & Sons, Inc.
New York London Sydney Toronto

Library of Congress Cataloging in Publication Data:

DePuy, Charles H
Introduction to organic chemistry.

Includes index.
1. Chemistry, Organic. I. Rinehart, Kenneth L., joint author. II. Title.
QD253.D36 1975 547 74-19442
ISBN 0-471-20350-5

Printed in the United States of America

10-9 8 7 6 5 4 3 2 1

preface

This new edition is a brief introduction to organic chemistry. It is primarily for students who need, or would like to learn, the fundamentals of the subject but who are mainly interested in its applications to other areas of science or to the world at large. The present edition resembles the first one in many ways, and yet it is an extensive revision. These revisions, as much as possible, consist of rewriting to improve clarity and to update the sections that contain new information. Also changes in organization now allow a quicker entry into what is, for nonmajors at least, the most interesting and useful part of the subject: its applications.

We use the basic concepts of reaction mechanisms and theoretical organic chemistry, particularly resonance theory, as a model to help the student understand relationships in reactivity among the various functional groups. But we were not satisfied with our own results in the classroom when we merely presented fundamental theories and then left it to the student to make the jump from theory to practical applications in biology or other areas. Thus, in this revision, we maintain our emphasis on introducing the basics succinctly and applying them to examples from petroleum chemistry, polymer chemistry, biochemistry, or other areas of general interest.

The number of problems at the end of each chapter is greatly increased and, in response to requests, we answer only a portion of them. We retain the programmed nomenclature problems but they are now included at the end of the chapters to which they refer. We also have retained the legends to the figures, which allow the student to review a chapter easily. Every essential concept is summarized concisely in these figure legends. Finally, we continue to use a bright red color to emphasize the groups undergoing change in a reaction.

As in the first edition, we emphasize the three-dimensional aspects of organic chemistry, introduce stereochemistry at an early stage, and continue to employ it throughout. We do not stress synthesis, but for students who wish to study this area in more detail, a number of synthetic problems are introduced. We discuss fully biochemical pathways (particularly protein biosynthesis and the metabolism of carbohydrates, lipids, and amino acids). Our goal is to emphasize how organic reactions occurring in biological systems can be recognized as the same reactions that occur in the laboratory.

Most of the material can be covered in one semester of three one-hour lectures a week. Even in a shorter course, the student will have been exposed to a great many industrial and biological applications of organic chemistry. Many of the problems of modern civilization and their solutions involve organic chemistry. We hope that this text will help students to recognize both the problems and their solutions.

Charles H. DePuy
Kenneth L. Rinehart

contents

Introduction to Organic Chemistry

1
stereochemistry, structure, and saturated hydrocarbons

In an age when atomic physics and interplanetary rocketry are preeminent, many persons contend that mathematics is the universal language of science. This is not entirely true. A vast area of science, containing the giant petroleum and plastics industries and pharmaceutical research, and stretching to the border of life itself, requires a second language: organic chemistry.

Organic chemistry is a complex science. In the hundred-year period since it became recognized as a distinct branch of chemistry, more than one million separate organic compounds have been prepared, and these are only a small portion of the possibilities. Fortunately, in spite of its complexity, organic chemistry is highly systematic and may be discussed in terms of classes of molecules and a few types of reactions. The existence of an infinity of numbers does not make arithmetic impossible, since the operations of addition, subtraction, multiplication, and division learned with small numbers may be applied without change to large ones; similarly, in organic chemistry an understanding of the reactions of simple molecules may be applied substantially without change to large ones. We shall therefore follow a pattern: We shall introduce organic chemistry by discussing simple molecules; we shall then often apply the principles learned to a discussion of complex molecules, in order to demonstrate the generality of the subject.

The application of the adjective "organic" to the chemistry of carbon compounds is in some ways misleading, since it implies a direct association between organic chemistry and living organisms. Indeed, it was first thought that all organic compounds had to originate in living matter. Organic chemistry as a separate science is often said to date from 1828, when the German chemist Friedrich Wöhler synthesized an organic molecule (urea) from an inorganic chemical (ammonium isocyanate). In the century following Wöhler's discovery, the complete separation of organic chemistry from biological systems became popular among many organic chemists, and the synthesis of complex molecules from "coal, air, and water" was often a goal in itself. More recently, enormous advances in our understanding of the chemistry of biological systems have completed the circle; today more and more time and effort in organic chemistry are being devoted to reactions and processes that bear directly upon organic systems arising from living matter. Consequently, at every opportunity we shall attempt to illustrate organic reactions and structure among naturally occurring molecules, and shall discuss the organic chemistry of biological systems. For our purposes, organic chemistry will be defined as the chemistry of carbon compounds wherever they occur, and an organic chemist as one who studies the structure and reactions of these carbon compounds.

Because they are the source of so much of our energy—natural gas for heating, bottled gas for stoves, gasoline for automobiles—hydrocarbons play an extraordinary role in all human affairs. We have never been more aware of our dependence upon these relatively simple organic molecules to run the machinery of our civilization, nor more aware of the limits of their supply. Thus, from the standpoint of practical importance as well as structural simplicity, hydrocarbons make the logical starting point for a study of organic chemistry.

Methane and Its Three-Dimensional Structure

Methane, the simplest hydrocarbon, is a gas of molecular formula CH_4 which occurs naturally in underground pockets in the petroleum-producing areas of the world. Colorless and odorless (the odor associated with natural gas in our homes is caused by the deliberate addition of traces of skunk oil), methane is an ultimate product of the decay of organic matter in the absence of air; it rises in bubbles from beneath the surface of swamps and marshes, hence its name "marsh gas." It burns cleanly and with the release of a great deal of heat (Fig. 1-1), and is the least polluting of all our common sources of energy. Despite the ease with which it burns, methane is, by the standards of organic chemistry, *inert*. By this we mean that at ordinary temperatures methane reacts only slowly or not at all with most chemical reagents. For example, it can be bubbled unchanged through concentrated acid or alkali

$$CH_4 + 2O_2 \longrightarrow CO_2 + 2H_2O + \text{Energy}$$

Figure 1-1. *The combustion of methane is one of the principal sources of energy in modern civilization. The combustion of 16 g of methane (1 mole) releases 213,000 cal of heat, enough to raise the temperature of 2.13 liters of water from 0 to 100°C.*

and heated without effect with most oxidizing and reducing agents. Because the carbon in methane is combined with the maximum number of hydrogens possible, four, we say that it is a *saturated hydrocarbon.*

Under sufficiently vigorous conditions, for instance, in the heat of a match, saturated hydrocarbons will react. One of their most useful reactions is with the halogens, especially chlorine or bromine. If a sample of methane is mixed with chlorine gas, no reaction occurs, but in the presence of sunlight a reaction does take place and a mixture of organic compounds results (Fig. 1-2). Gaseous hydrogen chloride is also formed. The various products of the

$$CH_4 \xrightarrow[\text{light}]{Cl_2} \underset{\text{methyl chloride}}{CH_3Cl} + \underset{\text{methylene chloride}}{CH_2Cl_2} + \underset{\text{chloroform}}{CHCl_3} + \underset{\text{carbon tetrachloride}}{CCl_4} + HCl$$

Figure 1-2. *Methane reacts with chlorine in sunlight to form a mixture of chloro-, dichloro-, trichloro-, and tetrachloromethanes. Only one compound of each formula is known.*

reaction all have different boiling points and may be readily separated from one another and identified. As their molecular formulas indicate, the products obviously correspond to compounds in which successively one, two, three, and four hydrogen atoms of methane are replaced by chlorine atoms. These compounds may be named *chloromethane* (or, alternatively, methyl chloride), *dichloromethane* (or methylene chloride), *trichloromethane* (chloroform), and *tetrachloromethane* (or carbon tetrachloride).

Next, consider the implications for the structure of methane that only *one compound of each structure exists,* i.e., one methyl chloride, one methylene chloride, etc. For example, the simplest explanation for the fact that only one

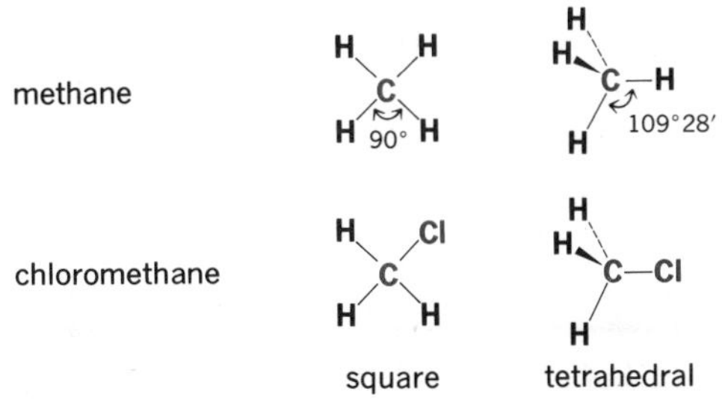

methane
90°
109°28′
chloromethane
square
tetrahedral

Figure 1-3. *Two possible structures for methane and chloromethane which correctly predict that only one compound of the formula CH_3Cl should exist. In the square model all four bonds from carbon are in the same plane (that of the page). In the tetrahedral model, the solid lines indicate bonds in the plane of the page, the dotted line, behind the page, and the heavy line, in front.*

molecule CH_3Cl exists is that the hydrogens in CH_4 are equivalent, so that no matter which is replaced by chlorine, the same molecule results. Such would be the case, for instance, if methane were square (Fig. 1-3); but other possibilities exist, such as the tetrahedral model also shown in Fig. 1-3.

Although both square and tetrahedral spatial formulas predict only one chloromethane, the square arrangement predicts that two dichloromethanes, CH_2Cl_2, should exist (Fig. 1-4). In one of these molecules the two chlorine

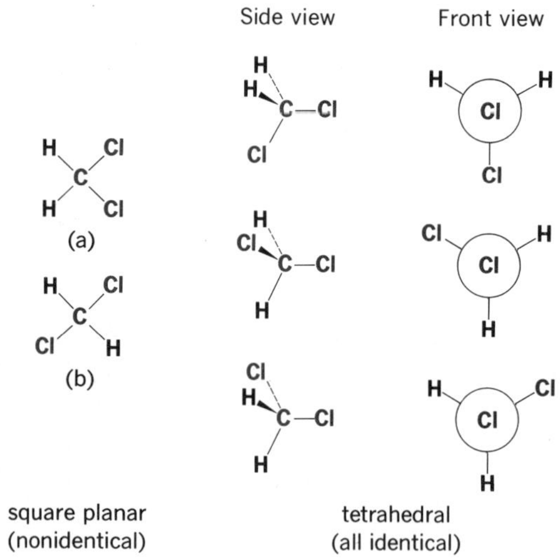

Figure 1-4. *If methane were square planar, two dichloromethanes, (a) and (b), should exist. The tetrahedral structure correctly predicts the existence of only one molecule with the formula CH_2Cl_2.*

atoms would be adjacent to each other, and in the other they would be on opposite sides of the carbon atom. Note that the distance between the two chlorine atoms would differ for these two compounds. The tetrahedral model, on the other hand, correctly predicts the existence of only a single dichloromethane. Some practice may be required until this can be clearly recognized. *All aspects of the stereochemistry of organic molecules may be explained only if it is assumed that whenever a carbon atom is surrounded by four atoms, these atoms are tetrahedrally arranged.* This conclusion, first reached on strictly organic chemical grounds in 1874, has been fully confirmed by more recent physical methods.

Molecular Models

The term tetrahedral is used to describe the geometry of methane because a regular tetrahedron is formed when the centers of the hydrogen atoms of the molecule are connected. If the carbon atom is regarded as being at the center of the tetrahedron, the lines linking it to the hydrogen atoms point to the corners of the tetrahedron. Organic chemists frequently rely on molecular models so that they can more easily visualize the stereochemistry of molecules (Fig. 1-5). For illustrations in this text we shall occasionally use

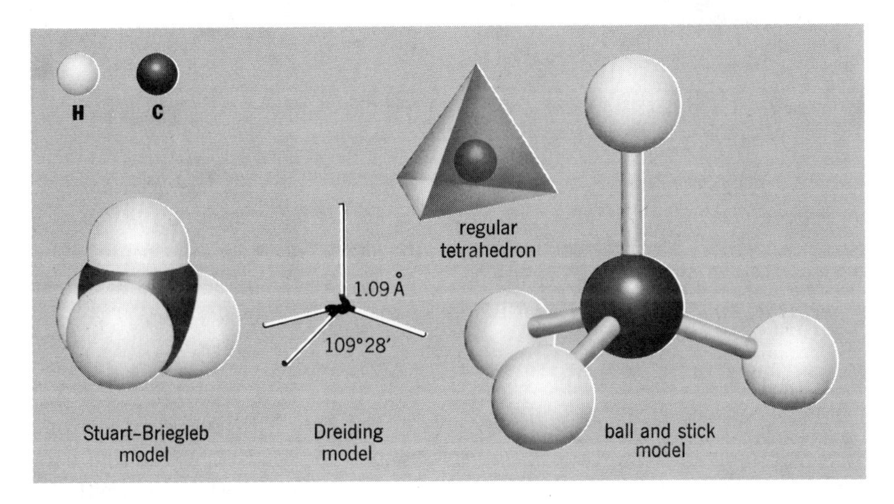

Figure 1-5. *Stuart-Briegleb models show the relative sizes of the atoms, while Dreiding models make bond lengths and angles more obvious; both types are accurately constructed to scale. The cheaper ball and stick models are helpful in understanding stereochemistry.*

two types of model: the type designed by Stuart and Briegleb, and that designed by Dreiding. The first type is the "space-filling" model. It represents to scale the space occupied by the electrons of the atoms involved. The other type, the Dreiding model, accurately represents only the positions of the atomic nuclei, yet clearly shows bond lengths and bond angles. For methane the C—H bond length is 1.09 Å (1 Å or Ångström unit = 10^{-8} cm) and the bond angle is 109°28′. Dreiding models provide a more open representation of the molecule and therefore permit a clearer view of the relationships among atoms. The Stuart–Briegleb models allow one to visualize the actual volume occupied by atoms.

Ethane

Organic chemistry exists as a separate science because of its complexity. The main reason for this complexity is the ability of carbon atoms to form strong bonds with other carbon atoms so that very large molecules containing many carbon atoms exist and are stable.

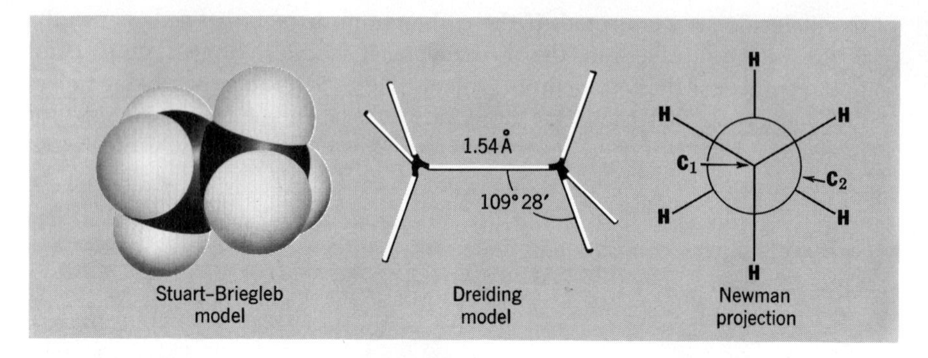

Figure 1-6. *Two tetrahedral carbon atoms are joined with six hydrogen atoms in the ethane molecule. This figure gives three representations. Note that in the front view, which looks along the carbon–carbon axis and is called a* Newman projection, *the rear carbon atom is represented by a circle. The C—C bond length is 1.54 Å. As in methane, the C—H bond length is 1.09 Å, and the bond angles are 109°28′.*

Two tetrahedral carbon atoms may join together with six hydrogen atoms to form ethane, C_2H_6, whose structure is shown in Fig. 1-6. Notice that each carbon is attached tetrahedrally to four other atoms, just as in methane, although now one of the four is another carbon atom. The bond angles, 109°28′, are also the same as those in methane; this is the so-called *tetrahedral angle,* a constant for a regular tetrahedron.

Because carbon–carbon bonds are strong and difficult to break, methane and ethane are much alike in chemical properties. Ethane is *saturated,* i.e., it does not combine with more hydrogen nor is it easily attacked chemically. It burns readily to carbon dioxide and water (Prob. 1-1), and, like methane, it reacts with chlorine in the presence of sunlight. In this way from one to six of the hydrogens of ethane may be replaced by chlorine. Let us see what our three-dimensional tetrahedral structure for the carbons in ethane predicts for these compounds.

Prob. 1-1. Write a balanced equation for the combustion of ethane.

No matter how it is formed, there is only *one* monochloroethane, C_2H_5Cl (Fig. 1-7). This is certainly consistent with the structure given for ethane because all six hydrogens in this molecule are equivalent. It is found, however, that there are *two* dichloroethanes, $C_2H_4Cl_2$, each having chemical and

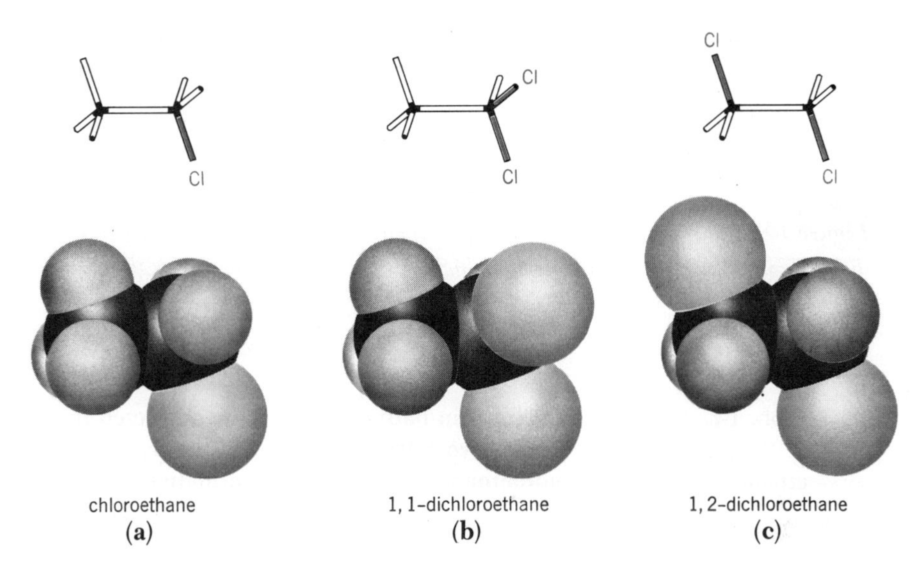

Figure 1-7. *Models of chloroethane and of the two structural* isomers *of dichloroethane. The latter two compounds have the same molecular formula, $C_2H_4Cl_2$, but different structural formulas.*

physical properties slightly different from the other. This is the first time we have encountered what will turn out to be a common situation: two molecules having the same molecular formula but different structures. Such molecules are called structural *isomers* of each other. The existence of two isomers having the formula $C_2H_4Cl_2$ is readily explained. In one, the two chlorine atoms are attached to the same carbon atom; in the other, a chlorine is attached to each carbon. To give each of these compounds a distinctive name, we number the carbon atoms and speak of the first as 1,1-dichloroethane and the second as 1,2-dichloroethane.

Although organic chemistry is a three-dimensional science, convenience and the limitations of paper and blackboard require two-dimensional representations. Up to this point, we have used projection formulas to represent the third dimension. When formulas are condensed, the third-dimensional representation is lost, but for our discussions of many reactions this loss is not important, and what is lost in precision is gained in convenience. Compare, for example, the two representations of the chlorination of ethane. In the upper line of Fig. 1-8 we see *extended structural formulas,* in which each bond is represented by a straight line. The lower line contains *contracted structural formulas,* in which all atoms attached to each carbon atom are grouped together on a single line, usually following the carbon atom.

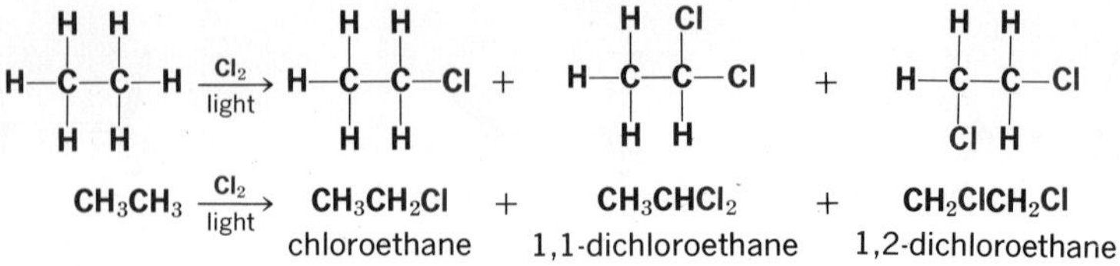

$$CH_3CH_3 \xrightarrow[\text{light}]{Cl_2} CH_3CH_2Cl + CH_3CHCl_2 + CH_2ClCH_2Cl$$

chloroethane, 1,1-dichloroethane, 1,2-dichloroethane

Figure 1-8. *Chlorination of ethane leads to a mixture of one monochloroethane, two dichloroethanes, and a number of more highly chlorinated ethanes. Here the reaction is written with both extended and contracted structural formulas.*

Prob. 1-2. Draw the structures in both extended and contracted forms of the two trichloroethanes, two tetrachloroethanes, one pentachloroethane, and one hexachloroethane, formed in the chlorination of ethane.

The existence of only a single 1,2-dichloroethane is not predicted by the formula for ethane as we have drawn it in Fig. 1-6; this can be seen most clearly in the Newman projection formula, where the two chlorines might replace hydrogens that are close to one another or on opposite sides (Fig. 1-9). In fact, such isomers do exist, but at room temperature there is extremely rapid rotation about the single bond between the two carbon atoms so that the various isomeric forms are in rapid equilibrium with one another and cannot be separated. If 1,2-dichloroethane were cooled down nearly to absolute zero ($-273°C$), it should be possible to separate these isomers, and this has been done for somewhat related compounds. Isomers like these, which are rapidly interconverted by rotation, are called *conformers,* or *conformational isomers.* This freedom of rotation simplifies many problems of isomers; for all practical purposes, only one isomer of 1,2-dichloroethane needs to be considered. In

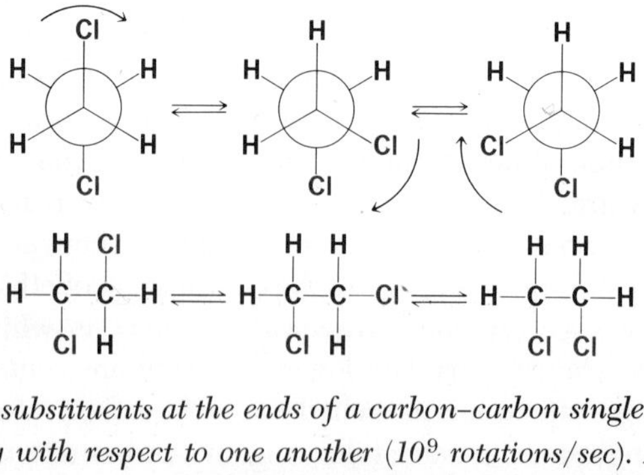

Figure 1-9. *The substituents at the ends of a carbon–carbon single bond are rotating extremely rapidly with respect to one another (10^9 rotations/sec). As a consequence of this rotation, conformational isomers cannot be separated at room temperatures.*

writing structural formulas for molecules that are rapidly rotating, there is a possibility of confusion for the unwary. For example, the extended structural formula for 1,2-dichloroethane can be written in several different ways (Fig. 1-9), yet all represent the same molecule.

Propane

When three carbon atoms are joined with eight hydrogen atoms, the resulting product is propane, C_3H_8 (Fig. 1-10). Notice that in the series methane, CH_4, ethane, C_2H_6, and propane, C_3H_8, each molecule differs from the previous one by an additional unit of CH_2. The series can be extended indefinitely in steps of CH_2 so that each member fits the formula C_nH_{2n+2}. A group of molecules of this type, each differing from the next by the structural unit CH_2, is known as a *homologous series*. Methane, ethane, and propane constitute the first three members of the homologous series known as the *continu-*

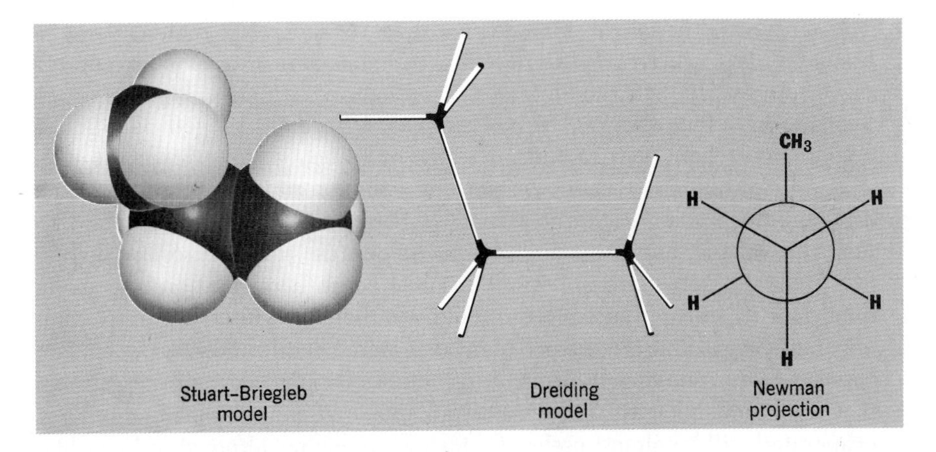

Figure 1-10. *The structure of propane, C_3H_8, in two side views and in a Newman projection looking down the C-1, C-2 bond. Because there is free rotation about this and the other C—C bond in the molecule, several other conformations could be given.*

ous-chain saturated hydrocarbons, or *normal alkanes.* Since members of a homologous series have quite similar chemical and physical properties, one can predict with confidence that propane, for example, will react with chlorine in sunlight in a manner similar to the reaction between either methane or ethane and chlorine.

One physical property of propane of great practical importance is the *critical temperature.* This is the temperature above which the gas will not liquefy no matter what the external pressure. For methane the critical temperature is $-82°$, and liquid methane cannot exist at room temperature. For propane this temperature is $97°$; the application of a modest amount of pressure causes propane to liquefy, despite the fact that its boiling point at atmospheric pressure is $-42°$. Since a liter of liquid contains a great deal more of a substance than a liter of gas, and since propane vaporizes when exposed to atmospheric pressure, it is the fuel of choice in areas that are not served by a methane pipeline because it is easily transported and stored in liquid form.

Prob. 1-3. There is free rotation throughout all the bonds of the propane molecule. The chlorination of propane gives two different monochloropropanes and four different dichloropropanes. Draw the structures of these six compounds in both extended and contracted forms.

Butane

The next member of the homologous series of saturated hydrocarbons is butane, C_4H_{10}. There are two different compounds of this formula, and we meet here for the first time hydrocarbons that are structural isomers of one another. In butane the four carbon atoms may be joined together either in a continuous chain or with one carbon atom branched from the others (Fig. 1-11). To distinguish the two isomers by common names, we call the first *n*-butane (for *normal*), and the second, isobutane. The two butanes are, like methane, ethane, and propane, good fuels. Because they are so low boiling (see Table 1-1), they are important gasoline constituents that are especially important in starting an engine in cold weather. In winter the proportion of butanes in gasoline is increased to make it more volatile and so make starting easier, while in summer the amount of butanes is reduced. A can of gasoline that has stood open to the air until all the low-boiling constituents have evaporated will be almost useless in starting an engine, although it looks like perfectly good gasoline.

We shall use *n*-butane to examine the ideas of *conformational isomers* in somewhat greater detail, for despite the fact that these isomers cannot ordinarily be separated, they can have profound effects on the physical and chemical properties of molecules. If we draw a Newman projection down the central C—C bond of *n*-butane, we have a situation exactly like that in 1,2-dichloroethane, namely, two tetrahedral carbons each substituted with one group other than hydrogen. Let us start out with the two methyl groups next to one another and see how the energy of the molecule changes as rotation about the C—C bond takes place.

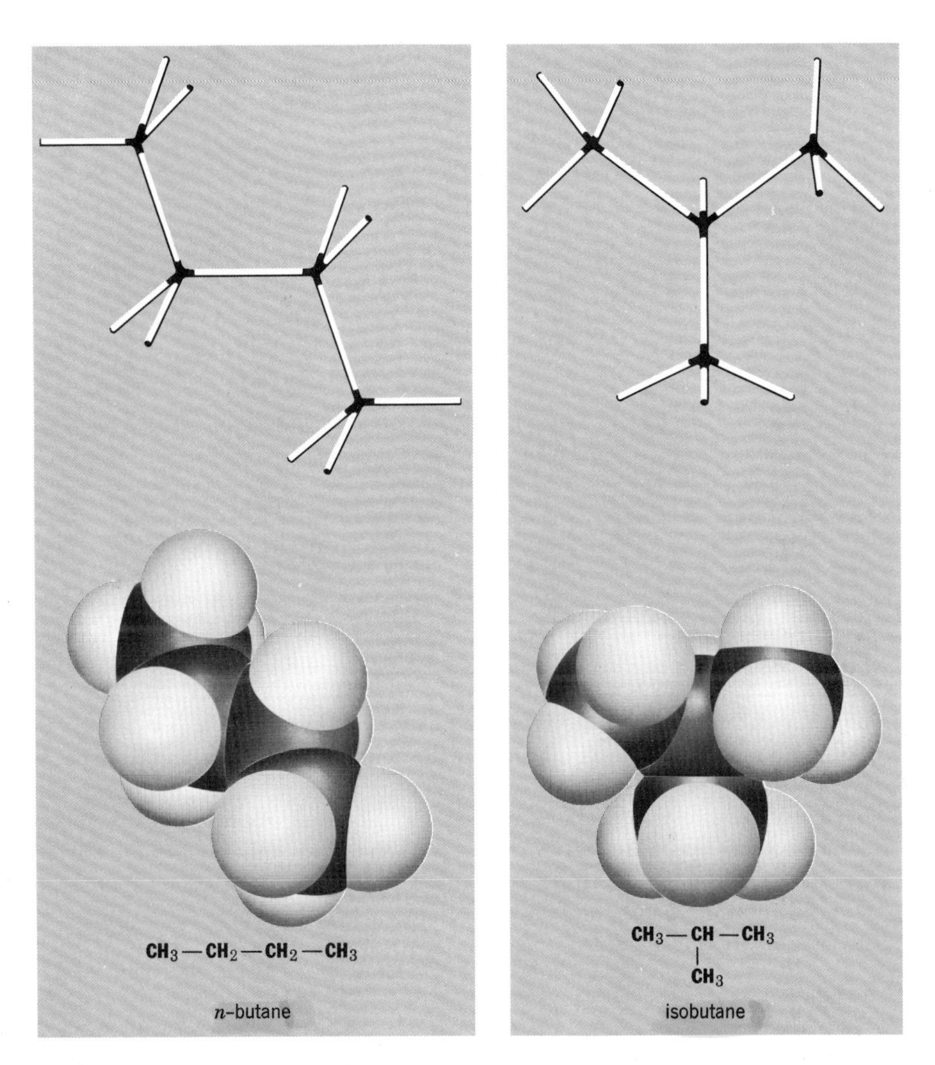

Figure 1-11. *Two isomeric butanes (C_4H_{10}) exist;* n-*butane has a continuous chain of carbon atoms, while isobutane has a branched chain. These isomers have somewhat different physical and chemical properties,* n-*butane boiling at* $-0.5°$ *and isobutane at* $-12°$.

Conformer A (Fig. 1-12), with the two methyl groups adjacent, has the advantage that each substituent on the front carbon (C-2 of butane) falls between two substituents on the back carbon; we say that the substituents are *staggered*, which is a relatively favorable arrangement. However, conformer A is not the lowest energy conformer of butane, for the two methyl

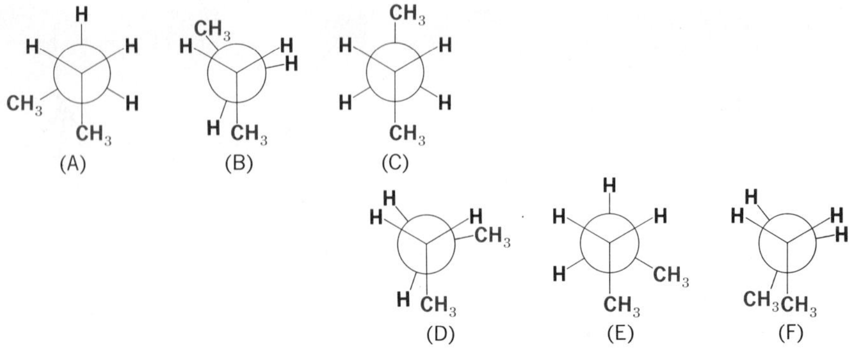

Figure 1-12. *As rotation about the C-2, C-3 bond of butane occurs, the molecule passes from one* conformation *to another. The most stable form (C) has the two methyl substituents* anti *to one another, the least stable form (F) has them* eclipsed. *The* staggered *forms, A, C, and E, are of lower energy than the* eclipsed *forms, B, D, F.*

groups are rather large and bump into and repel one another. We call conformer A a *gauche* form, from the English word meaning awkward. As rotation of the back carbon (C-3) occurs, the methyl groups get farther apart, but now the substituents on the two carbons *eclipse* one another (conformer B) and the energy rises. In conformer C the methyl groups are as far apart as it is possible for them to be, and all groups are staggered. This, the *anti* form, is clearly the most favorable structure for *n*-butane. As rotation continues, conformers D and E repeat A and B, and the least stable conformer of all is F in which the two methyl groups are eclipsed. Despite the fact that there is rapid rotation about the bonds in *n*-butane, the molecule spends most of its time in the most stable conformation, conformer C.

Hydrocarbon Polymers: Crystalline Compounds Versus Rubbers

As a practical application of the ideas of conformational isomers, and as a further introduction to hydrocarbons, let us consider two polymers, polyethylene and butyl rubber, which will be familiar to everyone, the former in plastic squeeze bottles, the latter in bicycle and automobile inner tubes. Both polymers, whose chemistry of formation will be extensively discussed in Chapter 4, are long-chain saturated hydrocarbons, containing as many as 1000 carbon atoms. Their structural formulas, shown in Fig. 1-13, indicate that polyethylene is related in structure to *n*-butane, and butyl rubber to isobutane.

We now consider a Newman projection of polyethylene. Like *n*-butane,

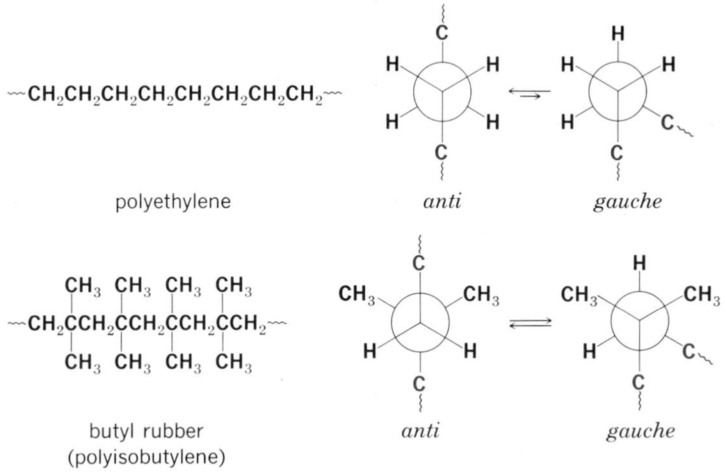

$\sim CH_2CH_2CH_2CH_2CH_2CH_2CH_2CH_2 \sim$

polyethylene

anti ⇌ *gauche*

$$\sim CH_2C(CH_3)_2CH_2C(CH_3)_2CH_2C(CH_3)_2CH_2C(CH_3)_2 \sim$$

butyl rubber
(polyisobutylene)

anti ⇌ *gauche*

Figure 1-13. *The structural formula and Newman projection along the backbone of polyethylene and polyisobutylene. A typical molecule contains 1000 or more carbon atoms.*

the anti form is the most stable; if each bond in the molecule assumes this form, a long, rodlike molecule will result, which can pack together into a high-melting solid form with a great deal of crystallinity. In butyl rubber (polyisobutylene), on the other hand, the anti and gauche forms are of nearly identical energy because of the presence of the methyl substituents. Simply on the basis of chance, therefore, $\frac{2}{3}$ of the C—C backbone bonds are in the gauche conformation, which leads to coiling of the molecule. When stretched, the molecules straighten out into the anti form, but when released they again coil up. This gives polyisobutylene rubberlike properties. Here, then, we have an example of how the population of conformations can influence the physical properties of important organic substances.

Pentanes

As the number of carbon atoms increases, so does the number of possible isomers. The next member of the alkane series, pentane, C_5H_{12}, has three isomers. The five carbons may be arranged in a continuous chain to give *n*-pentane, with one branch to give isopentane, or with two branches to give neopentane (Fig. 1-14). Each of these pentanes is a well-known organic species, differing slightly in physical and chemical properties from the other two.

These three isomers furnish another example of the effect of the structure

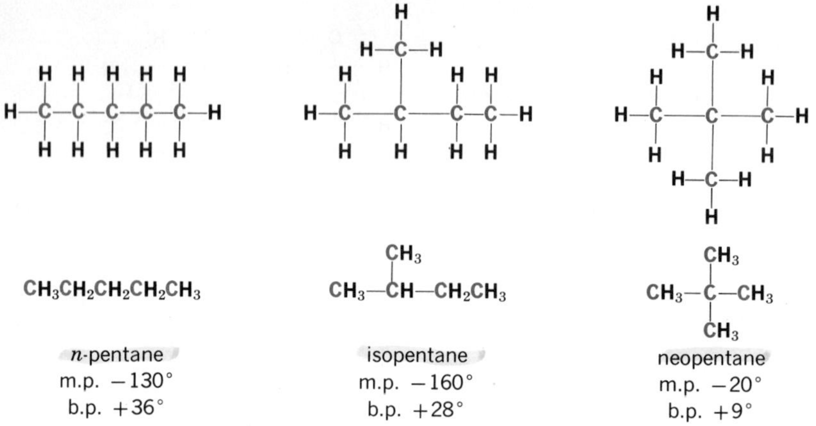

$CH_3CH_2CH_2CH_2CH_3$	$CH_3—CH(CH_3)—CH_2CH_3$	$CH_3—C(CH_3)_2—CH_3$
n-pentane	isopentane	neopentane
m.p. −130°	m.p. −160°	m.p. −20°
b.p. +36°	b.p. +28°	b.p. +9°

Figure 1-14. *The structures of the three isomeric pentanes,* C_5H_{12}.

of a molecule on its physical properties. We look first at the melting points. The *n*-pentane melts higher than isopentane because it can assume a rodlike shape in which adjacent molecules can pack closely together. But neopentane melts more than 100° higher than either of its isomers. A Stuart–Briegleb space filling model of neopentane will show it to be an almost spherical molecule, which can easily pack into an almost perfect crystal. The boiling points of the three isomers are nearly in the opposite order. The long *n*-pentane molecules fit rather neatly together; they present the greatest opportunity for intermolecular attraction. Consequently, evaporation occurs more slowly than from its branched-chain isomers. The spherical neopentane molecules exert relatively little attraction on each other and so have a low boiling point. As a general rule, the more symmetrical the molecule, the higher its melting point and the lower its boiling point relative to its isomers.

Higher Homologs

Table 1-1 contains the names and structures of the continuous-chain isomers of the first ten saturated hydrocarbons and the number of possible isomers for each molecular formula. The names in the table should be memorized, because they will be used frequently. Notice how quickly the number of isomers mounts and how impossible it would be to memorize a different name for each. Obviously, a more systematic nomenclature is necessary; the standard one will be described in the next section.

Table 1-1
Continuous-Chain Saturated Hydrocarbons and the Number of Their Possible Isomers

Compound	Structural Formula	Boiling Point, °C	Melting Point, °C	Total Number of Hydrocarbons Having the Same Molecular Formula
methane	CH_4	−162	−182	1
ethane	CH_3CH_3	−89	−183	1
propane	$CH_3CH_2CH_3$	−45	−90	1
n-butane	$CH_3(CH_2)_2CH_3$	−0.5	−138	2
n-pentane	$CH_3(CH_2)_3CH_3$	36	−130	3
n-hexane	$CH_3(CH_2)_4CH_3$	68	−95	5
n-heptane	$CH_3(CH_2)_5CH_3$	98	−91	9
n-octane	$CH_3(CH_2)_6CH_3$	125	−57	18
n-nonane	$CH_3(CH_2)_7CH_3$	151	−51	35
n-decane	$CH_3(CH_2)_8CH_3$	174	−30	75

Prob. 1-4. Draw condensed structural formulas for all five isomeric hexanes.

Systematic Nomenclature

If there can be an infinite number of saturated hydrocarbons, there must be an infinite number of names to identify them. A simple, unequivocal system of nomenclature is therefore a necessity. Such a system was devised by the International Union of Chemistry at Geneva in 1892, and the task of keeping it up to date has fallen to its successor, the International Union of Pure and Applied Chemistry (IUPAC). The IUPAC system will be applied extensively throughout this book. Its basic rules can be easily illustrated for saturated hydrocarbons. Briefly, the longest *continuous chain* of carbon atoms in the molecule is considered to be the parent chain from which the molecule is derived through the replacement of hydrogen atoms. If we take isobutane as an example, its longest continuous chain of carbon atoms is three. In the IUPAC system, this compound would be considered to be a derivative of propane, in which one hydrogen atom has been replaced by a CH_3 group. (See Fig. 1-15.)

If we are to name branched-chain compounds as derivatives of continuous-chain hydrocarbons, we shall need names for fragments like CH_3—, which we have called *methyl* in Fig. 1-15. These fragments are not molecules but combining groups, which may be considered as being derived from the

Figure 1-15. *In the IUPAC nomenclature system the hydrocarbon whose common name is isobutane is named as a derivative of propane, since three carbons is the longest continuous chain of carbons in isobutane.*

corresponding hydrocarbon by loss of a hydrogen atom. The group CH_3CH_2— is called *ethyl,* since it is derived from ethane. Two different combining groups may be formed from propane by the loss of hydrogen from either the 1- or 2-carbon; four butyl groups are possible, two from *n*-butane and two from isobutane. As a class these combining groups, which may be thought to arise from the loss of a hydrogen atom from an alkane, are known as *alkyl groups.* In this book we shall make extensive use in nomenclature of six of these alkyl groups; their names and structures, as given in Table 1-2, should be committed to memory.

Prob. 1-5. Draw structures for the other two butyl (C_4H_9—) combining groups which are not given in Table 1-2.

It is now possible to state systematically the rules for IUPAC nomenclature for saturated hydrocarbons and to illustrate them with a more complicated molecule. We shall do this for the molecule in Fig. 1-16. On the basis of its molecular formula, C_7H_{16}, the molecule belongs to the family of isomeric

Table 1-2
Names and Structures of Six Common Alkyl Groups

Group	Name	Group	Name
CH_3—	methyl	$CH_3CH_2CH_2CH_2$—	*n*-butyl
CH_3CH_2—	ethyl		
$CH_3CH_2CH_2$—	*n*-propyl	CH_3—C(CH_3)(CH_3)—	*t*-butyl
CH_3CH(CH_3)—	isopropyl		

```
CH3—CH—(CH3)
       |
(CH3)—CH—CH2
          |
          CH3
```

2,3-dimethylpentane

Figure 1-16. *In the IUPAC nomenclature of an alkane, the longest continuous chain of carbon atoms in the molecule is located (even though it may not be written in a straight line), and named as the parent. The substituent groups (circled) are named, and the chain is numbered from the end that gives these groups the smallest numbers. The name is punctuated as shown.*

heptanes, since it has seven carbon atoms. However, the IUPAC name is derived in a different manner. First, the longest continuous chain of carbon atoms is located and named. In this illustration the longest chain (shown in red) is five. Hence the parent from which the name is to be derived is *pentane,* taken from the list of Table 1-1. Second, the substituent groups (circled) are named. In this case, two *methyl* groups are attached to the longest chain. Third, the parent chain is numbered from the end that gives the substituents their smallest numbers. Here, numbering from the left gives the full name *2,3-dimethylpentane,* while numbering from the right would give the name *3,4-dimethylpentane.* Since the former has the smaller numbers, it is the correct name.

When the IUPAC system is to be used in naming saturated hydrocarbons, these common errors should especially be watched for:

1. Failing to locate the longest continuous chain of carbon atoms, which may not be written in a straight line.
2. Numbering from the wrong end.
3. Failing to write the same number twice, when two identical substituents are on the same carbon atom (e.g. 2,2-dimethylpropane).
4. Confusing the *n*-propyl and isopropyl groups.

In Fig. 1-17 two complicated hydrocarbons are named by the IUPAC

```
                  CH3
                  |
                  CH—CH3
                  |
CH3CH2CH2—C—CH3
                  |
                  CH—CH2—CH2—CH2
                  |                  |
          CH3—CH2                   CH3
```

4-methyl-5-ethyl-4-isopropylnonane

```
                                CH2—CH3
                                |
CH3CH2CH2CH—CH2—CH—CH3
              |
              CH—C(CH3)3
              |
              CH2
              |
              CH2—CH2—CH3
```

3-methyl-5-*n*-propyl-6-*t*-butyldecane

Figure 1-17. *Two examples of the IUPAC nomenclature of hydrocarbons. Note especially the difference between the isopropyl and* n*-propyl groups.*

system, and in the problems at the end of the chapter there is a set of graded questions designed to emphasize the fundamentals of the system.

One further point of nomenclature needs to be introduced, not so much for the naming of the hydrocarbons themselves as for naming the derivatives to be considered in future chapters. A fairly complicated molecule will ordinarily contain various types of carbon–hydrogen groups. At the ends of the molecule are CH_3 groups; along the chain are CH_2 and CH groups, and there may even be an occasional carbon atom that carries no hydrogen atoms at all and is attached to four other carbon atoms. A carbon atom attached to only one other carbon atom, like those of the methyl groups, is called a *primary carbon*. The carbon atom that is attached to two other carbon atoms, like those of the CH_2 groups, is said to be a *secondary carbon*. Carbon atoms attached to three and four other carbon atoms are called *tertiary* and *quaternary*, respectively (see Fig. 1-18).

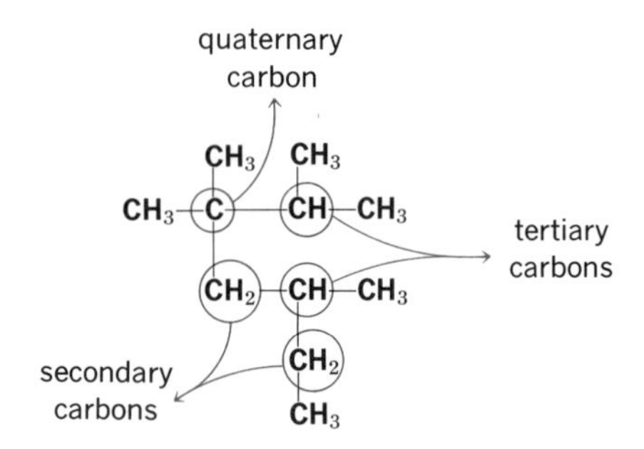

Figure 1-18. *It is convenient to designate carbon atoms as primary (1°), secondary (2°), tertiary (3°), or quaternary (4°), depending upon whether they are bonded to one, two, three, or four other carbons. All the methyl groups contain primary carbon atoms.*

Cyclic Saturated Hydrocarbons

The molecules mentioned thus far all have been composed of carbon-atom chains, often having one or more branches. The ends of these chains can be joined together to form a ring or a cycle of carbon atoms. When all the carbon atoms of the ring are saturated, the molecules are said to be *cycloalkanes*. In most respects cycloalkanes resemble their noncyclic counterparts, both chemically and physically. However, certain of their chemical features merit special attention.

Cyclopentane

In many ways the least complicated alicyclic hydrocarbon is cyclopentane. If the carbon atoms in this molecule lie in a plane, they form a nearly perfect pentagon (Fig. 1-19); in fact, the usual symbol for cyclopentane is a pentagon. The internal angles of a pentagon are 108°, almost the normal tetrahedral angle (109°28′).

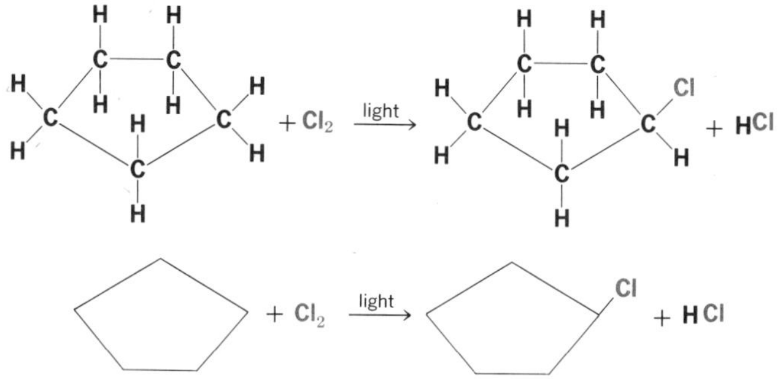

Figure 1-19. *Cyclopentane hardly differs chemically from* n-pentane, *as shown here by its reaction with chlorine and light. The symbol for a cyclic hydrocarbon is a regular polygon with as many corners as there are atoms in the ring. Unless otherwise specified, enough hydrogen atoms are understood to be attached to these carbon atoms so that each carbon has a valence of four.*

Cyclopentane and *n*-pentane are similar, except that the cyclic compound has two fewer hydrogen atoms. Although both are nearly inert chemically, they react with halogens in the presence of light through the substitution of one or more hydrogen atoms. Since the cyclic structure confers a high degree of symmetry on the molecule, only one monochlorocyclopentane is possible. Here cyclopentane differs from *n*-pentane, which has three monochloro derivatives.

Cyclic molecules give rise to a new type of isomerism. This is indicated by an examination of the possible isomers of dichlorocyclopentane. Both of the chlorine atoms may become attached to the same carbon atom to give 1,1-dichlorocyclopentane, or they may be on adjacent carbon atoms as in 1,2-dichlorocyclopentane, or they may be in the 1,3- position. *Two isomers of both 1,2- and 1,3-dichlorocyclopentanes exist.* In one of the two isomeric 1,2-dichloro compounds, both chlorine atoms are attached to the same side of the ring to give *cis*-1,2-dichlorocyclopentane; in the other isomer, the two chlorine atoms are attached to opposite sides of the ring to give *trans*-1,2-dichlorocyclopentane. The prefix *cis* signifies that two substituents are on the same side of a ring; *trans,* that they are on opposite sides. Similarly, there are *cis*- and *trans*-1,3-dichlorocyclopentanes (see Fig. 1-20).

Notice the difference between the *cis* and *trans* isomers: Both have chlorines bonded to the same carbon atoms, but their arrangements in space differ. These isomers, which differ in their three-dimensional structure, are

1,1-dichlorocyclopentane

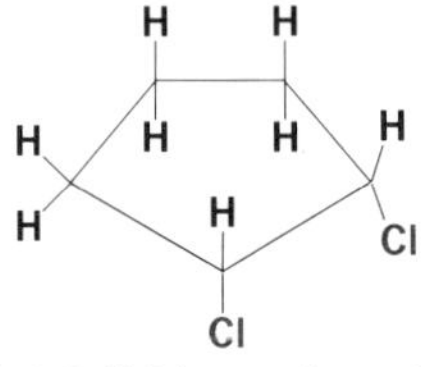

cis-1,2-dichlorocyclopentane

trans-1,2-dichlorocyclopentane

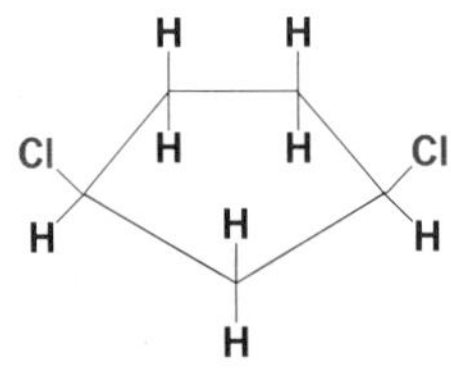

cis-1,3-dichlorocyclopentane

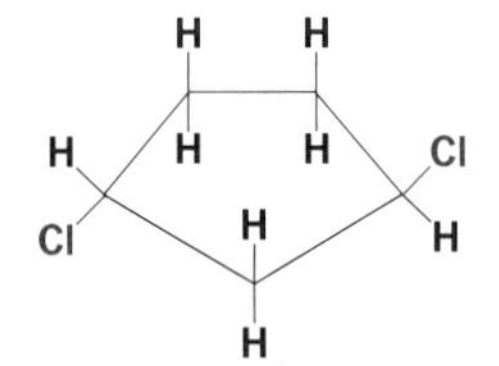

trans-1,3-dichlorocyclopentane

Figure 1-20. *As well as structural isomerism, the dichlorocyclopentanes exhibit* geometrical isomerism, *since both the 1,2- and 1,3-isomers can exist in a* cis *and a* trans *form.*

called *stereoisomers.* In the present example, in which the chlorine atoms are either on the same or on opposite sides of the ring, they are more precisely called *geometrical isomers.* This type of isomerism should be clearly distinguished from structural isomerism, in which the isomers differ from one another in the order in which the atoms are joined together. For example, 1,1-dichlorocyclopentane is a structural isomer of any of the other four dichlorocyclopentanes.

Figure 1-21 contains Newman projections of *cis-* and *trans-*1,2-dichloro-

cis

trans

Figure 1-21. *Newman projections along the 1,2- bond of* cis- *and* trans-*1,2-dichlorocyclopentanes. In order to rotate around the 1,2- bond the ring would have to be broken; since this could only happen at extremely high temperatures, these isomers are distinct species, which may be separated from one another.*

cyclopentanes. Compare these structures carefully with those in Fig. 1-12. Which conformer of *n*-butane corresponds most closely to the cyclopentane ring, as shown? Notice that if the carbon atoms in cyclopentane were completely in one plane there would be complete *eclipsing* of all the bonds in a form corresponding to structure F of Fig. 1-12. To relieve some of this eclipsing, the cyclopentane molecule is slightly twisted so that one of its carbon atoms is bent out of the plane of the other four.

Cyclohexane

Rings containing six carbons, the cyclohexanes, form readily and are widely distributed in nature. Because they play such an important role in all of organic chemistry, it is important that we examine their structure in some detail. Despite its seeming complexity, cyclohexane can be analyzed by the conformational model we have developed for *n*-butane in Fig. 1-12.

First, would we expect all six carbons of cyclohexane to lie in one plane? Not likely. In the first place the internal angles of a regular hexagon are 120°, rather far from the normal 109.5° tetrahedral angles of saturated carbons. So we would have to bend the bonds to make cyclohexane planar. Second, we have just seen with cyclopentane that a planar cyclic molecule has all of its C—C bonds eclipsed; in cyclopentane, with nearly unstrained 110° bond angles in a planar form, some of these angles are bent to escape eclipsing. So, on two accounts, bond angles and eclipsing, we predict that cyclohexane will not be planar.

If we make a molecular model of cyclohexane using only the ring carbon atoms and ignoring for the moment the hydrogen atoms, we find that three nearly strainless conformations exist, two identical *chair* forms and one *boat* form (Fig. 1-22).

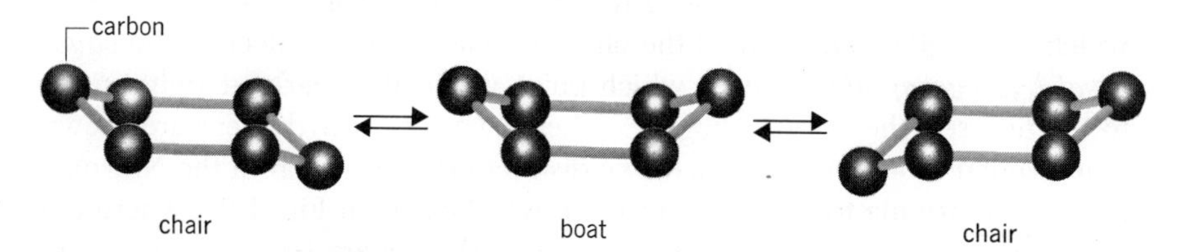

Figure 1-22. *The carbons of cyclohexane can adopt three distinct conformations, two low-energy chair forms and one higher-energy boat form. These conformations are in rapid equilibrium with one another.*

These conformations can easily be transformed into one another simply by rotation around carbon–carbon bonds; hence they are rapidly interconverting at room temperature, just like the various conformations of *n*-butane. For reasons discussed in the following paragraph, the two chair forms, which are identical to one another in an unsubstituted cyclohexane, are of lower energy than the boat form. Cyclohexane, therefore, spends most of its time in one or the other of the two chair forms, which interconvert by way of the boat conformation.

To see why the chair form of cyclohexane is more stable than the boat form, and why cyclohexanes in general are so common and so stable, let us examine a front view of cyclohexane in the chair and boat form (Fig. 1-23).

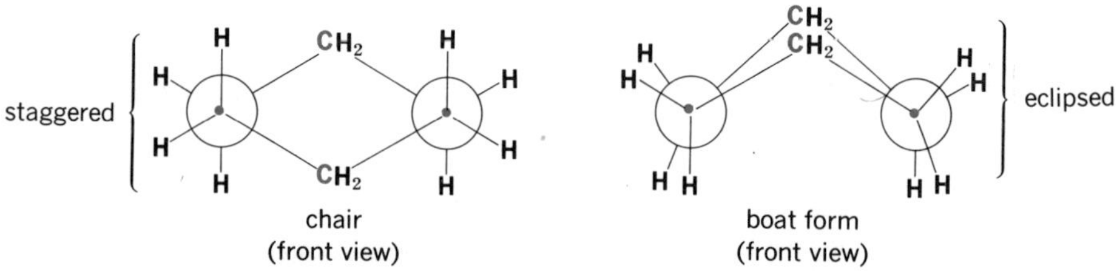

Figure 1-23. *In the chair form of cyclohexane all C—C bonds are in the energetically favorable staggered form; in the boat form there are eclipsed conformations that are energetically unfavorable.*

Notice that in the boat form the two CH_2 groups along each side, the "gunwales" of a boat, are in the energetically unfavorable *eclipsed* conformation, in which the hydrogen atoms are directly opposite one another. Two carbon atoms are also eclipsed. *In either chair form all carbons and hydrogens are staggered;* as a consequence these chair forms are of lower energy than the boat form.

Next we examine the hydrogen atoms carefully in one of the chair forms of cyclohexane; they will be seen to fall into two distinct groups, those six which run horizontally around the circumference of the molecule (the *equatorial hydrogens*) and those six which point vertically, nearly at right angles to the first set (the *axial hydrogens*). These two sets of hydrogens are shown in molecular models in Fig. 1-24. The distinction is also clear in the Newman projection formula for the chair form of cyclohexane in Fig. 1-23, where the axial hydrogens point straight up and down, the equatorial hydrogens more nearly right and left. *When one chair form of cyclohexane converts to the other, those six hydrogens which were axial in the first chair form become equatorial in the second,* and those six which were equatorial become axial. Since the two chair forms are of equal energy and in rapid equilibrium with one another, each hydrogen spends half of its time in an axial position, half in an equatorial.

The distinction between an axial and an equatorial position becomes

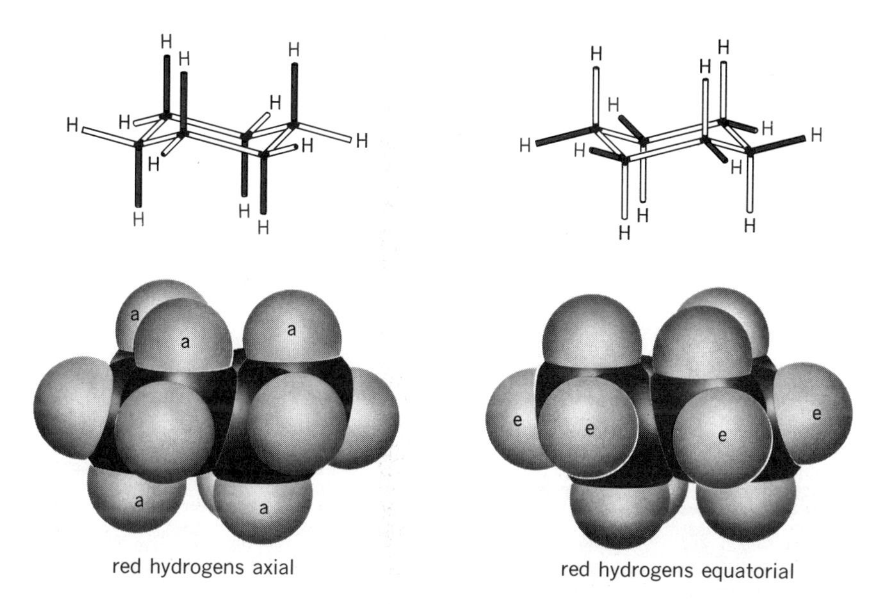

Figure 1-24. *The hydrogen atoms on a cyclohexane ring fall into two classes,* axial *and* equatorial. *Conversion of one chair form of cyclohexane to the other converts all axial hydrogens to equatorial, and* vice versa.

important in a substituted cyclohexane because a substituent is more stable in an equatorial position than in an axial position. This is so because axial substituents bump into the hydrogen atoms in the other axial positions, leading to what is called "*steric repulsion.*" This interference is illustrated in the molecular model of *axial*-methylcyclohexane in Fig. 1-25. A substituent in an equatorial position is not as crowded and does not experience as much steric repulsion. This makes that chair form with the substituent equatorial relatively more stable than the form with the substituent axial, and the molecule spends more of its time in the more stable conformation. For example, methylcyclohexane spends 90% of its time in that chair form with the methyl equatorial, and only 10% of its time in the other chair form. During the transformation from one chair form to the other, it passes briefly through a boat conformation. As the size of the substituent increases, so does the tendency to remain with the substituent equatorial. In *t*-butylcyclohexane there is no detectable amount of axial conformer present.

Large and Small Rings

Cyclic hydrocarbons containing rings of fewer than five carbon atoms must have internal bond angles that are much smaller than the normal 109°28′

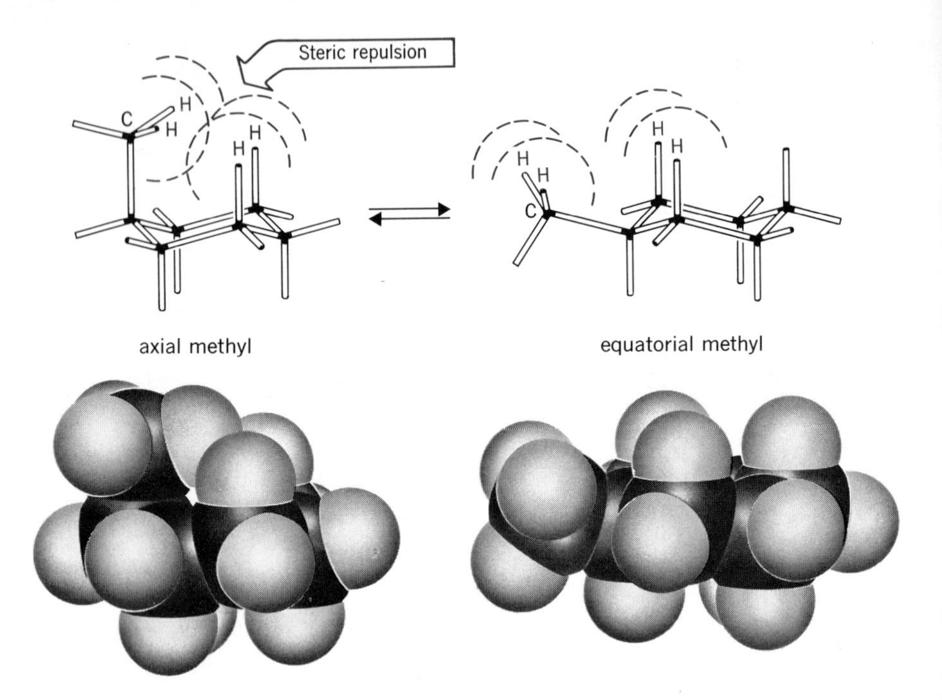

Figure 1-25. An axial substituent is sterically hindered *by hydrogens in the other axial positions, while an equatorial group is not so hindered. The equilibrium between the two chair forms is thus displaced toward the form with the equatorial substituent.*

of a tetrahedron. Simple geometry shows that these angles would be 90° in cyclobutane and 60° in cyclopropane; a square and an equilateral triangle are the usual symbols for these molecules (Fig. 1-26). Both of these hydrocarbons are stable and easy to form, despite the fact that their bonds are bent greatly from normal. Compounds with rings larger than six carbon atoms are also well known.

An example is muscone (Fig. 1-27), which contains a ring of 15 carbon atoms. This compound is obtained from a small gland of the Himalayan male musk deer and is an attractant for the female of the species. It is also the

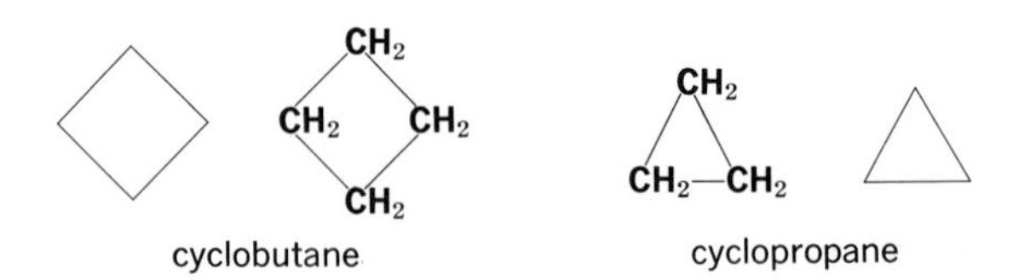

***Figure 1-26.** Cyclobutane and cyclopropane are both stable molecules, despite the fact that the carbon–carbon bonds must be highly strained to bend them from the normal tetrahedral angle of 109°28′ to 90° and 60°, respectively.*

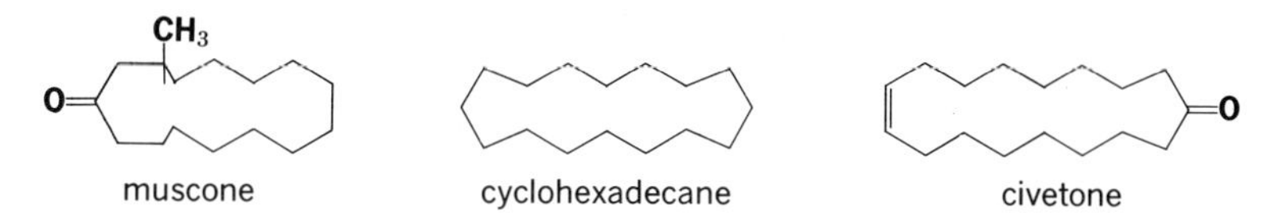

Figure 1-27. *Examples of compounds containing large rings. The groups C=C and C=O will be discussed in Chapters 3 and 8, respectively. Civetone is from the civet cat and, like muscone, is used in perfumes.*

base of many expensive perfumes. In large amounts it has a powerful, disagreeable odor, but when sufficiently dilute it appears to elicit the same response from the human male toward the human female that it does from the female musk deer toward the male of her species. Stimulated by the economic advantage of obtaining an improved and less-expensive substitute for muscone, many chemists have prepared a number of large-ring compounds and studied their properties, and large ring compounds like muscone and civetone can now be prepared more cheaply synthetically than through the slaughter of animals.

Nomenclature of Cyclic Compounds

The IUPAC nomenclature of cyclic molecules is not very different from that of noncyclic compounds; it does, however, incorporate extensions of the rules. The ring is generally considered to be the parent molecule, and the prefix "cyclo" is used. The number 1 is assigned to the ring carbon atom that gives the smallest numbers to the substituents, which are usually arranged in the order of increasing complexity (Fig. 1-28).

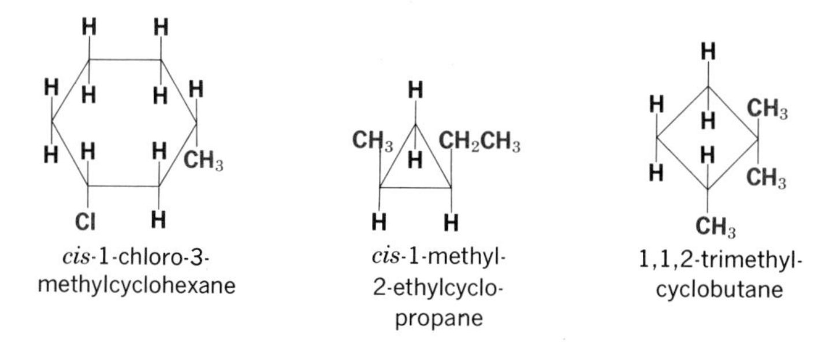

Figure 1-28. *In the IUPAC nomenclature of cyclic hydrocarbons, numbering is begun with the carbon atom that gives the smallest numbers for the location of substituents.*

Group names may also be used for cyclic compounds. The cyclohexyl group, for example, is a cyclohexane ring less one hydrogen. Additional instruction on the nomenclature of cyclic hydrocarbons will be found among the problems at the end of the chapter.

Fused-Ring Compounds

If adjacent carbon atoms are shared by two rings, the molecule is said to contain *fused rings*. The most important examples occur in the cyclohexane series; cyclohexanes may be fused together in their stable chair form. Two cyclohexane rings fused together give *decalin*, as shown in Fig. 1-29. If further

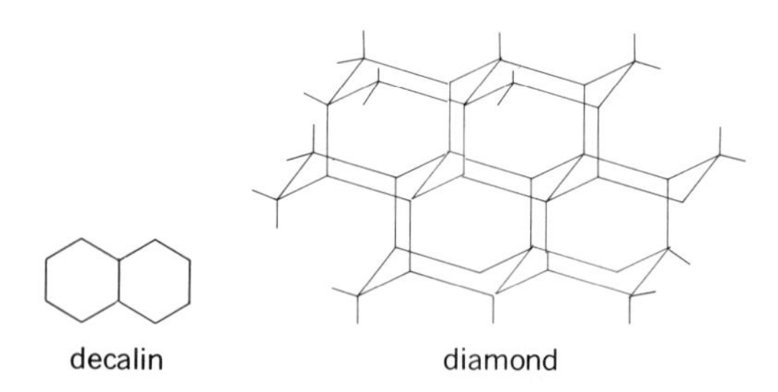

Figure 1-29. *Six-membered rings may be fused together. In diamond, fused six-membered rings continue in all directions.*

six-membered rings are fused onto this configuration progressively fewer and fewer hydrogens are left; eventually, pure carbon, in the form of *diamond*, results. The great hardness and stability of diamond is accounted for by the fusing of carbon atoms into an infinite number of cyclohexane rings, all in their stable chair form.

Reaction Mechanisms

The alkanes provide excellent examples of a general principle: More is required to make a chemical reaction proceed than favorable reaction energy. A mixture of methane and oxygen is indefinitely stable, despite the extremely *exothermic* (energy releasing) character of its reaction to give carbon dioxide and water. Similarly, a mixture of an alkane and chlorine can be kept indefinitely in the dark without reacting. However, if a spark is supplied to the former mixture or a burst of light to the latter, reactions rapidly ensue. To occur at an appreciable rate, a reaction must have a favorable pathway. The study of such reaction paths, called *reaction mechanisms*, has been very helpful in understanding how organic reactions take place and in correlating seemingly unrelated reactions.

It is difficult to bring about reactions of alkanes, because all their electrons (except the tightly bound 1*s* electrons of carbon) are involved in the formation of strong carbon–carbon and carbon–hydrogen bonds. The first step in the chlorination of an alkane has been shown to be the absorption of light by a chlorine molecule, which dissociates into chlorine atoms. A chlorine atom then attacks an alkane molecule, and, by removing a hydrogen atom, forms hydrogen chloride. This reaction occurs readily because the energy of the hydrogen–chlorine bond formed in the process more than compensates for the energy lost in breaking the carbon–hydrogen bond. In this way, a highly reactive site is created in the alkane molecule: a carbon atom with only three of its four valences used, and with one additional electron not employed in bonding. Fragments of this type are known as *free radicals* and occur often as short-lived intermediates in chemical reactions.

In the specific reaction for the chlorination of methane (Fig. 1-30), the

$$Cl_2 \xrightarrow{\text{light}} 2Cl\cdot$$

$$\cdot Cl\cdot + \underset{\text{methane}}{CH_4} \longrightarrow HCl + \underset{\text{methyl radical}}{\cdot CH_3}$$

$$CH_3 + Cl_2 \longrightarrow CH_3Cl + Cl\cdot$$

Figure 1-30. *The* mechanism *of the chlorination of methane. A chlorine atom reacts in the second step, while another one is generated in the third step, so that the sequence of reactions constitutes a* chain reaction.

methyl free radical ($\cdot CH_3$) is formed by the attack of a chlorine atom on methane. Under these conditions, the methyl free radical has a lifetime of approximately one-billionth of a second, before it reacts with a chlorine molecule to form methyl chloride and a new chlorine atom. After a comparably short time, the new atom in turn reacts with more methane to form a new, equally short-lived, methyl free radical. These steps are repeated many times in what is known as a *chain reaction* so that the dissociation of a single chlorine molecule can lead to the chlorination of many molecules of methane.

Summary

The material in this and all subsequent chapters can easily be reviewed by reading the legends under each figure. All important reactions and concepts are given concisely there, and the legends serve as a study guide and outline.

PROBLEMS

The first group of problems (1–15) is designed to teach IUPAC nomenclature.

1. Name each of the following hydrocarbons by the IUPAC system:
   (a) CH_4 (b) $CH_3CH_2CH_2CH_2CH_3$ (c) $CH_3CH_2CH_3$ (d) $CH_3(CH_2)_5CH_3$
   (Ans. Table 1-1)

2. Draw the structure of each of the following:
   (a) ethane (b) octane (c) butane
   (d) decane (e) nonane (f) propane
   (g) hexane (h) heptane
   (Ans. Table 1-1)

3. Name each of the following groups:
   (a) $CH_3CH_2—$ (b) $CH_3CH_2CH_2$ (c) $(CH_3)_3C—$
   Draw the structure of each of the following groups:
   (d) methyl (e) isopropyl (f) *n*-butyl
   (Ans. Table 1-2)

4. Among alkanes the longest continuous chain is considered to be the parent. Find and name this chain in the following compounds.
   (a) $CH_3CH—CHCH_2CH_3$ (with CH_3 and CH_3 branches) (b) $(CH_3)_2CHCHCH_2CH_2CH_3$ (with $CH_2CH_2CH_3$ branch) (c) $CH_3CH_2CHCH_3$ (with CH_2CH_3 branch)

5. Draw the structures for:
   (a) 4-isopropylheptane (b) 2,3-dimethylbutane
   (c) 2,2-dimethyl-3-ethyloctane (d) 4-*t*-butylnonane
   (e) 4-*n*-propylheptane

6. The chain is numbered from the end that gives the substituent the lower numbers. Name the following compounds:
   (a) $CH_3CH_2CHCH_3$ (with CH_3 branch) (b) $CH_3CHCH—CHCH_3$ (with CH_3, CH_2CH_3 and $CH_2CH_2CH_3$ branches) (c) $CH_3CH_2CHCH_2CHCH_3$ (with CH_2CH_3 and CH_3 branches)

7. Name each of the following compounds by the IUPAC system:
   (a) $CH_3CHCHCHCH_3$ (with CH_3, CH_3 and CH_2CHCH_3 branches; CH_3 on the last) (b) $CH_3CCH_2CHCHCH_3$ (with CH_3, CH_3, CH_2CH_3 and CH_2CH_3 branches)
   (c) $CH_3CH_2CCHCH_2CH$ (with CH_2CH_3, $CHCH_3$–CH_3, CH_3C–CH_3, and $CH_3CH_2CH_2CH_3$ branches)

8. Halogen atoms are treated as substituent groups: F(fluoro), Cl(chloro), Br(bromo), and I(iodo). Give structures for the following:
   (a) 2-bromopropane (b) 2-bromo-2-methyloctane
   (c) 1-chloro-2-bromobutane

9. The cyclic compounds are given a parent name corresponding to the number of carbons in their ring with a prefix *cyclo*-. Draw the structures of:
   (a) cyclobutane   (b) cyclopentane
   (c) cyclopropane   (d) cyclohexane
   Name the following compounds:

(e) (f)

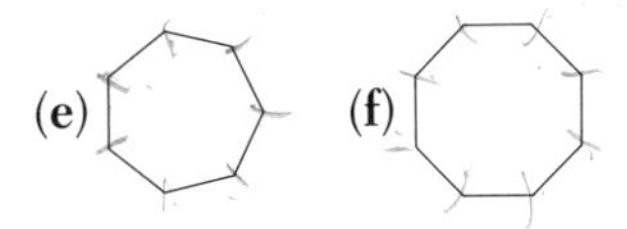

10. In cyclic systems the numbering is begun at the position that gives the substituents their smallest numbers. Give the structures of:
    (a) 1,1-dimethylcyclohexane
    (b) 1-bromo-2-chloro-4-isopropylcyclopentane
    (c) 1-*t*-butylcyclopropane
11. Name by the IUPAC system:

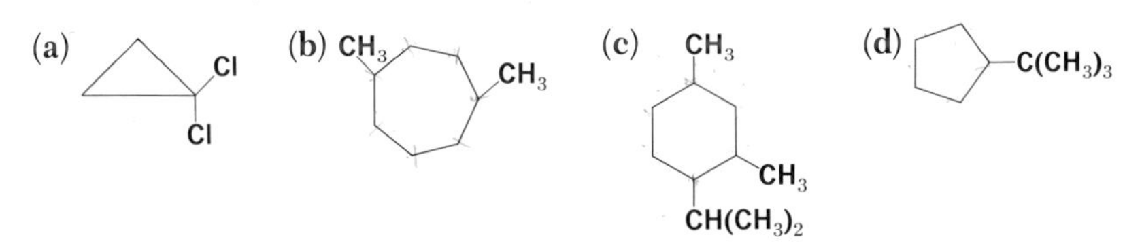

12. In cyclic systems containing two substituents, the two may be on the same side of the ring (designated *cis*) or on the opposite sides of the ring (*trans*). Name each of the following:

(a) CH_3, H, H, CH_3 (b) Cl, H, CH_2CH_3, H   (c) H, CH_3, $CH(CH_3)_2$, H

Draw structures for:
(a) *cis*-1,2-dimethylcyclobutane   (b) the two stereoisomers (geometrical isomers) of 1,3-dimethylcyclobutane

13. Draw the structures of the following compounds indicating the stereochemistry if called for by the name:
    (a) 2,3-dimethylhexane
    (b) 4-isopropylheptane
    (c) 4,5-dimethyl-3-ethyl-4-*n*-propylheptane
    (d) isopropyl chloride
    (e) 3-bromo-5-iodooctane

3 methyl, 4 isopropyl, 6, isopropyl, 7 ethyl nonane
3 ethyl, 4 6 di iso propyl, 7 methyl

(f) methylcyclobutane
(g) cyclooctane
(h) *trans*-1,2-dimethylcyclopentane
(i) *t*-butyl bromide
(j) *cis*-4-*n*-butylmethylcyclohexane
(k) cyclopropyl chloride

14. Name each of the following compounds according to the IUPAC system.

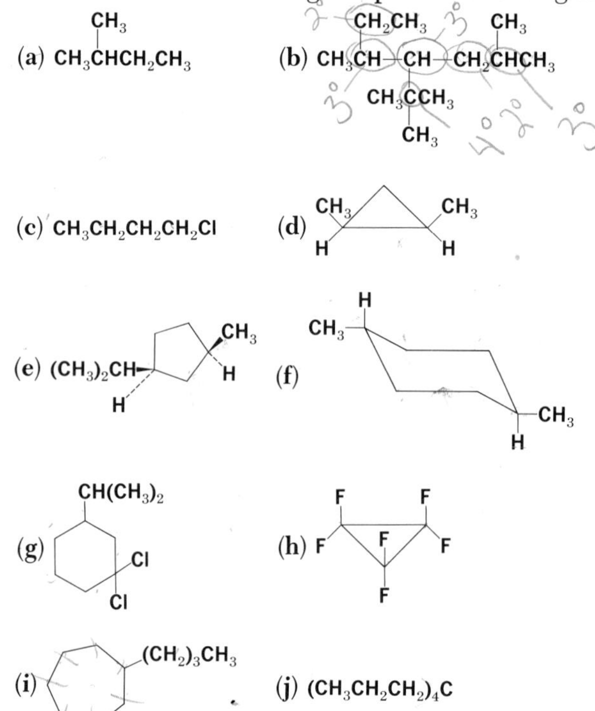

(i) (ring) $(CH_2)_3CH_3$ (j) $(CH_3CH_2CH_2)_4C$

15. Draw the structures of all nine isomers of heptane (C_7H_{16}). Name each according to the IUPAC system. (*Hint:* Draw the carbon skeletons first, starting with seven carbon atoms in a row, then with six, then with five, etc. Place carbon atoms at all possible positions, eliminate duplicates, and then add hydrogen atoms.)

16. In Prob. 14 (**b**), label each carbon as 1°, 2°, 3°, or 4°.

17. Write the structures of the following compounds in extended form. Be sure that each carbon is attached to four and only four other atoms.

(a) $(CH_3)_3CCl$
(b) $CH_3CH(CH_3)CH_2CH(CH_3)_2$
(c) $I_2C(CH_3)_2$
(d) $CH_3CH_2CH(CH(CH_3)_2)CH_3$
(e) $CH_3(CH_2)_6CH_3$
(f) $(CH_2)_3$

18. Using Newman projections

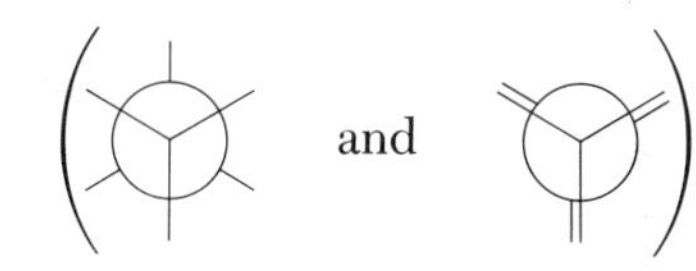

draw each of the following:

(**a**) the *anti* form of 1,2-dichloroethane
(**b**) an *eclipsed* form of ethane
(**c**) a view down any C—C bond of cyclopropane
(**d**) a view down any C—C bond of cyclohexane in the chair form
(**e**) a view down the 1,2-bond of methylcyclohexane with the methyl group in the equatorial position. (Use a model if possible.)
(**f**) The same with the methyl in the axial position.

19. Write a balanced equation for the complete combustion of octane (C_8H_{18}) to carbon dioxide and water.

20. Perform a "back of an envelope" calculation of the pounds of CO_2 produced if 10 gal of octanes (a principal constituent of gasoline) is burned in an automobile. By "back of an envelope" we mean an approximate calculation that will give you the right order of magnitude but which may be off by a factor of 2 or so. For instance, in this calculation assume 1 qt = 1 l, 1 l = 2 lb.

21. In a typical automobile engine, combustion is always incomplete. If 10% of the product (on a molecular basis) is CO rather than CO_2, how many lb of CO are produced per 10 gal of gasoline consumed?

2
petroleum and the analysis of organic compounds

The major source of hydrocarbons in the world today is petroleum. Formed during millions of years by decomposition of marine plants and animals, crude petroleum consists of a complex mixture of compounds, mainly hydrocarbons, but also of smaller amounts of organic molecules containing oxygen, sulfur, nitrogen, and even metals. The hydrocarbon content of petroleum may vary from as high as 97 to 98% in some Pennsylvania crude oils to as low as 50% for some Mexican and Southern oils. A recent compilation lists over 300 different hydrocarbons that have been isolated and characterized from one crude oil, and those comprise under 50% of the total amount of material present in the oil.

The usual first step in refining petroleum is to separate the crude oil into fractions on the basis of their boiling points. Fractions of a typical crude petroleum, arranged in the order of increasing boiling point, can be seen in Fig. 2-1. The gasoline obtained in this separation is known as *straight-run* gasoline and is of too low a quality to be used directly in today's automobiles. The naphthas yield kerosene, solvents, and diluents for paints, varnishes, and lacquers. Furnace and gas oils are burned in oil heaters and diesel engines or are used to make more gasoline. The residues furnish a great variety of common products, ranging from waxes, mineral oils, and paraffin to asphalt for paving.

Analysis of Hydrocarbons

If we wish to know something about the types and amounts of the various hydrocarbons present in a sample of petroleum, we shall need an analytical method for separating petroleum into its components and for measuring how

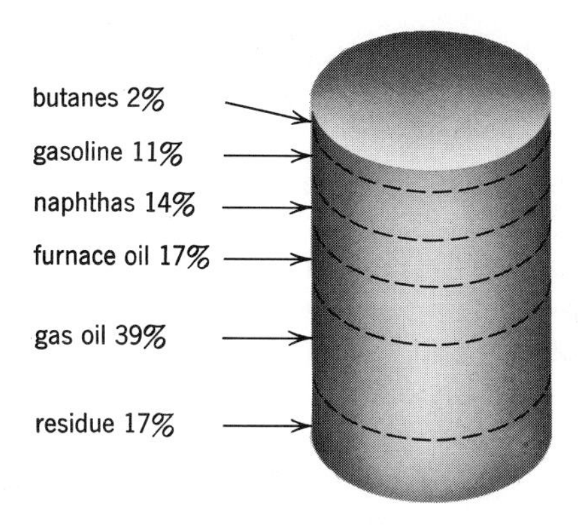

Figure 2-1. *The composition, in percentage of volume, of a West Texas crude oil.*

much of each compound is present in the mixture. This problem is by no means unique to hydrocarbon chemistry; nearly every organic reaction we shall study may give a mixture of isomeric products, and these products will usually have very similar physical properties. For instance, the bromination of propane in the presence of light gives a mixture of 1-bromopropane (b.p. 71°) and 2-bromopropane (b.p. 59°) and halogenation of more complex hydrocarbons gives mixtures of many isomeric compounds.

Mixtures of volatile organic compounds can be separated and analyzed quickly and efficiently by *gas chromatography*. A schematic version of such an instrument is shown in Fig. 2-2. A constant flow of helium is passed through

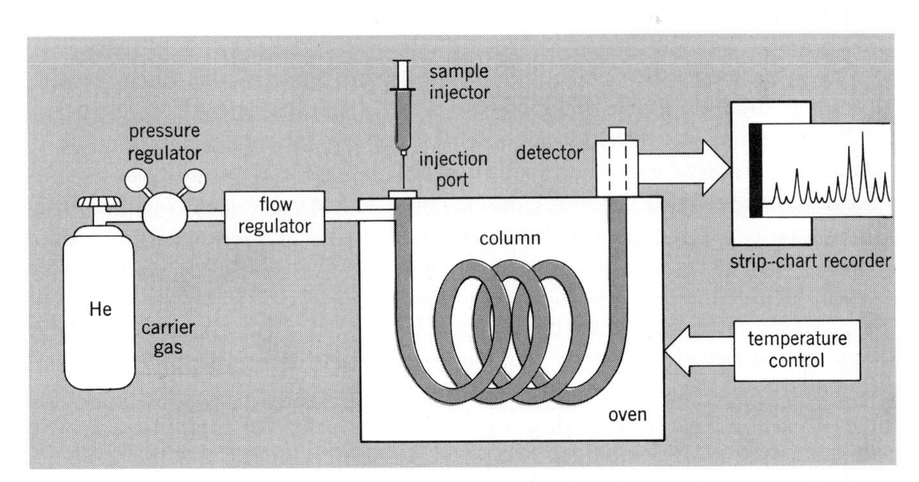

Figure 2-2. *A typical gas chromatograph. A tiny amount of a mixture of organic compounds is injected into a stream of helium. The mixture separates into its components on the column, and each component is detected separately and its quantity is recorded on a graph.*

a chromatography column, and a tiny amount of the mixture to be analyzed is injected into this stream through a rubber septum using a syringe. The mixture vaporizes and is carried through the column by the helium flow. The components pass through the column at different rates, some quickly, some more slowly. As they emerge from the column one by one, they are sensed by the detector, and the output of the detector is fed into a strip-chart recorder. A *gas chromatogram* of a hydrocarbon mixture is shown in Fig. 2-3.

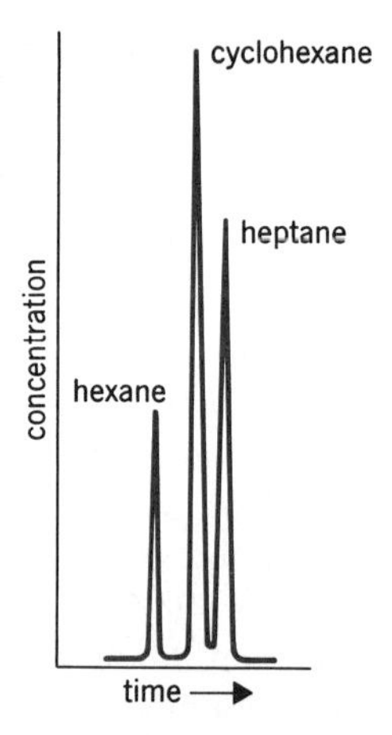

***Figure* 2-3.** *A gas chromatogram of a mixture of hydrocarbons; the area under each peak gives the relative concentration of each component present. Only 0.001 g of the mixture was used in this analysis.*

At a given temperature and helium flow rate, and for a given chromatographic column, the time it takes each component to reach the detector is a constant. Consequently, this *retention time* can be used to identify the compound. In addition the area under each chromatography peak is proportional to the quantity of material present. Consequently, in favorable cases one gas chromatogram, which can easily be carried out on 100 micrograms (0.0001 g) of material or even less, can give a qualitative and quantitative analysis of a complex mixture.

The chromatography column is the heart of a gas chromatograph. It usually consists of a piece of ⅛ in. copper or glass tubing about 6 ft long and coiled so as to fit into a small oven. This column is tightly packed with fine particles of finely crushed brick or other inert material whose surface has been coated with a high-boiling liquid. In the example of Fig. 2-3 this liquid is silicone oil. As the molecules of the mixture are carried through the column in the stream of helium, they dissolve in and reevaporate from these thin films of liquid on the surface of the tiny inert particles. Because of slight differences in the solubility of the gases in the liquid, accentuated by any slight differences in their boiling point, the various components of the mixture will require a longer or shorter time to pass through the column, and will be separated from one another. Hundreds of different high-boiling liquids have been used to coat the chromatography particles, and the coating can be chosen so as to accentuate differences in solubility among the components of a mixture. For especially difficult separations, like the gas chromatograms of petroleum and gasoline, the column can be made very long. By the proper choice of column and conditions, nearly any complex mixture can usually be separated as long as the components are even slightly volatile.

A gas chromatogram of a sample of saturated hydrocarbons from a North Slope Alaskan oil well is shown in Fig. 2-4. Notice the enormous complexity

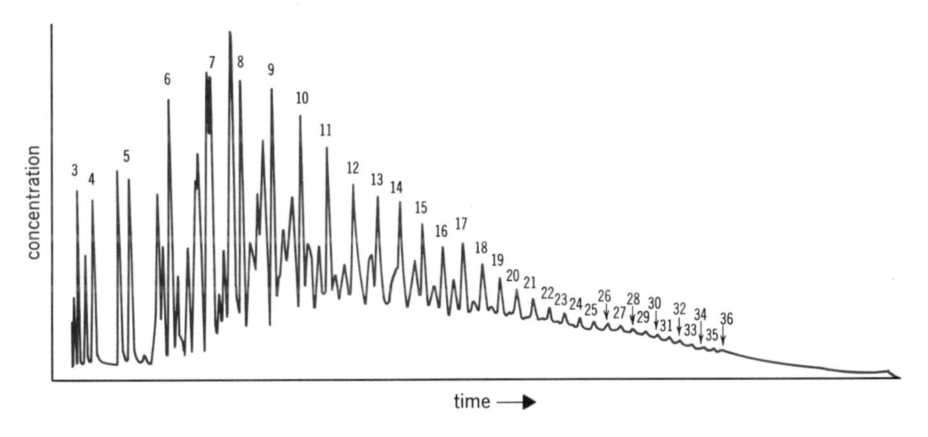

Figure 2-4. *A gas chromatograph of hydrocarbons from a North Slope Alaskan oil field. The numbers refer to peaks for the* n-*alkanes.*

of the mixture. Up through the hexanes (C_6) many individual components, both cyclic and noncyclic, can be identified, but beyond that only the relatively abundant normal hydrocarbons emerge from among the background of branched-chain isomers. This chromatogram will give some idea of both the power of gas chromatography and the complexity of petroleum.

Gasoline

Gasoline is a volatile mixture of hydrocarbons, cyclic and noncyclic, in the range of pentanes to decanes. The quality of a gasoline is very much dependent upon the structure of the hydrocarbons of which it is composed. Table 2-1 contains a comparison of the amounts of normal, branched, and cyclic alkanes in gasolines derived from some typical petroleum sources. The usual straight-run gasoline fraction encompasses hydrocarbons in the C_4 to C_{12} range. In this range there are 661 possible alkanes and 3839 possible alkenes (a class

Table 2-1
Per Cent (by Volume) of Various Hydrocarbon Types in a Fraction, Boiling from 40 to 102°, of Straight-Run Gasoline from Three Sources

Gasoline Source	Branched Alkanes	*n*-Alkanes	Cyclopentanes	Cyclohexanes
Oklahoma	36	21	23	20
Michigan	63	13	8	16
Texas	10	62	8	20

of compounds discussed in Chapter 3), as well as a large number of aromatic compounds (Chapter 5). Because of this possible mixture, it is understandable that no single formulation for gasoline is achievable.

The quality of a gasoline is measured by its octane number. In an internal combustion engine like that in a modern automobile, a mixture of air and vaporized gasoline is drawn into the cylinder on the downstroke of the piston; this mixture is compressed by the piston on its upstroke and is then ignited by a spark from the spark plug. A high-octane gasoline will burn quickly and smoothly, giving a surge of power to the piston, forcing it down and into its next stroke. A low-octane gasoline, on the other hand, will explode, setting up a shock wave, which batters against the piston and cylinder walls and gives rise to the familiar "knocking" sound.

Alkanes vary greatly in their ability to burn in an engine without knocking. The *octane ratings* of a number of alkanes are presented in Table 2-2. These are relative values, determined under carefully defined engine

Table 2-2
Octane Ratings of Some Hydrocarbons

Hydrocarbon	Octane Number
n-heptane	0
2-methylhexane	42
methylcyclohexane	75
2,3-dimethylpentane	91
2,2,4-trimethylpentane (isooctane)	100

conditions. It was discovered early that 2,2,4-trimethylpentane (called industrially "isooctane") caused little knocking, whereas *n*-heptane caused a great deal. A scale was therefore devised on which the octane rating of any gasoline or hydrocarbon was defined as being equal to the proportion of isooctane in an isooctane-*n*-heptane mixture that knocked under the same conditions. Methylcyclohexane, for example, knocks under the same conditions as a mixture of 75% isooctane and 25% *n*-heptane, and hence has an octane rating of 75.

Straight-run gasoline often contains a high percentage of *n*-alkanes, which makes its octane number low (typically around 70). In addition, the gasoline fraction amounts to only a relatively small percentage of the crude oil. Yet gasoline, along with fuel oil for heating, makes up the major market for petroleum. To solve the twin problems of low quality and low quantity, the petroleum industry began early in its development to study ways by which the higher-boiling components of petroleum could be converted to gasoline.

Cracking Petroleum into Gasoline

The industry found its solution in a process called "cracking." This process breaks large complex hydrocarbons into lighter ones through the application of heat and pressure. The first method used was *thermal cracking.* In this method the large alkane molecules are heated to temperatures near 700° until they split apart into smaller molecules. A great deal more gasoline thus becomes available, but, unfortunately, as thermal cracking leads primarily to *n*-alkanes, the octane number of gasoline from this source is not high. The low-octane problem was solved by the introduction of *catalytic cracking.* In this method, gas oils are mixed with a catalyst and heated to 400 to 550°, which cracks them to gasoline and at the same time causes the normal paraffins to rearrange to branched isomers. The octane rating of gasoline from this latter cracking is therefore higher than that of either straight-run gasoline or gasoline from thermal cracking (Table 2-3).

Table 2-3
The Composition of the Hexanes from Gasoline of Different Sources

Components	Catalytic Gasoline, %	Thermal Gasoline, %	Straight-Run Gasoline, %
n-hexanes	9	63	51
branched hexanes	91	37	49
	100	100	100

The process of catalytic cracking also produces hydrocarbons whose boiling points are too low for them to find much use in gasoline; these derivatives of butane, propane, and ethane are not wasted, however. Instead, some of them are *reformed* into higher molecular weight compounds (see Chapter 3) for inclusion in gasoline, while others are converted into a variety of commercial chemicals.

The octane number of a gasoline may be raised by increasing the quality of the hydrocarbons contained or by including any one of a number of antiknock agents. Tetraethyllead (TEL) is the antiknock agent most commonly employed (Fig. 2-5). The inclusion of as little as 3 ml of a TEL solution is

$$\underset{\text{ethyl chloride}}{4C_2H_5Cl} + 4PbNa \longrightarrow \underset{\text{tetraethyllead}}{Pb(C_2H_5)_4} + 4NaCl + 3Pb$$

***Figure* 2-5.** *Tetraethyllead (TEL) is produced from ethyl chloride and a lead-sodium alloy.*

sufficient to increase the octane rating of a gallon of typical straight-run gasoline from 70 to 84. The ethyl fluid used in modern gasoline contains about 61% tetraethyllead, 37% of a mixture of 1,2-dibromo- and 1,2-dichloroethanes, and 2% of a dye and impurities. The additive TEL moderates the combustion of the gasoline-air mixture and helps prevent knocking. In the process, the TEL burns and lead salts are formed. The chloro- and bromoethanes furnish halogen atoms for the formation of lead halides, which evaporate in the exhaust. In the absence of these additives, the cylinders of the engine would become coated with nonvolatile lead salts. The use of TEL is being discouraged to reduce soil and air pollution by lead. Since 1972 nonleaded and low-lead fuels have been advised for new autos by the manufacturers.

Gasoline is a carefully blended product. Although pure isooctane would make a 100-octane gasoline, it would be nearly useless in an automobile because the engine would never start without some very low-boiling components. The composition of gasoline must change with the season and with the altitude. Therefore, modern gasoline contains many additives—from "deicers" to dyes that improve the aesthetic appeal of the fluid.

Air Pollution by the Internal Combustion Engine

In an internal combustion engine, gasoline combustion is never the completely efficient reaction we would like it to be; among the materials leaving the cylinders are unreacted hydrocarbons and products produced by partial oxidation. One product of incomplete combustion is carbon monoxide. Many other, more complex, molecules containing double bonds, carbonyl groups and other functional groups, are among the products of partial combustion. These are swept into the exhaust system and, to a large extent, emitted to the air.

All these hydrocarbons and their partial oxidation products, plus lead salts from tetraethyl lead, would cause air pollution enough, but the problem is made worse by the fact that the cylinder temperature during combustion is high enough to bring about the oxidation of nitrogen in the air (Fig. 2-6). As a result nitric oxide (NO) is formed and this quickly oxidizes to the brown gas nitrogen dioxide (NO_2) in the atmosphere. In a smoggy city NO_2 is responsible for the brown color that hangs over the city. With its unappetizing

$$N_2 + O_2 \longrightarrow 2NO$$

$$2NO + O_2 \longrightarrow 2NO_2$$

Figure 2-6. *Nitrous oxide is formed in the engine cylinders and is given out in the exhaust. In the air it is oxidized to the brown gas NO_2.*

color, NO_2 is a bad pollutant by itself, but it leads to still more serious pollution. The mere fact that it is brown tells us that it absorbs light from the sun. The energy that it absorbs in this way leads to its dissociation into NO and an oxygen atom (Fig. 2-7). The oxygen atom formed is highly reactive

$$NO_2 + \text{sunlight} \longrightarrow NO + O\cdot \quad (O_2 \text{ returns NO to } NO_2)$$

$$O\cdot + \text{hydrocarbons} \longrightarrow \text{eye irritants}$$

$$O\cdot + O_2 \longrightarrow O_3$$

Figure 2-7. *The basic reactions which produce Los Angeles–type photochemical smog. One NO_2 molecule can serve as a source of many oxygen atoms.*

(analogous to the chlorine atoms formed by irradiation of Cl_2, Fig. 1-30) and rapidly attacks the hydrocarbons that are emitted by automobile engines. The products of this reaction between hydrocarbons and oxygen atoms are often eye irritants and other noxious pollutants. Ozone is also formed. Since the NO is reoxidized to NO_2, it can serve as a catalyst for the conversion of oxygen molecules into oxygen atoms; thus one has *photochemical smog.*

What steps are possible to reduce photochemical smog? In the first place the exhaust can be recycled through the cylinder to burn the exhaust products more efficiently, and cylinders have been redesigned so that combustion occurs at lower temperatures with less oxidation of nitrogen. At the present time, however, the long-run solution appears to be the so-called *catalytic converter,* which is placed in the exhaust system. It consists of a can of metal oxide mixed with finely divided clay to give a large surface area. In this highly efficient catalytic system complete combustion of all organic material to CO_2 and H_2O takes place. With rigid auto exhaust inspection, it may be possible to eliminate automobile air pollution with this device.

PROBLEMS

1. Draw the structural formulas for the hydrocarbons in Table 2-2.
2. In the next chapter we shall see that "isooctane" can be made from two molecules with the carbon skeleton of isobutane. Identify two such skeletons in isooctane.
3. If we assume that they appeared at the detector in the order of increasing boiling point, what would a gas chromatogram of a mixture of $\frac{1}{3}$ *n*-pentane (b.p. 36°) and $\frac{2}{3}$ neopentane (b.p. 9°) look like? To confirm the peak assignments, a sample of the mixture can be "spiked" by adding a sample of known compound. For instance, what would the chromatogram look like if about $\frac{1}{3}$ more *n*-pentane were added?

See p. 308

4. An oxygen atom has two unpaired electrons. It could react with a hydrocarbon to give an alkyl radical and a hydroxyl radical. These could combine to give an alcohol. Write equations by which methane could be converted to methyl alcohol (CH_3OH) by an oxygen atom.
5. There are five isomeric hexanes, C_6H_{14}, and they have boiling points of 50, 58, 60, 63, and 69°. Review what has been said about the variation of boiling point with structure and then guess which structure corresponds to each boiling point.
6. Draw Newman projections down the 1,2-bonds of *cis*- and *trans*-1,2-dimethylcyclopentanes (use Fig. 1-21 and replace the appropriate chlorine atoms by $-CH_3$ groups). At high temperature and over a catalyst an equilibrium can be established between these two isomers (bonds must be broken and reformed to do this). Which isomer would you expect to be more stable (which would predominate at equilibrium) and why?
7. Given here are the two chair forms of cyclohexane with all of their axial and equatorial bonds. Practice drawing them so that you can easily

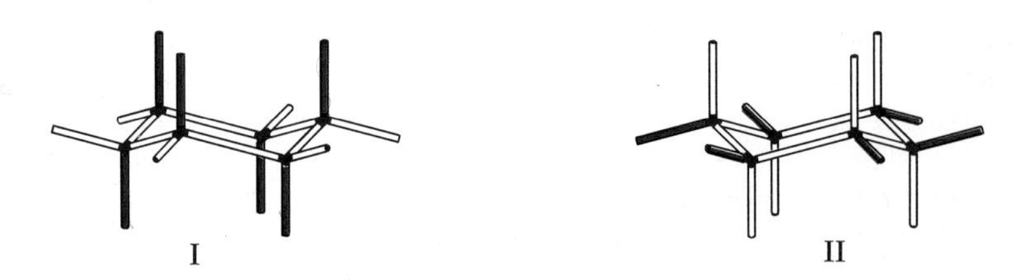

reproduce them. Use these drawings to answer the following questions:
   (a) Identify the axial and equatorial bonds in each structure. (Ans.: Fig. 1-24.)
   (b) Using structure I, draw *axial*-methylcyclohexane. Convert I to II; now does the methyl group occupy an axial or an equatorial position? Which form, I or II, will be more stable for methylcyclohexane?
   (c) Starting with *axial*-methylcyclohexane, place another methyl group

in the *axial*-4-position to give *diaxial*-1,4-dimethylcyclohexane. Is this *cis*- or *trans*-1,4-dimethylcyclohexane?

(d) Now convert to II; what position does each of the methyl groups occupy? Which form is more stable?

(e) Using structure I, draw *cis*-1,4-dimethylcyclohexane. Label each methyl *a* (axial) or *e* (equatorial). Convert to II and relabel. Which is more stable, I or II, in this isomer?

(f) At high temperature over a catalyst, *cis*- and *trans*-1,4-dimethylcyclohexane can be brought into equilibrium. Which isomer will be the more stable and why?

(g) Repeat parts (c) to (f) for the 1,2- and 1,3-dimethycyclohexane isomers.

3

unsaturated hydrocarbons: alkenes, alkynes, dienes

Hydrocarbons may be saturated or unsaturated. In saturated hydrocarbons there are a sufficient number of hydrogen atoms so that each carbon may be surrounded by four other atoms. Many hydrocarbons do not have enough hydrogens for this purpose; we say that they are *unsaturated.* Ethene, C_2H_4, and ethyne, C_2H_2, (common names *ethylene* and *acetylene*) are simple examples of unsaturated hydrocarbons. In the former each carbon is surrounded by three, and in the latter only two, other atoms. Such unsaturated molecules are, compared to alkanes, highly reactive and serve as sources for most of our industrial organic chemicals, including many plastics like polyethylene. In this chapter we shall introduce the basic nomenclature and reactions of unsaturated molecules, and in the next we shall discuss their use in the formation of polymers.

Ethylene and Acetylene

Ethylene (C_2H_4), the simplest of the alkenes, is a gas with a lower boiling point than ethane. It is formed in large amounts during the cracking of petroleum and is such an important and common industrial chemical that in many parts of the United States a company desiring to use it as a raw material simply hooks up to an ethylene pipeline, much as a householder hooks up to a natural gas line. Nearly 23 billion lb of ethylene were consumed in the United States in 1972. As its formula tells us, ethylene contains two fewer hydrogens than ethane; each carbon is bonded to only three atoms, one carbon and two hydrogens. It is convenient to maintain the *tetravalence* of carbon throughout organic chemistry, and so we write the structure of ethylene with a *double bond* between the two carbons. In keeping with its unsaturated character, ethylene will react with hydrogen gas in the presence of a catalyst to form ethane (Fig. 3-1).

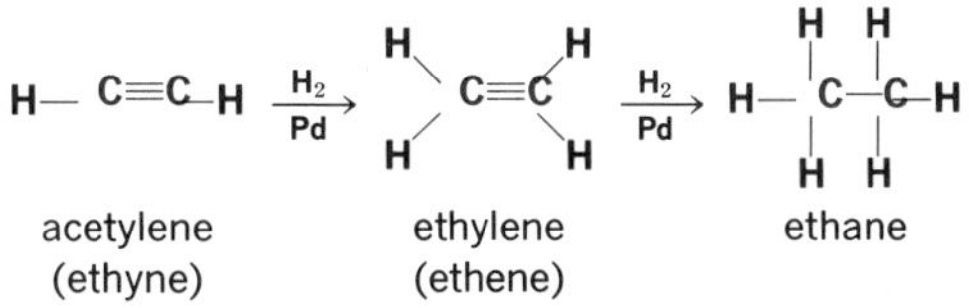

$$H{-}C{\equiv}C{-}H \xrightarrow[Pd]{H_2} H_2C{=}CH_2 \xrightarrow[Pd]{H_2} H_3C{-}CH_3$$

acetylene (ethyne) — ethylene (ethene) — ethane

Figure 3-1. *Alkynes (like acetylene) and alkenes (like ethylene) are both* unsaturated *toward hydrogen, while alkanes are saturated hydrocarbons. Note that each carbon atom maintains its tetravalence throughout these transformations.*

Acetylene (C_2H_2) is a gas used widely in industry both as a fuel and as a chemical for conversion to other compounds. Acetylene is familiar to some as the fuel in a welder's torch and to others as the illuminating gas once widely used in miners' and campers' lanterns. Today it is principally used as an intermediate in the syntheses of many industrially important compounds. Acetylene is made commercially by the action of water on calcium carbide, or by heating methane or ethane to very high temperatures. The maintenance of the tetravalence of carbon requires that acetylene be written with a triple bond between the carbon atoms. As Fig. 3-1 shows, ethylene may be produced from acetylene by the addition of one molecule of hydrogen in the presence of a catalyst.

Nomenclature of Alkenes and Alkynes

In Table 3-1 are the structures and names of a representative sample of simple molecules containing either a double or triple bond. The first thing to notice

Table 3-1

Examples of Nomenclature of Alkenes and Alkynes

Compound	Common Name	IUPAC Name
$H_2C{=}CH_2$	ethylene	eth ene
$CH_3{-}CH{=}CH_2$	propylene	prop ene
$CH_3{-}CH_2{-}CH{=}CH_2$	α-butylene	1-but ene
$CH_3{-}CH{=}CH{-}CH_3$	β-butylene	2-but ene
$CH_3{-}C(CH_3){=}CH_2$	isobutylene	2-methylprop ene
$CH_3{-}CH(CH_3){-}CH{=}CH_2$	isopropylethylene	3-methyl-1-but ene
$HC{\equiv}CH$	acetylene	eth yne
$CH_3{-}CH_2{-}C{\equiv}C{-}CH_3$	methylethylacetylene	2-pent yne

is that each has both a common and a systematic name. Unfortunately, it is true that most simple, and many complex, organic compounds have both a common and a systematic name, because most of them were first isolated and named before the IUPAC system of nomenclature existed, and common names persist, especially in industry. Although we emphasize systematic names here, some acquaintance with common names is necessary in order to read industrial and popular chemical articles.

In the IUPAC system the presence of unsaturation in the molecule is indicated by replacing the *ane* ending of the saturated hydrocarbon by *ene* if the molecule contains a double bond or *yne* if a triple bond is present. We thus have the series *ethane, ethene, ethyne.* And because, as we shall soon see, a double or triple bond is the source of most reactivity in an unsaturated molecule, we choose the parent chain for naming *so as always to include both unsaturated carbons* and number the chain from the end which will give the *smallest number to the double or triple bond.* With these few extensions, the IUPAC rules for naming alkenes and alkynes are very similar to those for alkanes. Three examples of the systematic nomenclature of alkenes and alkynes are given in Fig. 3-2. Note in the first example that, although the

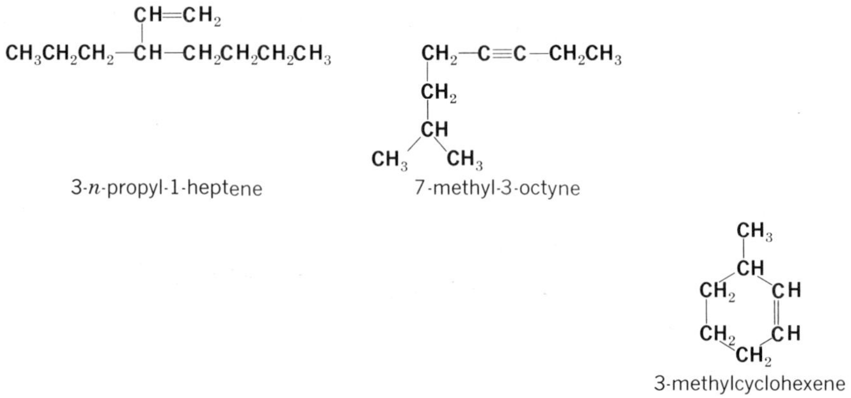

$CH=CH_2$
$CH_3CH_2CH_2-CH-CH_2CH_2CH_2CH_3$

3-*n*-propyl-1-heptene

$CH_2-C\equiv C-CH_2CH_3$
CH_2
CH
CH_3 CH_3

7-methyl-3-octyne

CH_3
CH
CH_2 CH
CH_2 CH
CH_2

3-methylcyclohexene

Figure 3-2. *Three examples of the systematic nomenclature of unsaturated hydrocarbons.*

longest continuous chain of carbon atoms in the molecule is eight, the longest that contains the double bond is seven, and so *heptene* is chosen as the parent name. In the second example numbering from the right gives the triple bond its smaller number. Note that only the number of the first carbon of a double or triple bond is given; the second carbon is assumed to have the next higher number. Therefore, in the third example, one carbon of the double bond in the cyclohexene ring is numbered 1, the second carbon of the double bond is assumed to be 2, and the methyl group is on C-3. Additional examples of the nomenclature of unsaturated hydrocarbons are given in the problems at the end of the chapter.

Structure of Alkenes

We were able to draw some useful conclusions about the structure of saturated hydrocarbons by considering the number of isomers they formed upon replacement of hydrogen by chlorine. Let us apply the same technique to ethylene. If we do, we find, as we might expect, only a single *monochloroethylene*, but three *dichloroethylenes* exist and can be separated from one another (Fig. 3-3). In contrast to the various conformers of 1,2-dichloroethane,

1-chloroethene (vinyl chloride) | 1,1-dichloroethene (vinylidene chloride) | *trans*-1,2-dichloroethene | *cis*-1,2-dichloroethene

Figure 3-3. *The one mono- and three dichloroethenes. The existence of two geometrical isomers of 1,2-dichloroethene indicates that there is no free rotation about a double bond.*

which cannot be separated from one another except possibly near absolute zero, *two stable noninterconverting geometrical isomers of 1,2-dichloroethene exist.* The existence of these two isomers shows that there is no freedom of rotation around a carbon–carbon double bond. In conformity with the names used in cyclic systems, that 1,2-isomer with the two substituents on the same side of the double bond is called *cis*-1,2-dichloroethene, and that 1,2-isomer with the substituents on opposite sides of the double bond is called *trans*-1,2-dichloroethene.

Many compounds that contain double bonds can exhibit geometrical isomerism. For example, there are four butylenes: isobutylene (2-methylpropene), 1-butene, and *cis*- and *trans*-2-butenes (Fig. 3-4). The two latter com-

isobutylene b.p. −7°C | 1-butene b.p. −6°C | *cis*-2-butene b.p. +4°C | *trans*-2-butene b.p. +1°C

Figure 3-4. *The four isomeric butylenes. Isobutylene and 1-butene are* structural isomers *of the* cis- *and* trans-2-*butenes, which are* geometrical isomers *of one another.*

pounds are *geometrical isomers* of each other, and both are structural isomers of isobutylene and 1-butene. Like structural isomers, geometrical isomers have different physical and chemical properties.

Prob. 3-1. Draw all six structurally and geometrically isomeric alkenes having the molecular formula C_5H_{10}.

Molecular models of ethylene and acetylene are shown in Fig. 3-5. Note especially that ethylene is planar, i.e., that the two carbons and the four

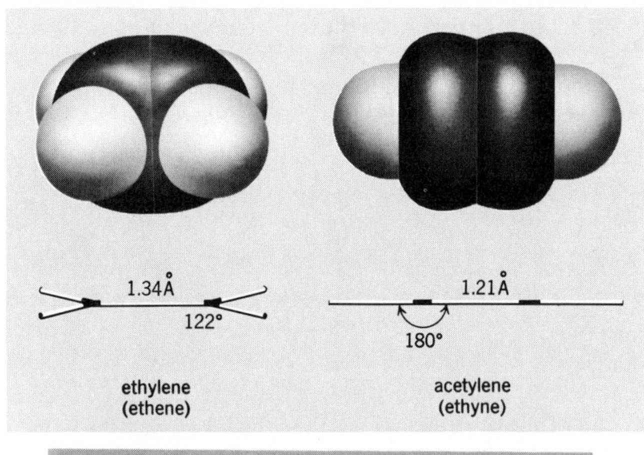

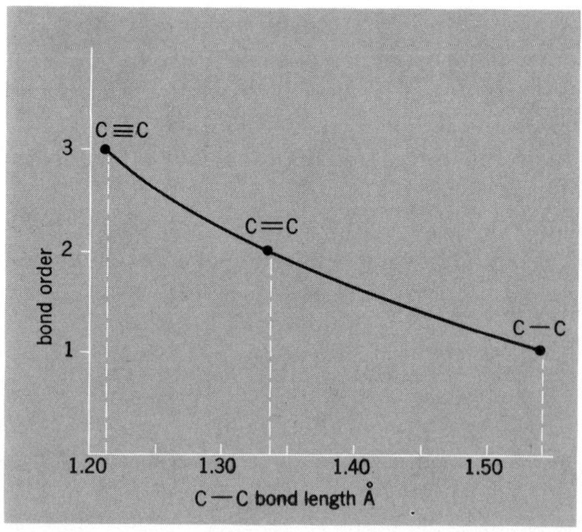

Figure 3-5. *Molecular models of ethylene and acetylene, and a graph of carbon-carbon bond length vs. the degree of unsaturation (bond order).*

atoms to which they are bonded all lie in the same plane. Acetylenes, on the other hand, are *linear.* Compare these structures carefully with that for ethane, Fig. 1-6. Note how the distance between the two carbons decreases as the number of bonds holding them together increases. This is shown graphically in Fig. 3-5, where the bond order (1 = single, 2 = double, 3 = triple bond) is plotted against the carbon–carbon bond length.

Prob. 3-2. Making use of the geometry of double and triple bonds given in Fig. 3-5, suggest an explanation for the fact that double bonds often occur within stable six-membered rings, but triple bonds never do.

Bonding in Organic Molecules

We now know enough about the structure and chemistry of simple organic molecules to make us wonder about the fundamental way in which these molecules are joined together. Why, for instance, is there free rotation in ethane but not in ethylene? Why are unsaturated molecules so reactive, as compared to saturated ones? Is there any simple way we can rationalize the fact that acetylene is linear, ethylene flat, and ethane tetrahedral? In short, we should like to apply bonding theory to these molecules in the hope that it will make a number of isolated facts fit together, and so become easier to remember. In what follows it is assumed that the reader has studied atomic and molecular structure in an earlier course and is familiar with the ideas of orbitals, ionic and covalent bonds, and a few other basic concepts. It may be helpful to review these topics in any general chemistry textbook.

Here we shall adopt an extremely practical attitude; we are not interested in the ultimate nature of matter, nor the truth or falsehood of any particular theory of bonding, or of atomic theory. What we want is as simple a visual picture as possible of why organic molecules react as they do, and why they have the particular shapes and properties they do. If we can develop such a picture, using concepts that are derived from atomic structure and inorganic chemistry, so much the better.

Atomic Orbitals

Recall that electrons in individual atoms occupy discrete *orbitals,* each capable of holding two, and no more than two, electrons of opposite spin. The hydrogen atom, for example, has one electron in the 1*s* orbital, while helium has two electrons (with opposite spins) in this same orbital. Continuing down the periodic table, we find that the next element, lithium, has three electrons. Two of them occupy the 1*s* orbital, which is now filled; the third enters the orbital of next higher energy, the 2*s* orbital. In beryllium the 1*s* and 2*s* orbitals are both filled, and beginning with boron the three 2*p* orbitals begin filling until, with neon, the first row of the periodic table is completed. The orbital structure of the first ten elements is given in Fig. 3-6.

1*s*	2*s*	$2p_x$	$2p_y$	$2p_z$	
↑					**H**
↑↓					**He**
↑↓	↑				**Li**
↑↓	↑↓				**Be**
↑↓	↑↓	↑			**B**
↑↓	↑↓	↑	↑		**C**
↑↓	↑↓	↑	↑	↑	**N**
↑↓	↑↓	↑↓	↑	↑	**O**
↑↓	↑↓	↑↓	↑↓	↑	**F**
↑↓	↑↓	↑↓	↑↓	↑↓	**Ne**

Figure 3-6. *The electronic configuration of the first ten elements. Each arrow indicates one electron present in the orbital. Arrows in opposite directions indicate electrons of opposite spins.*

An electron in a 1*s* orbital is constantly moving with respect to the nucleus within the confines of that orbital. Although, according to the Uncertainty Principle, we cannot locate an electron exactly, we can determine the *probability* of finding an electron in a given volume of space. If we plot such a probability for a 1*s* electron against the distance from the nucleus, we get the graph in Fig. 3-7. Extending such a graph to three dimensions and chopping it off so that we get a region of space in which the electron spends 90% of its time gives us a pictorial representation of an atomic 1*s* orbital. An atomic 2*s* orbital is represented by a similar sphere but with a larger radius. An electron in a 1*s* orbital is of lower energy than one in a 2*s* orbital because it spends more of its time close to the atomic nucleus.

Constructed in a similar way, the 2*p* orbitals have shapes resembling dumbbells. There are three of these atomic 2*p* orbitals, of identical energy, and only differing from one another by their orientation in space. Placed in an *xyz* coordinate system, one points along the *x* axis, one along the *y* axis and one along the *z* axis. Drawings of these three orbitals are given in Fig. 3-8. Although it is not easy to show this in a drawing, if the three 2*p* orbitals

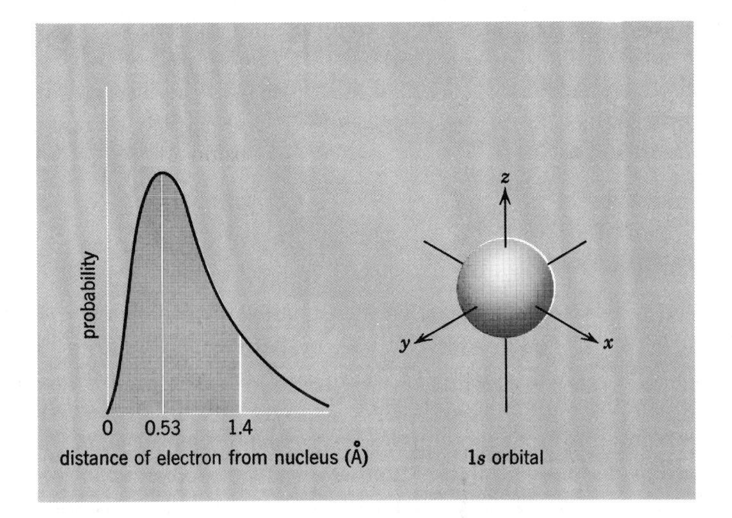

Figure 3-7. *The graph represents the relative probability of finding an electron at various distances from the nucleus of a hydrogen atom. The most probable distance is 0.53 Å (0.53×10^{-8} cm). The probability is 90% of finding it within the gray area (1.4 Å). A 1s orbital is defined as a region in space within which a 1s electron spends 90% of its time. This is a sphere with a radius of 1.4 Å.*

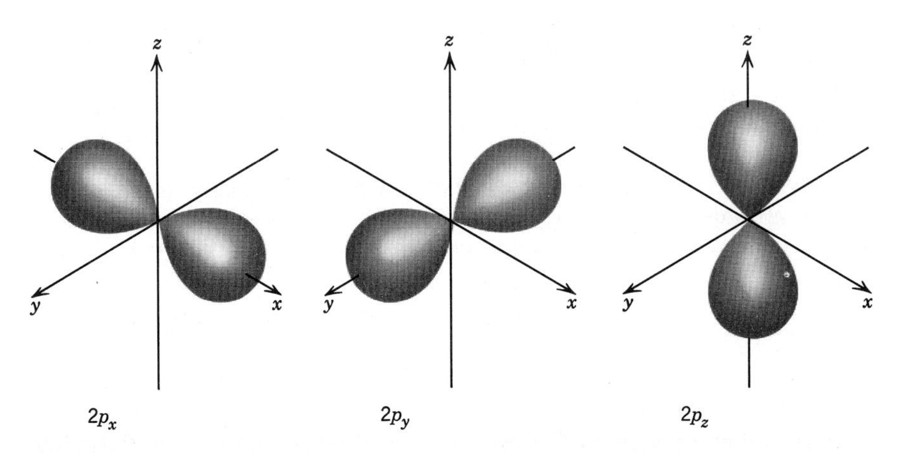

Figure 3-8. *The three 2p orbitals are dumb-bell shaped and arranged at right angles to one another. Each can hold two electrons. Superimposed upon one another, the three orbitals would make a sphere.*

were all drawn on the same set of axes, they would add together to form a perfect sphere. The electron distribution in neon then, in which each $2p$ orbital has two electrons, is spherical, just as it is in the nitrogen atom with one electron in each $2p$ orbital.

Molecular Orbitals

We have been discussing orbitals that have been widely used in understanding atomic structure. It would be convenient and satisfying if we could use these same orbitals to explain molecular structure; in fact, this is just what has been done. For simple inorganic molecules the extension of atomic orbitals to molecular orbitals is exceedingly easy. We postulate that covalent bonds arise by the overlap of atomic orbitals to give a *molecular orbital,* capable of holding two electrons of opposite spin. Molecular hydrogen, H_2, arises, for instance, by the overlap of two $1s$ atomic orbitals into a sigma (σ) molecular orbital encompassing both hydrogen nuclei. The bond in HF may be thought of as arising from the overlap of a $1s$ hydrogen atomic orbital with a $2p_z$ atomic orbital from fluorine (Fig. 3-9). This scheme also accounts for the formula

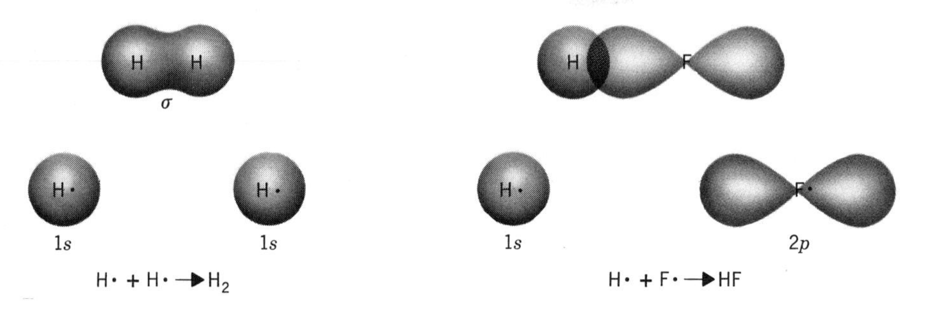

Figure 3-9. *Molecular orbitals may be thought of as arising from the overlap of atomic orbitals. The orbitals in H_2 and HF are called sigma (σ) because they are cylindrically symmetrical along the axis of the bond. Fluorine contains eight other electrons in filled* s *and* p *atomic orbitals.*

and, at least approximately, the structure of water, since oxygen has two orbitals, p_y and p_z, capable of forming a bond with hydrogen atoms. Similarly, the formula and structure of ammonia, NH_3, follows naturally from the electronic configuration of the nitrogen atom.

Bonding in Methane

It is this relatively simple picture of the construction of molecular orbitals from *s* and *p* atomic orbitals that we wish to apply to organic molecules, but we already run into problems with the simplest example, methane. If we look at the electronic configuration of the carbon atom (Fig. 3-6), we see that it has one electron each in its $2p_x$ and $2p_y$ orbitals; we might expect it to form a bond to hydrogen with each of these to give CH_2 with a 90° bond angle, rather far from the CH_4 with 109°28′ bond angles that actually forms. Obviously, some modification is necessary.

Instead of picturing methane as arising from a carbon atom in its lowest energy state, we picture first adding energy to the atom to excite one of the $2s$ electrons to a $2p$ orbital (Fig. 3-10). The considerable energy required to do this is called the *promotion energy*. Once this is done, we have a carbon atom with four valence electrons, each in its own orbital and each able to combine with a hydrogen atom to form a C—H bond. The additional energy gained by forming four C—H bonds instead of only two more than compensates for the promotion energy so that the formation of CH_4 is more favorable than the formation of CH_2.

Hybridization

The electronic distribution within a carbon atom with one of its electrons "promoted" is spherical, with one electron in the spherical $2s$ orbital and one each in the $2p_x$, $2p_y$, and $2p_z$ orbital. We must remember that these *s* and *p* orbitals are mathematical devices, and that dividing a spherical electron density into an *s* and three *p* orbitals is not the only possible way the division can be made. To describe the bonding in methane, we divide the spherical electron density into four equal parts, each pointing at the corners of a tetrahedron. In terms of our original *s* and *p* orbitals, each of these new *hybrid* orbitals will contain one-quarter of the $2s$ orbital and one quarter each of the $2p_x$, $2p_y$, and $2p_z$ orbital. To indicate this relationship, we call them sp^3 *hybrids* to show that they contain one part *s* to three parts *p*.

Bonding in Ethylene

In ethylene each carbon is surrounded by three other atoms in the same plane. Let us imagine this to be the plane containing the *x* and *y* axes of Fig. 3-8. We see from that figure that the $2p_z$ orbital has zero density in that plane; it cannot contribute to a hybrid orbital in the *xy* plane. The other three

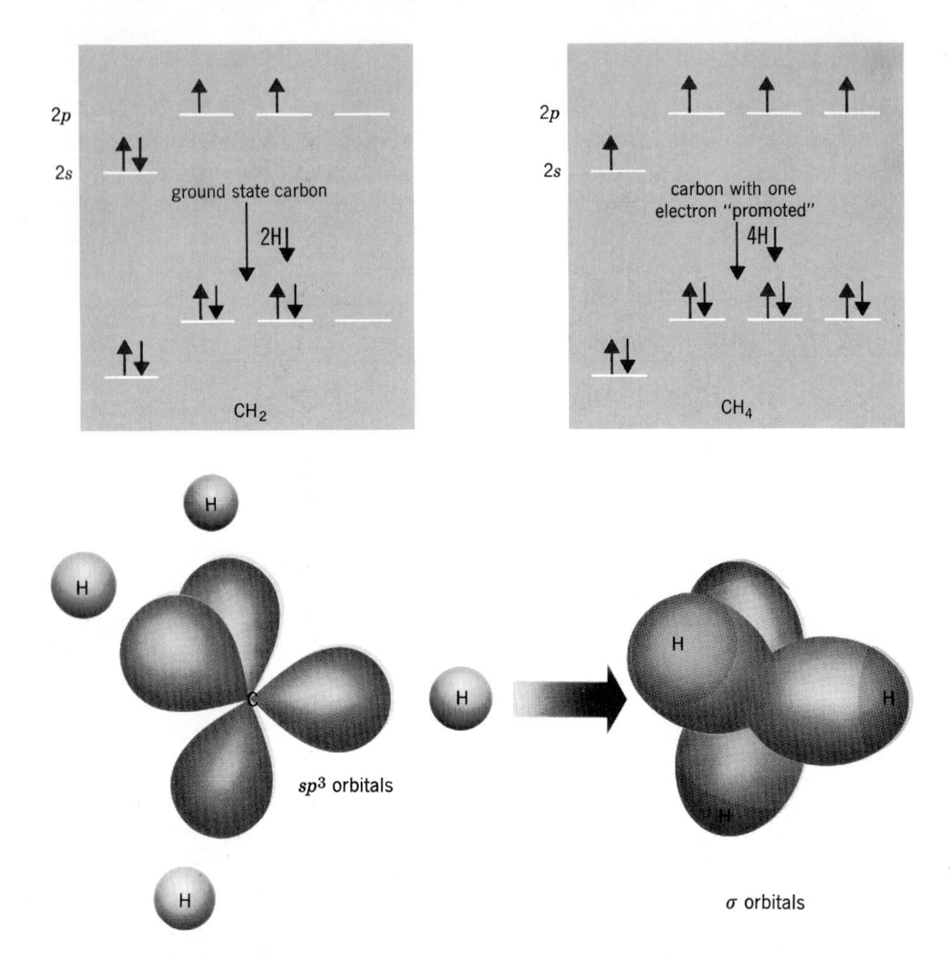

Figure 3-10. *To account for the tetrahedral bonding in methane, the carbon atom is said to "promote" one electron so that it may form four bonds instead of two. The resulting "promoted" atom has a spherical electron distribution. If this were divided into four equal parts each resulting orbital would be* ¼ s *and* ¾ p, *labeled* sp^3; *these* hybrid orbitals *point at the corners of a tetrahedron.*

orbitals, $2s$, $2p_x$, and $2p_y$, can. From these three orbitals three new hybrid orbitals may be constructed, called sp^2, each ⅓ *s* and ⅔ *p* in character. In addition to these hybrid orbitals, we have an unchanged $2p_z$ orbital. This is represented in Fig. 3-11.

Two such *trigonally hybridized* carbon atoms can form a strong *sigma bond* through end-on overlap of one of their sp^2 hybrid orbitals. The other two hybrid orbitals can form sigma bonds to hydrogen. Each carbon retains a p_z orbital. These two p_z orbitals may overlap side to side to form the second

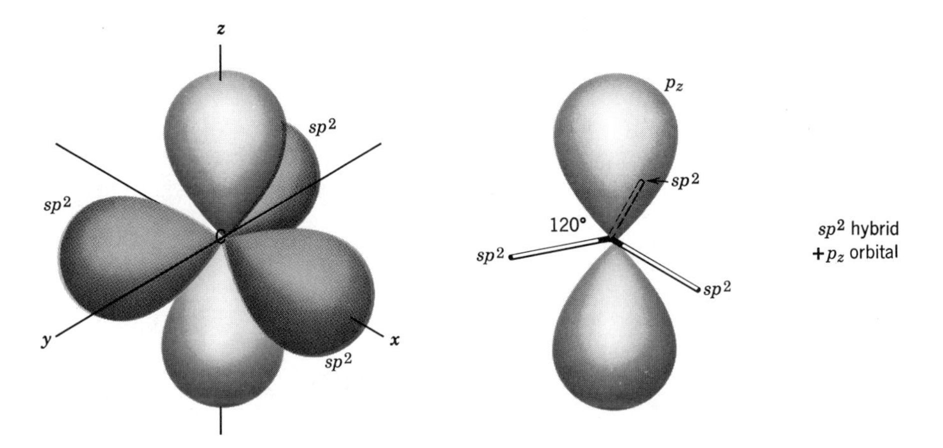

Figure 3-11. *Each carbon in ethylene is considered to be* trigonally hybridized, *i.e., to have three* sp^2 *hybrid orbitals pointing to the corners of a triangle. This leaves the* p_z *orbital unhybridized.*

of the two bonds of the carbon–carbon double bond in ethylene. Such a bond is called a pi (π) bond (Fig. 3-12).

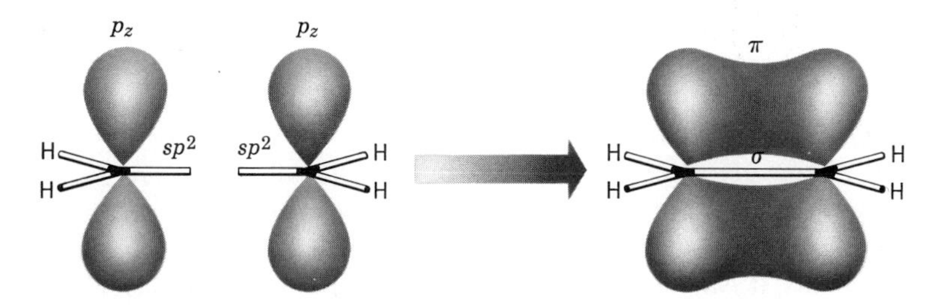

Figure 3-12. *In ethylene, one bond of the carbon–carbon double bond is a* σ *bond formed by the end-on overlap of* sp^2 *orbitals and the other is a* π *bond formed by the edgewise overlap of* p_z *orbitals.*

This bonding picture for ethylene has the advantage of making certain important properties of alkenes easily understood. First, the greater reactivity of alkenes as compared with alkanes can be ascribed to the pair of electrons in the π orbital. These electrons, which are relatively far from the nuclei, have no counterpart in the structure of alkanes. Second, the concept of a π bond also accounts for the existence of *cis* and *trans* isomers among alkenes. Notice in Fig. 3-13, that overlap of two *p* orbitals to form a π bond can occur only if the orbitals are parallel. Rotating one CH_2 group 90° completely destroys this overlap. Alkenes resist this rotation, and *cis* and *trans* forms have

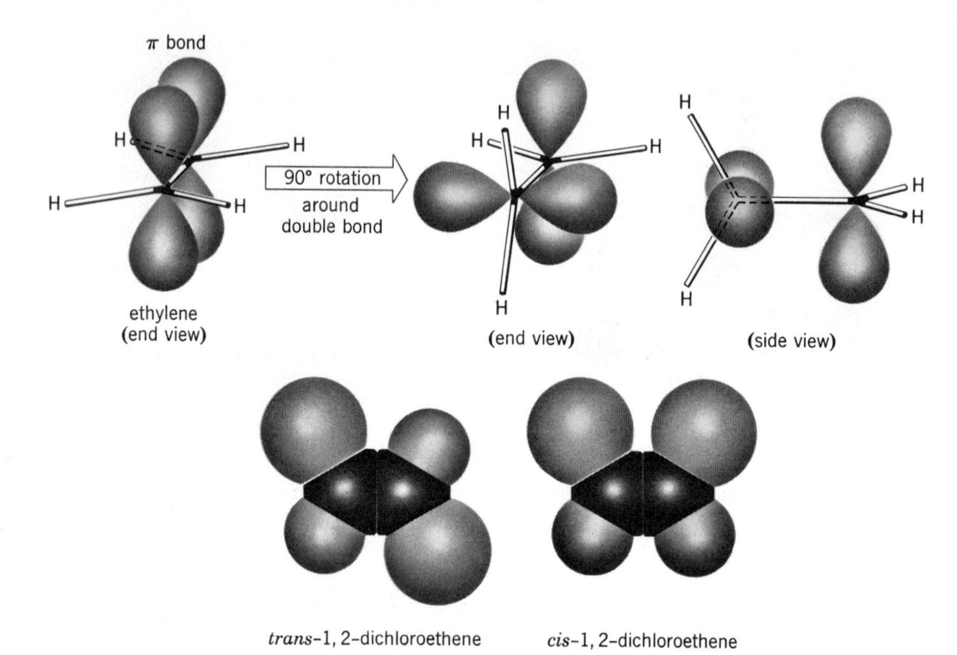

Figure 3-13. *The π bond is destroyed if the two* p *orbitals are twisted about the C—C axis. As a consequence, there is no freedom of rotation about double bonds, and* cis- *and* trans-*isomers have separate existence.*

separate existences. It is possible, by using sufficient energy (for instance, by heating) to convert a *cis* form to a *trans* form, or vice versa. A great deal of energy is required, but not so much as would be needed to break the whole molecule apart; only the π bond need be broken. From this and other types of experiments, it can be shown that a carbon–carbon π bond is weaker than a carbon–carbon σ bond.

Bonding in Acetylene

In acetylene we have two carbon atoms linearly bonded to two other atoms. Suppose these atoms are placed along the x axis in Fig. 3-8. Only the $2s$ orbital and the $2p_x$ orbital have electron density along this axis; $2p_y$ and $2p_z$ are both zero there. From the $2s$ and $2p_x$ orbitals two *sp hybrid orbitals* may be constructed, leaving the $2p_y$ and $2p_z$ orbitals unchanged. Such a *linearly hybridized* carbon atom is shown in Fig. 3-14.

Two *sp* hybridized carbons can be bonded end-on through an *sp* hybrid orbital, and each in turn can form a σ bond with a hydrogen. In addition

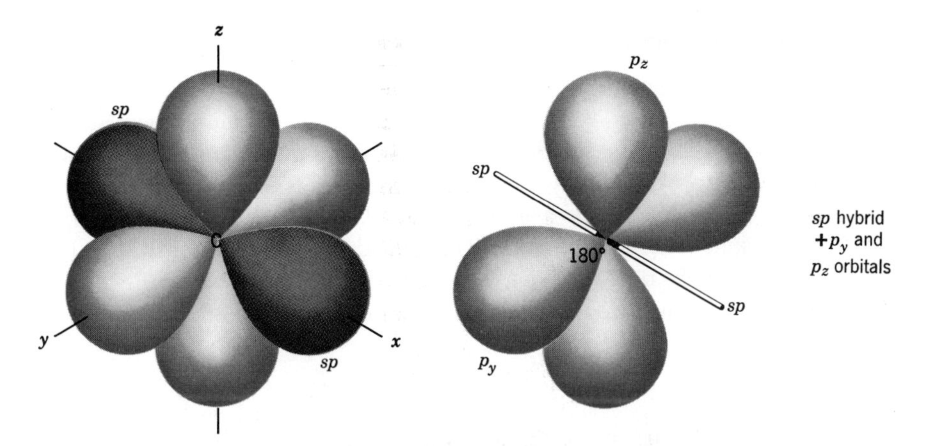

Figure 3-14. *In acetylene each carbon is* linearly hybridized, sp. *The* p_y *and* p_z *orbitals are not used in making this hybrid.*

Two *sp* hybridized carbons can be bonded end-on through an *sp* hybrid orbital, and each in turn can form a σ bond with a hydrogen. In addition two π bonds can form from the overlap of the two perpendicular unhybridized *p* orbitals (Fig. 3-15). In this picture the carbon–carbon triple bond in an alkyne is composed of two π bonds from the overlap of the p_y and p_z orbitals plus one σ bond from the overlap of *sp* hybrid orbitals.

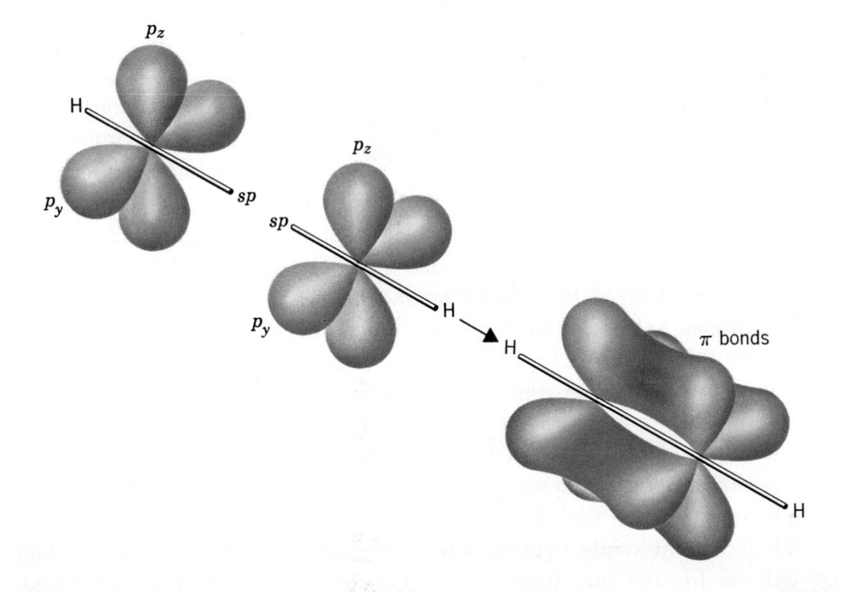

Figure 3-15. *In acetylene the triple bond is formed by the end-on overlap of* sp *hybrid orbitals to form a σ bond plus the edgewise overlap of both* p_y *and* p_z *orbitals to form two π bonds.*

Reactions of Alkenes and Alkynes

The most characteristic way in which alkenes and alkynes react is by *addition* to their multiple bonds. We have already seen that hydrogen, in the presence of a catalyst, will add to a double or a triple bond, yielding a saturated hydrocarbon. More commonly, the attaching reagent is an *acid*, either a proton acid like HCl or H_2SO_4, or a more general *Lewis acid*, in the sense of an electron-seeking reagent like Br_2. These acids add across the π bond of ethylene extremely rapidly, even in the absence of any catalyst (Fig. 3-16). By contrast, there is little tendency for a double or triple bond to react with a *base* like NaOH.

$$CH_2{=}CH_2 + HCl \longrightarrow \underset{\text{chloroethane}}{H{-}CH_2{-}CH_2{-}Cl}$$

$$CH_2{=}CH_2 + \underset{(H_2SO_4)}{HOSO_3H} \longrightarrow \underset{\text{ethyl sulfate}}{H{-}CH_2{-}CH_2{-}OSO_3H}$$

$$CH_2{=}CH_2 + Br{-}Br \longrightarrow \underset{\text{1,2-dibromoethane}}{Br{-}CH_2{-}CH_2{-}Br}$$

$$CH_2{=}CH_2 + NaOH \longrightarrow \text{No Reaction}$$

Figure 3-16. *Three addition reactions of ethylene. Basic reagents like sodium hydroxide do not usually react with double bonds.*

The addition of bromine to propylene occurs in an exactly analogous way, producing 1,2-dibromopropane, but two different products *could* form in the addition of an unsymmetrical reagent like HCl to an unsymmetrical double bond like the one in propylene. These two possible products are shown in Fig. 3-17. Actually, only one of the two products forms, that predicted by Markovnikov's rule, named after an eminent Russian organic chemist of the nineteenth century. The rule says that *the positive part of the reagent adds to that carbon atom of the double bond which has the greater number of hydrogen atoms.* As applied to the addition of hydrogen chloride to propylene, the positive part of the reagent (H^+) adds to the carbon atom of the double bond having two hydrogen atoms (C-1). The negative part of the reagent (Cl^-) adds to the other carbon. (Note that we count only the hydrogens on the carbons of the *double bond.* Of course, the methyl carbon has the most hydrogens of any carbon in propylene, but no addition can take place to the $-CH_3$ group, for this carbon is already "saturated.")

Markovnikov's rule is extremely important, for it tells us the product that will be formed any time an unsymmetrical acid [X^+Y^-] adds to an unsymmetrical alkene. Some additional examples of the application of this rule are given in Fig. 3-18. Consider the reaction of HBr with 2-methyl-2-

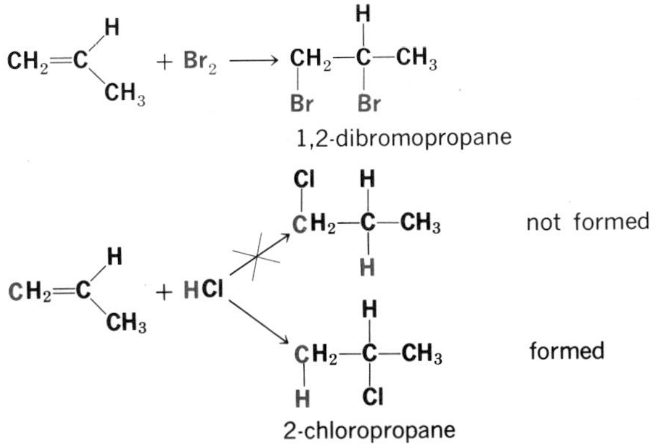

$CH_2{=}C(H)CH_3 + Br_2 \longrightarrow CH_2Br{-}CHBr{-}CH_3$

1,2-dibromopropane

$CH_2{=}C(H)CH_3 + HCl \longrightarrow$ $CH_2Cl{-}CH_2{-}CH_3$ not formed

$CH_2{=}C(H)CH_3 + HCl \longrightarrow$ $CH_3{-}CHCl{-}CH_3$ formed

2-chloropropane

Figure 3-17. *Addition of bromine to the double bond of propylene can occur in only one way, to give the dibromo compound shown. Addition of HCl could occur in two directions; in fact the positive part of the reagent (in red) adds only to carbon 1, which already has two hydrogens, rather than to carbon 2, which already has only one hydrogen.*

butene. We must first recognize that the only reactive site in the molecule is at the unsaturated carbon–carbon bond. Of the two carbons that are joined by this double bond, one is directly attached to a hydrogen atom (C-3); the other (C-2) is attached only to other carbons. When HBr adds to this double bond, the H^+ will become attached to C-3, the Cl^- to C-2. If both carbons of a double bond have the same number of hydrogens directly attached, then addition will occur equally well in either direction.

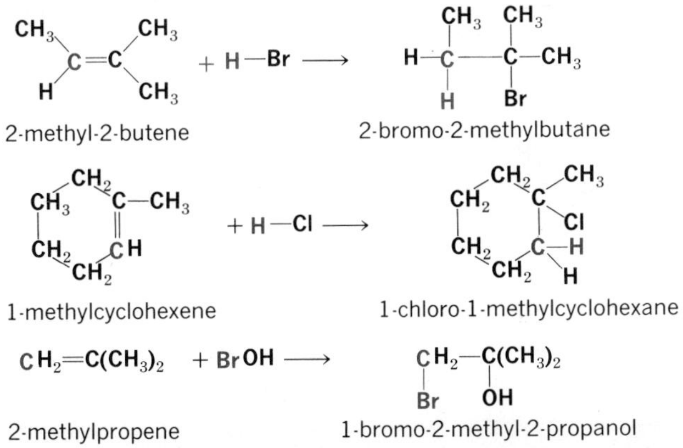

$(CH_3)HC{=}C(CH_3)_2 + H{-}Br \longrightarrow CH_3CH_2{-}CBr(CH_3){-}CH_3$

2-methyl-2-butene → 2-bromo-2-methylbutane

1-methylcyclohexene $+ H{-}Cl \longrightarrow$ 1-chloro-1-methylcyclohexane

$CH_2{=}C(CH_3)_2 + BrOH \longrightarrow CH_2Br{-}C(OH)(CH_3)_2$

2-methylpropene → 1-bromo-2-methyl-2-propanol

Figure 3-18. *Three more examples of the application of Markovnikov's rule. In the hypohalous acids (HOBr, HOCl, HOI) the halogen is the positive part of the reagent.*

Prob. 3-3. Use Markovnikov's rule to predict the product expected to be formed by the addition of HCl to 1-pentene, 2-methyl-2-pentene, 3-methyl-cyclopentene. The addition of hypochlorous acid (HOCl) to each of these compounds.

Markovnikov's rule also applies to the addition of acidic reagents across triple bonds. In the first addition an alkene forms, which may then react further to yield a saturated product (Fig. 3-19).

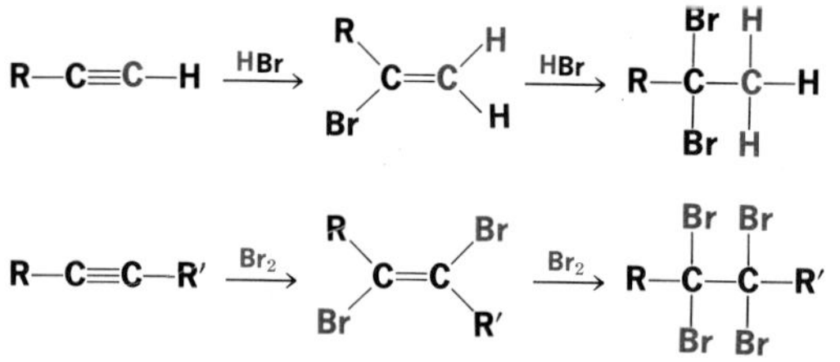

$$R{-}C{\equiv}C{-}H \xrightarrow{HBr} (R)(Br)C{=}C(H)(H) \xrightarrow{HBr} R{-}CBr_2{-}CH_2{-}H$$

$$R{-}C{\equiv}C{-}R' \xrightarrow{Br_2} (R)(Br)C{=}C(Br)(R') \xrightarrow{Br_2} R{-}CBr_2{-}CBr_2{-}R'$$

Figure 3-19. *Acidic reagents add to alkynes, obeying Markovnikov's rule.*

Mechanism of Addition to a Double Bond

Addition of an acid to a double bond occurs in steps, with the positive part of the reagent adding first, followed by addition of the negative part. A very short-lived *carbon cation* is formed as an intermediate (see Fig. 3-20). Although it will ordinarily exist for only a very short time before reacting with an anion to form a neutral product, these carbon cations (sometimes called *carbonium ions*), nevertheless, have a real existence; under special circumstances some

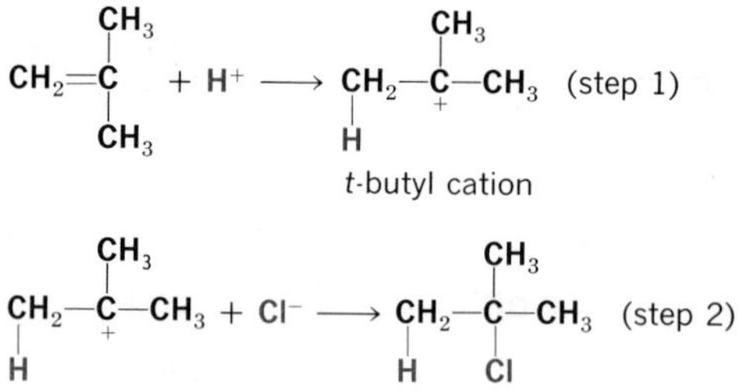

$$CH_2{=}C(CH_3)_2 + H^+ \longrightarrow H{-}CH_2{-}\overset{+}{C}(CH_3){-}CH_3 \quad \text{(step 1)}$$

t-butyl cation

$$H{-}CH_2{-}\overset{+}{C}(CH_3){-}CH_3 + Cl^- \longrightarrow H{-}CH_2{-}C(CH_3)(Cl){-}CH_3 \quad \text{(step 2)}$$

Figure 3-20. *Addition of an acid to a double bond occurs in two steps. In the first step the positive part of the reagent adds, forming a carbon cation. This highly reactive, short-lived intermediate quickly reacts with the anion to form a neutral molecule.*

carbonium ion salts can even be isolated as high-melting crystalline solids, similar in many ways to typical inorganic salts.

The role of carbon cations as intermediates in addition reactions may become clearer if we examine the reactions of Fig. 3-20 in more detail. First, we should emphasize the *Lewis* definition of an acid as an *electron-seeking reagent.* According to this definition, H^+ is an acid because it seeks an electron pair to complete its first shell of electrons. A base, in the Lewis sense, is an *electron-donating reagent.* Because a carbon–carbon unsaturated bond is rich in electrons, it is not surprising that an electron-seeking acid should react with it; in fact step 1 of the addition is a Lewis acid–Lewis base reaction, with the π electrons of the double bond playing the role of the Lewis base, the proton that of the Lewis acid. Since the electrons of the π bond are rather loosely held, those are most readily attacked. The result is that the electron pair of the π bond becomes converted to a C—H sigma bond, leaving C-2 attached to only three groups and bearing a positive charge. These changes are shown pictorially in Fig. 3-21.

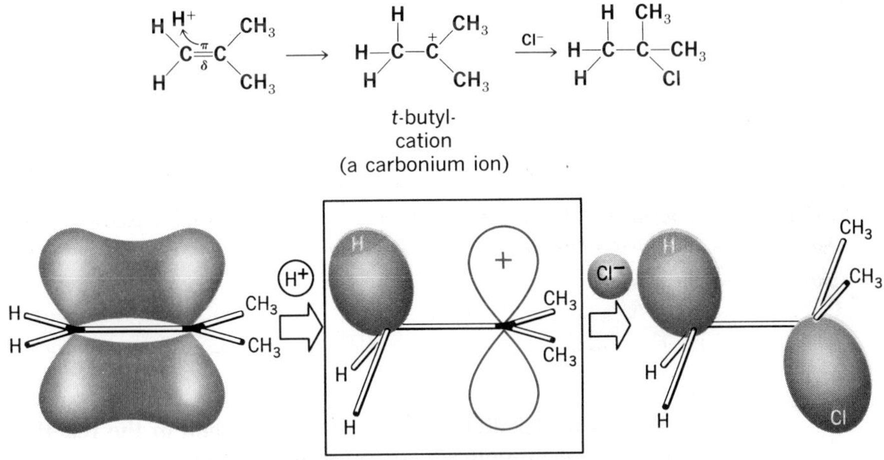

Figure 3-21. *An orbital representation of the reaction of hydrogen chloride and isobutylene. Filled orbitals are shaded, empty orbitals are unshaded.*

The carbonium ion formed as a product of step 1 is itself a Lewis acid, seeking to complete the octet around carbon. In fact, a carbon cation is one of the strongest Lewis acids known, for carbon compounds show very little tendency to exist in ionic form. The *t*-butyl cation immediately begins to search for a pair of electrons and it finds them in the chloride ion. This second step may also be thought of as an acid-base reaction, with the carbon cation as the acid (electron-seeking) and the chloride ion as the base (electron-donating).

Addition to a double bond is then a combination of two acid-base reactions.

Addition of bromine to a double or triple bond proceeds in exactly the same stepwise fashion. The bromine molecule may be thought of as Br^+Br^-, with one of the bromines as the positive, acidic part of the reagent. The reaction of Br_2 with a double bond again gives an intermediate carbonium ion, which quickly combines with bromide ion to form a neutral molecule (Fig. 3-22).

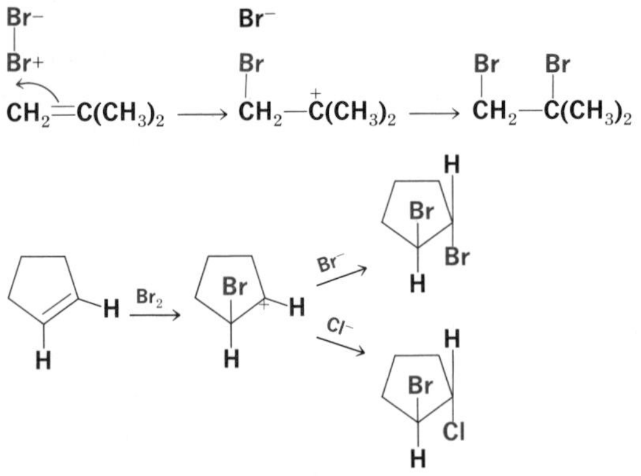

$$\text{Br}^-\text{—Br}^+ + CH_2{=}C(CH_3)_2 \longrightarrow \underset{\text{Br}}{CH_2}\text{—}\overset{+}{C}(CH_3)_2 \;\; \text{Br}^- \longrightarrow \underset{\text{Br}}{CH_2}\text{—}\underset{\text{Br}}{C}(CH_3)_2$$

Cyclopentene $\xrightarrow{Br_2}$ bromocyclopentyl carbonium ion; $\xrightarrow{Br^-}$ trans-1,2-dibromocyclopentane; $\xrightarrow{Cl^-}$ trans-1-bromo-2-chlorocyclopentane

Figure **3-22.** *Addition of halogen to a double bond also proceeds in two steps, by way of an intermediate carbonium ion. Some of the evidence for this intermediate is that addition occurs to give a* trans *product, and that in the presence of sodium chloride some bromochloro product is formed. Neither result would be expected if the two bromines were to add simultaneously.*

Prob. 3-4. Why is no 1,2-dichlorocyclopentane formed in the reaction of cyclopentene with bromine in the presence of sodium chloride?

Prob. 3-5. If propylene were treated with bromine in the presence of sodium chloride, would you expect to get 1-bromo-2-chloropropane or 1-chloro-2-bromopropane? Why?

Once we recognize that addition to a double bond is a two-step reaction, and that carbonium ions are intermediates, it becomes easier to understand Markovnikov's rule. We see that the initial addition of the positive part of the reagent determines the direction of addition, for once this occurs, the position of the anion follows automatically. Consider the two positions to which a proton might add to isobutylene (Fig. 3-23). Two different carbon

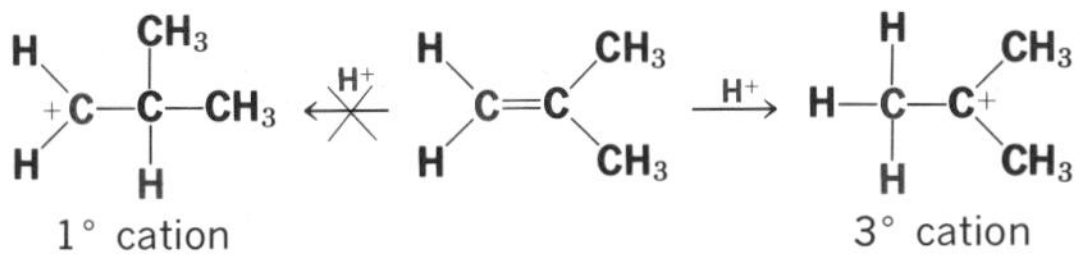

$$H_2\overset{+}{C}-CH(CH_3)-CH_3 \xleftarrow{H^+} (H)_2C{=}C(CH_3)_2 \xrightarrow{H^+} H-CH_2-\overset{+}{C}(CH_3)_2$$

1° cation 3° cation

Figure 3-23. *An acidic reagent adds to a double bond so as to form the less unstable of the two possible carbon cations. In a 3° cation the attached alkyl groups help disperse the positive charge, while in a 1° cation the attached hydrogens cannot do this.*

cation intermediates would be formed. The one actually formed, that one in accord with Markovnikov's rule, is a *tertiary cation*, i.e., has the positive charge on a carbon attached to three other carbons. The cation that would be formed by addition in the opposite, non-Markovnikov direction, would be a *primary cation*. All available experimental evidence indicates that although all simple carbon cations are highly reactive species, *tertiary (3°) cations are much more stable and easier to form than secondary (2°) cations, which in turn are more stable and easier to form than primary (1°) cations.* It is therefore much easier for the proton to add to C-1 of isobutylene to form the 3° cation than it is to add to C-2 to form the still higher energy 1° cation.

The concept of the relative stability of carbon cations can be a confusing one, and its importance merits further emphasis. Both primary and tertiary carbon cations are of high energy and are extremely reactive species. When we say a 3° cation is "more stable" than a 1° one, we are only speaking in relative terms, in the same sense that we might say that dynamite is more stable than nitroglycerine. The energy relationships that underlie Markovnikov's rule are often expressed graphically by means of an *energy diagram* (Fig. 3-24). From this diagram we can see that a 3° cation is of lower energy than a 1° cation (even though both are of high energy relative to neutral molecules) and so less energy is required to form a 3° than a 1° cation, starting from the alkene. Nevertheless, once formed, even a 3° cation reacts very quickly to form products.

Why should a 3° cation be more stable than a 1° cation? Because the methyl groups that surround a 3° cation have electrons (those of their C—H bonds) which they can share, in part, with the highly electron-deficient cation. To some extent the positive charge in a 3° cation is dispersed or *delocalized*, i.e., spread out over a larger volume; the delocalized charge is lower in energy

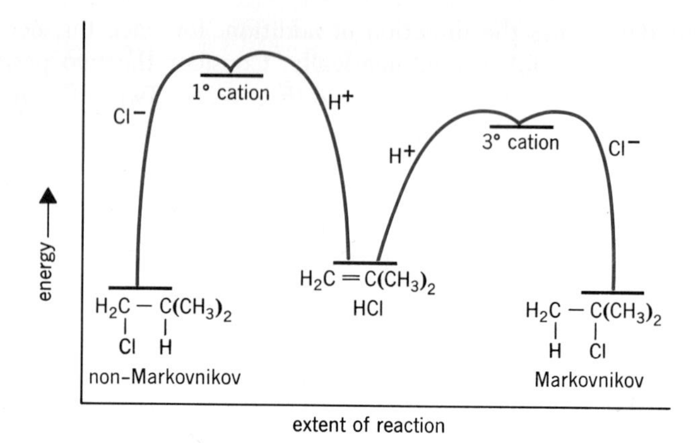

Figure 3-24. An energy diagram *which shows that less input of energy is required to form the Markovnikov addition product by way of a 3° cation than the non-Markovnikov addition product by way of a 1° cation. The products themselves do not differ appreciably in energy, but the intermediates through which they are formed do.*

than the concentrated or *localized* charge. In the *t*-butyl cation, for example, each of the three methyl groups can share some of the electron density of their C—H bonds with the central carbon, and so take on some of the positive charge. The two hydrogen atoms attached to a primary cation, by contrast, have no electrons to share and are unable to disperse charge from a neighboring carbon cation. The charge on a 1° cation is more localized, and hence of higher energy, than the charge on a 3° cation. A 2° cation, with two alkyl groups and one hydrogen, is of intermediate energy. Indeed, stable salts of 1° cations have never been observed, and those of 2° cations are exceedingly difficult to obtain compared to those of 3° cations. A great deal of experimental evidence confirms that the attachment of alkyl groups stabilizes adjacent carbon cations, and so their order of stability is $3° > 2° > 1°$.

It should be emphasized that Markovnikov's rule applies only to simple alkynes and alkenes, in which the double bond is substituted only with hydrogens and alkyl groups. If we attach atoms other than carbon or hydrogen, Markovnikov's rule may no longer apply. Perhaps an example will serve to emphasize the role of positively charged carbon intermediates in these additions. If we add HCl to ethylene substituted with a derivative of the ammonium ion, as shown in Figure 3-25, addition occurs in the opposite direction from that predicted by Markovnikov's rule. The reason for this becomes obvious if we draw the structure of the carbonium ions that would be formed by addition of a proton to C-1 and C-2. Addition of a proton in the direction predicted by Markovnikov's rule would generate a carbon cation adjacent to the already positively charged nitrogen atom, and these two charges would

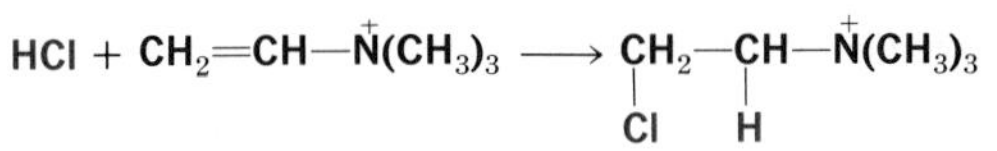

$$HCl + CH_2{=}CH{-}\overset{+}{N}(CH_3)_3 \longrightarrow \underset{\underset{Cl}{|}}{CH_2}{-}\underset{\underset{H}{|}}{CH}{-}\overset{+}{N}(CH_3)_3$$

Figure 3-25. *Addition of HCl to the substituted ethylene shown proceeds in the opposite direction to that predicted by Markovnikov's rule, since the positively charged nitrogen atom, in contrast to an alkyl group, repels an adjacent positive charge. This result emphasizes that Markovnikov's rule applies to double and triple bonds substituted only by alkyl groups.*

repel each other fiercely. Addition in the opposite direction, to give the anti-Markovnikov product, proceeds by way of an intermediate in which the two positive charges are farther apart, so that they repel one another less. This direction is favored, and anti-Markovnikov addition is observed. When a double bond is substituted by groups other than alkyl, their effect on an adjacent positive charge (either attractive, as halogen groups, or repulsive, as the positive nitrogen atom in Fig. 3-25) must be examined individually before the direction of addition can be predicted. We shall see how to make these predictions in future chapters.

Prob. 3-6. Draw the structures of the intermediate cations that would be formed by addition of H^+ in the two possible directions in Fig. 3-25.

Molecules with Two or More Double Bonds

Many molecules contain two or more double bonds and are accordingly called *dienes, trienes,* etc. Some examples of their nomenclature are given in Fig. 3-26; other examples are given among the problems at the end of the chapter. The longest chain containing as many double bonds as possible is chosen as the parent, with numbering begun at the end that gives the double bonds their smaller numbers. If the double bonds are separated by at least one

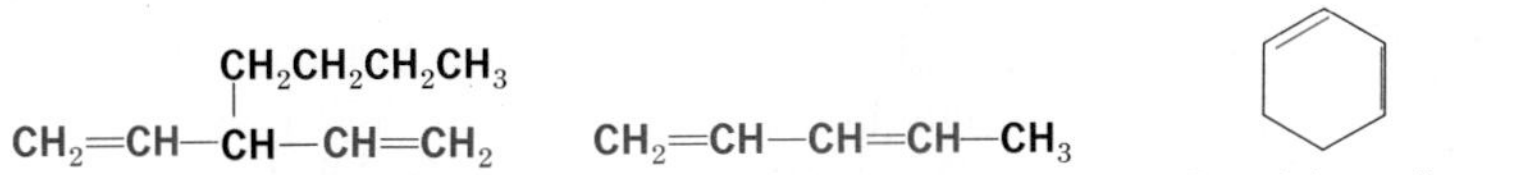

$$CH_2{=}CH{-}\overset{\overset{CH_2CH_2CH_2CH_3}{|}}{CH}{-}CH{=}CH_2$$

3-*n*-butyl-1,4-pentadiene
(unconjugated)

$$CH_2{=}CH{-}CH{=}CH{-}CH_3$$

1,3-pentadiene
(conjugated)

1,3-cyclohexadiene
(conjugated)

Figure 3-26. *If double bonds are separated by one or more saturated carbon atoms, they are said to be* unconjugated. *If they are adjacent, in the sense that they are separated by only one single bond, they are said to be* conjugated.

saturated carbon atom, as in 1,4-pentadiene, the two have little effect on one another and react independently. Bonds of this type are said to be *unconjugated double bonds* (Fig. 3-26). Each double bond in an unconjugated diene reacts exactly as it would if the molecule contained only a single double bond.

If the double bonds are adjacent, as in 1,3-pentadiene, they are said to be *conjugated*, and conjugated double bonds can sometimes modify the course of alkene reactions.[1] As an example, consider the addition of one molar equivalent of HCl to 1,3-butadiene; not only the expected product from Markovnikov addition across the 1,2 double bond is formed, but also an unexpected product of *1,4 addition*, in which the HCl adds across the ends of the conjugated double bonds and the remaining double bond moves to the middle (Fig. 3-27). Because systems of conjugated double bonds play such an important role in the chemistry of benzene, other aromatic hydrocarbons (Chapter 5), synthetic rubbers (Chapter 4), and biosynthesis (Chapter 16), we must examine this reaction more closely.

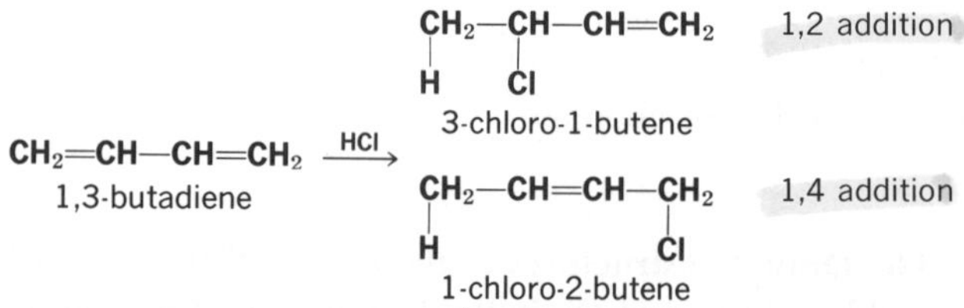

$CH_2{=}CH{-}CH{=}CH_2$ (1,3-butadiene) $\xrightarrow{HCl}$

- $CH_2(H){-}CH(Cl){-}CH{=}CH_2$ — 3-chloro-1-butene — 1,2 addition
- $CH_2(H){-}CH{=}CH{-}CH_2(Cl)$ — 1-chloro-2-butene — 1,4 addition

Figure 3-27. *Addition to conjugated double bonds may take place simultaneously in the normal manner and at the ends of the system.*

As with a simple double bond, addition of HCl to a conjugated diene proceeds in steps, by way of an intermediate carbon cation (Fig. 3-28), but now the cation is formed adjacent to a carbon–carbon double bond. Since a double bond is electron-rich, it can share its electrons with the adjacent electron-deficient carbon cation to an even greater extent than an alkyl group can. This electron-sharing is easy to see if we draw the molecular orbital (MO) picture for butadiene. As the proton uses the electron pair of one double bond to form a new C—H bond, electrons from the other π orbital flow into the now emptying p orbital on C-2, so that in the intermediate cation a two-electron π orbital is shared by three carbons and the positive charge is not localized on C-2 but is delocalized. We can obtain the same result just by using normal double and single bonds (valence-bond (VB) picture). We note that a second structure can be written for the intermediate cation just by moving an electron pair from the C-3,C-4 double bond. No movement of atoms is involved here, just of electrons. These two structures for the cation clearly

[1] Double bonds can be even closer together, as they are in allene $CH_2{=}C{=}CH_2$. Such *cumulative* double bonds are rare, and their reactions will not be discussed.

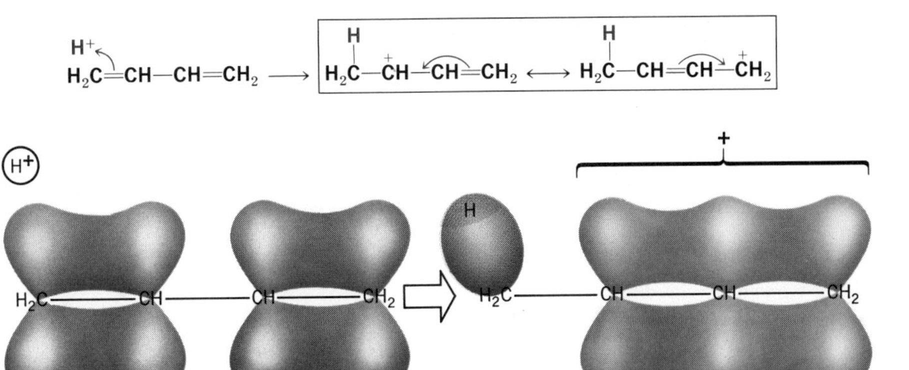

Figure 3-28. *Addition of an acid to the end of a conjugated diene gives an intermediate cation in which the charge is shared by both the C-2 and C-4–carbon atom, here shown in both a* valence-bond *(VB) and* molecular-orbital *(MO) picture. The VB picture shows clearly how only C-2 and C-4 (and not C-3) share the charge, while the MO picture makes the dispersal of charge by overlap of π orbitals very obvious.*

indicate that the positive charge is not localized on C-2 but is delocalized. Since C-2 and C-4 share the positive charge, it is easy to understand why the chloride ion might combine with either of them to form the two different products actually isolated in their reaction. Additional examples of this type of *charge delocalization* will be discussed in Chapter 5.

Prob. 3-7. What three products might you expect to obtain by the addition of one equivalent of hydrogen chloride to 1,3,5-hexatriene (CH_2=CH—CH=CH—CH=CH_2)? Draw both VB and MO structures for the intermediate cation.

Naturally Occurring Alkenes and Alkynes

Molecules that contain double and triple bonds occur very widely in nature. The simplest alkene, ethylene, is itself a plant hormone, and controls the ripening and color development of fruit. Farmers spray orchards with ethylene to control the ripening of fruit. Many plants and trees contain oily mixtures of substances known as *terpenes,* which are to a large extent unsaturated hydrocarbons (Fig. 3-29). *Turpentine* is a mixture of such oils from various trees; among the hydrocarbons it contains is *pinene.* Lemon oil contains the isomeric diene *limonene.* Vitamin A (*retinol*), which occurs in animals but not in plants, may be extracted from fish liver oils. A large number of acetylene derivatives also occur naturally.

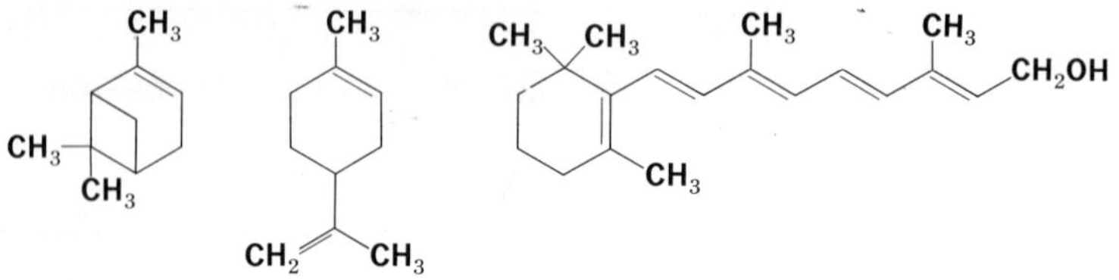

pinene limonene vitamin A

Figure 3-29. *A few examples of naturally occurring unsaturated compounds.*

Visual Chemical Tests for Alkenes and Alkynes

In many instances it is useful to have some simple chemical reaction which, by the generation or disappearance of a color, the formation of a precipitate, or other visual indication, will reveal the presence of a particular functional group in an organic molecule. Suppose, for example, you wanted to know if a clear liquid you found in an unlabeled bottle were hexane, 1-hexene, or 1-hexyne. Since all three have similar physical properties, this would be difficult to do simply by measuring the boiling points, for example. Chemically, it is less difficult. A drop of the hydrocarbon is dissolved in an inert solvent (CCl_4). Next a deep brown solution of bromine in CCl_4 is added dropwise. If the hydrocarbon is hexane, no reaction will occur and the hydrocarbon solution will immediately take on the brown color of the added bromine. Both hexene and hexyne react immediately with bromine, and *decolorize* it, because the addition products, dibromohexane and tetrabromohexane, are colorless. If the unknown hydrocarbon is unsaturated, as each drop of the brown bromine solution is added its color will fade and the hydrocarbon solution will remain clear. Eventually, when enough bromine has been added to saturate all double and/or triple bonds present, the brown color will remain. If the concentration of bromine in the initial solution is known, and if the addition is made from a buret, the method can be used for the quantitative analysis of alkenes and alkynes. In any event, the *decolorization of a solution of bromine in* CCl_4 *is a useful visual test for unsaturation* (double or triple bonds).

Since both 1-hexene and 1-hexyne decolorize bromine, a second visual test is necessary to distinguish them from one another. If an alkyne has a hydrogen atom attached to the triple bond, as it does in 1-hexyne or in any other *terminal acetylene* (one in which the triple bond is on the end of the hydrocarbon chain), it will form an insoluble precipitate when treated with certain metal ions, particularly Ag^+ in basic solution (Fig. 3-30). The appearance of a precipitate of a *silver acetylide* may be used as a diagnostic test for the presence of a terminal acetylene. There is, unfortunately, no simple visual test by which a nonterminal acetylene may be distinguished from an alkene.

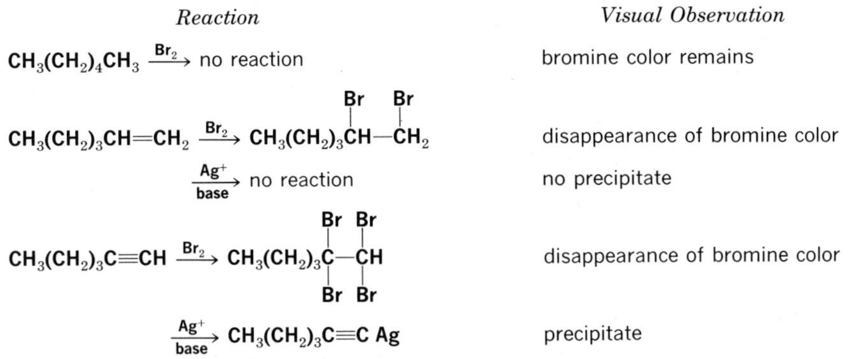

Reaction	*Visual Observation*
$CH_3(CH_2)_4CH_3 \xrightarrow{Br_2}$ no reaction	bromine color remains
$CH_3(CH_2)_3CH{=}CH_2 \xrightarrow{Br_2} CH_3(CH_2)_3CH(Br){-}CH_2(Br)$	disappearance of bromine color
$\xrightarrow[\text{base}]{Ag^+}$ no reaction	no precipitate
$CH_3(CH_2)_3C{\equiv}CH \xrightarrow{Br_2} CH_3(CH_2)_3CBr_2{-}CHBr_2$	disappearance of bromine color
$\xrightarrow[\text{base}]{Ag^+} CH_3(CH_2)_3C{\equiv}C\,Ag$	precipitate

Figure 3-30. *A series of visual chemical tests by which hexane, 1-hexene, and 1-hexyne may be distinguished.*

PROBLEMS

1. The IUPAC ending for a double bond is *ene* and for a triple bond is *yne*. Draw the structure of:
   (a) 2-butyne (b) 1-pentene
   (c) cyclohexene (d) 2-methyl-2-butene
   (e) 1-methylcyclopentene (f) cyclooctyne
   Give the IUPAC name for:
   (g) $CH_2{=}CH_2$ (h) $CH_3CH{=}CH_2$ (i) $HC{\equiv}CH$
   (j) 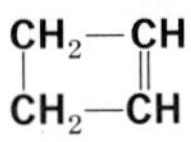 $CH_2{-}CH$ / $CH_2{-}CH$ (ring: $CH_2{-}CH{=}CH{-}CH_2$, four-membered) (k) $CH_3C{\equiv}CH$ (l) $(CH_3)_2CHC{\equiv}CH$
   (m) 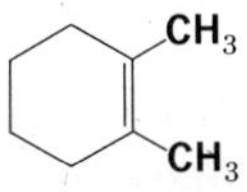 cyclohexene ring bearing CH_3 and CH_3 on the double-bond carbons (n) 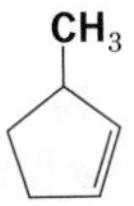 cyclopentene ring with CH_3

2. Always choose as the parent the longest chain containing the double or triple bond and number from the end which gives the first carbon of the functional group the smaller number.
   Draw the structure of:
   (a) 3-*n*-propyl-1-hexene (b) 2-isopropyl-1-pentene
   (c) 2,4-diethyl-1-octene (d) 1-methyl-5-ethylcyclohexene
   Name by the IUPAC system:
   (e) 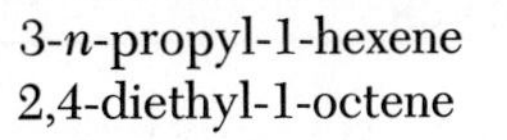 $CH_3CH_2CH(C{\equiv}CH)CH_2CH_2CH_2CH_3$ (f) 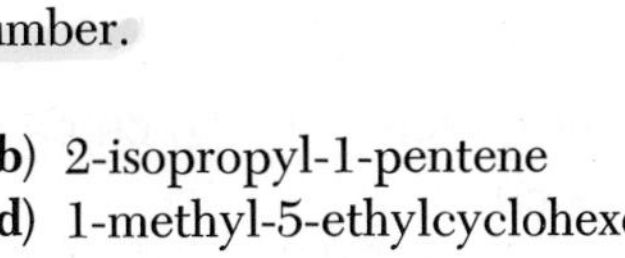 $CH_3CH(CH_2CH_3)CH_2{-}CH(CH_3){-}CH_2CH{=}CH_2$

(g)

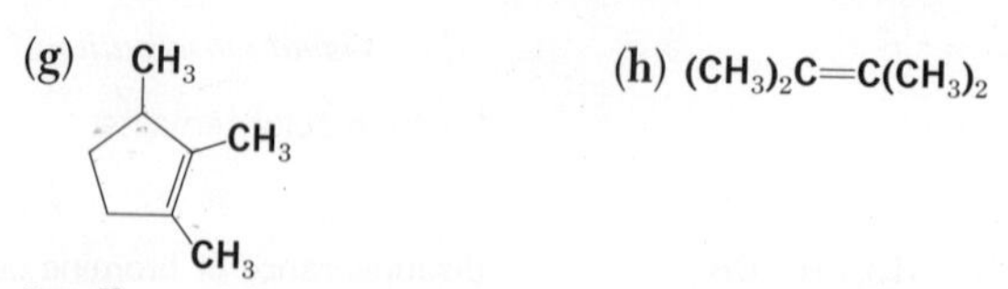

(h) $(CH_3)_2C{=}C(CH_3)_2$

3. In alkenes the substituents on a double bond may be arranged on the same side (*cis*) or on opposite sides (*trans*). Draw the structure of:
   (a) *cis*-3-hexene (b) *trans*-3-hexene
   (c) 2,3-dichloro-*cis*-2-butene (d) *cis*-1,2-diiodoethene
   (e) 4-ethyl-*cis*-2-octene (f) 2,4-dimethyl-*trans*-2-heptene
4. In compounds containing two functional groups, each must receive a number and the ending will be diene (two double bonds), triene (three double bonds), etc. Draw the structures for:
   (a) 1,3-butadiene (b) 2-methyl-2,4-hexadiene
   (c) 1,4-pentadiene (d) 3-ethyl-1,4,7-octatriene
   Give the IUPAC names for the following:
   (e) $CH_3CH{=}CHCH{=}CH_2$ (f) $CH_2{=}C(CH_2CH_3){-}CH(CH{=}CHCH_3)CH_2CH_3$
   (g) $HC{\equiv}CCH(CH_3)C{\equiv}CH$ (h) $HC{\equiv}CCH(C{\equiv}CH)CH_2CH_2CH_3$
   (i) (j) CH_3, CH_3
5. If a molecule contains both a double and a triple bond, both endings are used, the -ene preceding the -yne. Give the structures of:
   (a) 1-buten-3-yne (b) 1-cyclononen-5-yne
   (c) 2-methyl-3-ethyl-2-octen-4-yne
   The double bond receives the lower number in a molecule containing both a double and a triple bond. Name each of the following:
   (d) $CH_3CH{=}CHCH_2C{\equiv}CH$ (e) $CH_3CH(CH_3)C{\equiv}CCH{=}CHCH_3$
   (f) $HC{\equiv}CCH(CH_2CH_3)C(CH_2CH_3){=}CHCH_3$ (g) $CH_3CH_2CH{=}C(CH_3)C{\equiv}CH$
6. Give the product of the reaction of 1-butene with:
   (a) H_2/Pd (b) HCl (c) Br_2
   (d) HOCl (e) Cl_2 (f) O_2, match
   (g) H_2SO_4 (h) NaOH
7. What product would you expect for the addition of one molar equivalent

of each of the following reagents to 1-butyne? Of two molar equivalents?
(a) H_2/Pd (b) HCl (c) Br_2

8. Complete the following reactions:
   (a) $CH_3CH{=}C(CH_3)_2 + HCl \longrightarrow$
   (b) $CH_2{=}CH(CH_2)_3CH_3 + HOBr \longrightarrow$
   (c) 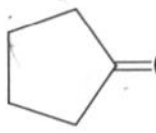 (cyclopentylidene)$=CH_2 + Br_2 \longrightarrow$
   (d) 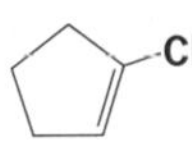 (cyclopentenyl)$-CH_3 + HCl \longrightarrow$
   (e) 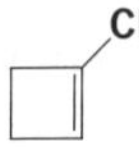 (cyclobutenyl)$CH_3 + Br_2 \longrightarrow$ Include stereochemistry of product.
   (f)  (cyclohexadiene) $+ H_2$ (excess) $\xrightarrow{Pd}$
   (g) $CH_3C{\equiv}CH + Ag^+ \xrightarrow{base}$
9. Which of the following compounds are conjugated and which unconjugated dienes? Draw the structure of a 1,4-addition product of bromine with each of the conjugated dienes. Neglect *cis* and *trans* isomerism.
   (a) 1,3-hexadiene (b) 1,4-hexadiene
   (c) 1,3-cyclohexadiene (d) 1,4-cyclohexadiene
10. Draw all three geometrical (*cis, trans*) isomers of 2,4-hexadiene.
11. Draw the structure of propene and label each bond from each carbon atom as *sp*, sp^2, sp^3. Repeat for propyne.
12. Describe simple visual chemical tests that would serve to distinguish among the following three compounds. Tell exactly what you would do and see or not see.
    $CH_3CH_2C{\equiv}CH$ $CH_3C{\equiv}CCH_3$ $CH_3CH_2CH_2CH_3$
13. Alkenes usually undergo addition reactions more readily than acetylenes. In the manufacture of vinyl chloride by addition of hydrogen chloride to acetylene, one worry is the further reaction of vinyl chloride with hydrogen chloride. In order to minimize this side reaction, should the reaction be carried out (a) with an excess of hydrogen chloride over acetylene, (b) with equal amounts of the two reagents, or (c) with an excess of acetylene?

4

addition polymers and free radicals

Under suitable conditions most alkenes that contain a terminal double bond can be induced to add together to form extremely long molecules known as *addition polymers*. As an example, in the presence of oxygen as a catalyst, and under high pressure and temperature, ethylene may be converted to *polyethylene*, in which 500 or more individual ethylene molecules form a straight-chain hydrocarbon whose average molecule will contain 1000 or more carbon atoms. The word *polymer* is derived from the Greek for "many parts"; in this example ethylene is the *monomer* that gives rise to polyethylene, the *polymer*. From a commercial standpoint, *polymerization*, as the process of forming polymers is called, is probably the most important reaction in organic chemistry, since polymers account for a high proportion of the sales of organic chemicals.

Acid-Catalyzed Polymerization

One of the most useful polymers is *butyl rubber*, made by the polymerization of isobutylene (Fig. 4-1). If a trace of acid (either a protonic acid like HCl or, better yet, a nonprotonic Lewis acid like BF_3) is added to a concentrated solution of isobutylene, an almost explosive polymerization reaction occurs, the isobutylene reacts completely, and a rubbery, rather gummy precipitate of polyisobutylene forms. We cannot speak of an exact formula for such a polymer, for its various molecules will have somewhat different chain lengths, depending upon the number of monomer units they contain, but we can speak of average molecular weights and average chain lengths. Polymers with average chain lengths of 1000 to 2000 isobutylene units (4000 to 8000 carbon atoms) can easily be produced.

The chemical reactions that lead to the formation of polyisobutylene from isobutylene are not hard to understand. Let us consider, for simplicity a vessel that contains only a large amount of isobutylene. To this vessel we add a tiny amount of gaseous HCl. We know that the proton of HCl will add to isobutylene, following Markovnikov's rule, to form the tertiary butyl

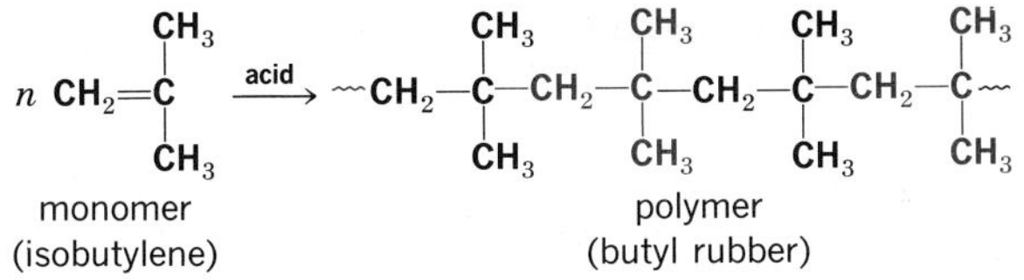

$$n\ CH_2{=}C(CH_3)_2 \xrightarrow{\text{acid}} \sim CH_2{-}C(CH_3)_2{-}CH_2{-}C(CH_3)_2{-}CH_2{-}C(CH_3)_2{-}CH_2{-}C(CH_3)_2 \sim$$

monomer (isobutylene) — polymer (butyl rubber)

Figure 4-1. *The polymerization of isobutylene to polyisobutylene (butyl rubber) is brought about by a trace of acid. The polymer may contain an average of 1400 isobutylene units and have an average molecular weight of 80,000.*

cation (Fig. 4-2, step 1). Under the conditions we discussed in the last chapter, in which isobutylene was reacted with an equimolar amount of HCl, this cation immediately combines with Cl^- to form the addition product, *t*-butyl chloride. But our new conditions are different, for there is only a tiny amount of Cl^- present and it is relatively improbable that the cation will find this anion to react with. Instead the cation adds, like any other powerful Lewis acid, to the carbon–carbon double bond of another isobutylene molecule (Fig. 4-2, step 2), for it is completely surrounded by these highly reactive molecules. The product of this self-addition is itself a tertiary cation that can add again to isobutylene; this reaction can be repeated over and over again (steps 3 to 1000). Finally, the polymerization comes to a halt when one of the cations loses a proton by a reaction that is the reverse of step 1 (step 1001) or combines

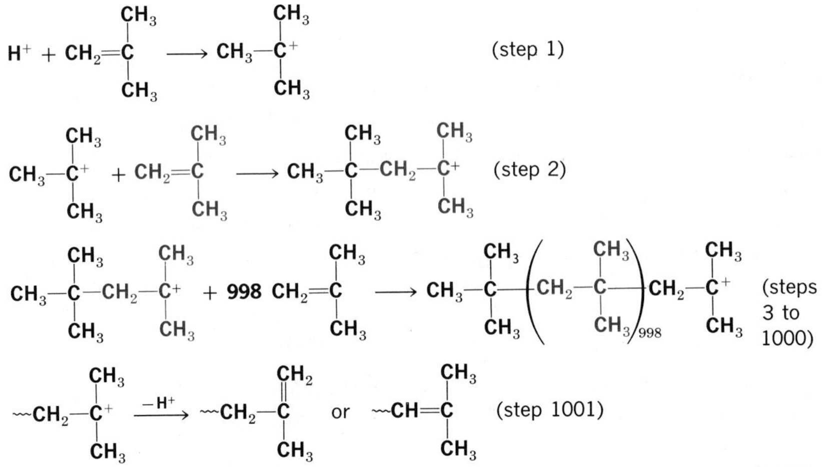

$$H^+ + CH_2{=}C(CH_3)_2 \longrightarrow CH_3{-}C^+(CH_3)_2 \quad \text{(step 1)}$$

$$CH_3{-}C^+(CH_3)_2 + CH_2{=}C(CH_3)_2 \longrightarrow CH_3{-}C(CH_3)_2{-}CH_2{-}C^+(CH_3)_2 \quad \text{(step 2)}$$

$$CH_3{-}C(CH_3)_2{-}CH_2{-}C^+(CH_3)_2 + 998\ CH_2{=}C(CH_3)_2 \longrightarrow CH_3{-}C(CH_3)_2\left(CH_2{-}C(CH_3)_2\right)_{998}CH_2{-}C^+(CH_3)_2 \quad \text{(steps 3 to 1000)}$$

$$\sim CH_2{-}C^+(CH_3)_2 \xrightarrow{-H^+} \sim CH_2{-}C({=}CH_2){-}CH_3 \ \text{ or } \ \sim CH{=}C(CH_3)_2 \quad \text{(step 1001)}$$

Figure 4-2. *The steps in the formation of a polyisobutylene molecule 1000 monomer units long by acid-catalyzed polymerization. The average chain length will depend upon the reaction conditions (temperature, time of reaction, and concentration of reagents).*

with a chloride ion. Although the very end of the chain may contain a double bond or a chlorine atom, its presence cannot affect the properties of a molecule of 4000 carbon atoms so that the resultant polymer is an inert hydrocarbon. The reason for its rubbery properties was discussed on p. 13.

Prob. 4-1. The very desirable hydrocarbon 2,2,4-trimethylpentane ("iso-octane"), octane number 100, may be formed by the *dimerization* of isobutylene, followed by reduction with H_2 and Pt. Write equations for this conversion in which polymerization stops at the end of step 2.

Free-Radical Polymerization

Even more important than acids in the polymerization of alkenes are chemical species known as *free radicals,* atoms or groups of atoms that contain a single, *unpaired electron.* The chlorine atom, $:\ddot{\underset{..}{Cl}}$, is such a free radical, and we have seen that it is reactive enough to remove a hydrogen from a saturated hydrocarbon (Fig. 1-30).

A free radical is usually a neutral species and may be considered related to a Lewis acid by the addition of a single electron. For example, addition of an electron to a proton would give the neutral hydrogen atom (H·), a free radical. Analogously, the bromine free radical $:\ddot{\underset{..}{Br}}\cdot$ is related to the Lewis acid $:\ddot{\underset{..}{Br}}^{\oplus}$. These relationships are illustrated in Fig. 4-3.

Like the corresponding acids, free radicals are highly reactive. Ordinarily, they occur only as short-lived intermediates in chemical reactions. One principal reason for their short lifetime is their quick combination into stable neutral molecules (Fig. 4-4). Despite their usual neutrality, free radicals each

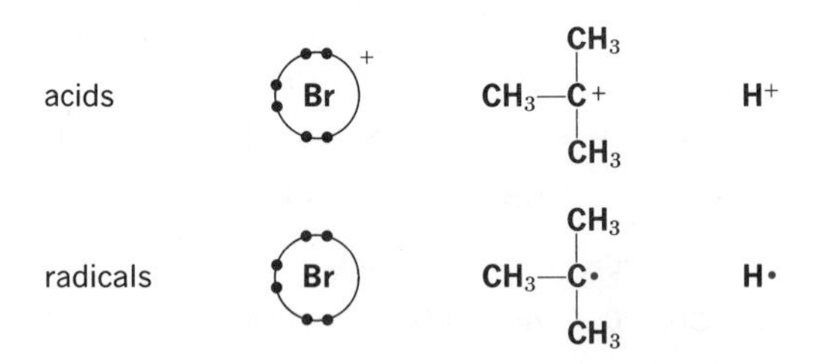

Figure 4-3. *The top three reagents are positively charged acids, each of which readily adds to a double bond. Note that each contains an even number of electrons. Below each is the corresponding free radical, a neutral species containing one more electron which is unpaired. Since free radicals are one electron short of having a complete shell, they too are reactive toward the electron-rich double bond.*

$$(CH_3)_3C\cdot \quad \cdot C(CH_3)_3 \longrightarrow (CH_3)_3C{-}C(CH_3)_3$$

Figure 4-4. *Whenever two free radicals come together, they may combine with the formation of a single bond. As a consequence it is seldom possible to have an appreciable concentration of free radicals, and they exist mainly as short-lived intermediates.*

lack one electron and therefore have incomplete outer shells. This makes them seek an electron and, like acids, add to carbon–carbon double bonds.

Because of their instability, free radicals are difficult to form. However, compounds of one class break down easily into free radicals and often serve to initiate organic free-radical reactions. These compounds, *peroxides,* contain oxygen–oxygen bonds. The bonds are so weak that when the peroxides are heated or irradiated with light, they split into two free radicals (Fig. 4-5).

$$RO{-}OR \xrightleftharpoons{\text{heat}} R{-}O\cdot \quad \cdot O{-}R$$

Figure 4-5. *Peroxides readily break down into free radicals; therefore they are convenient sources of small amounts of these species.*

One of the most significant industrial uses of free-radical reactions occurs in the formation of polymers. Polymerization induced by acids creates useful polymers from only a few alkenes. By using free radicals to touch off the reaction, the chemist may develop many useful products from substituted alkenes. The steps in the free-radical polymerization of an alkene closely resemble those in which an acid acts as catalyst, except that free radicals rather than carbonium ions serve as the agents. A few free radicals are first produced, usually through the breakdown of some peroxide molecules that have been added as an *initiator;* these free radicals then add to the alkene. If the object were to obtain a polymer like the one used in making vinyl tile, the chemist would employ *vinyl chloride* (IUPAC name: chloroethene) as the alkene. The two steps (a) and (b) in Fig. 4-6 that begin the polymerization are called the *initiation* reactions.

Next, the carbon free radical formed in the addition reacts with another molecule of vinyl chloride, step (c); this addition is repeated many times (n times in the example shown) until a long chain of atoms is produced. The repetitive steps in which the polymer is actually formed are said to be *propagation* reactions. In a typical polymerization sequence, more than 1000 molecules of vinyl chloride will be incorporated into the chain. Sooner or later, the free radical in step (c) may collide with another free radical, (d), which may be either the end of another chain, a radical from peroxide, or

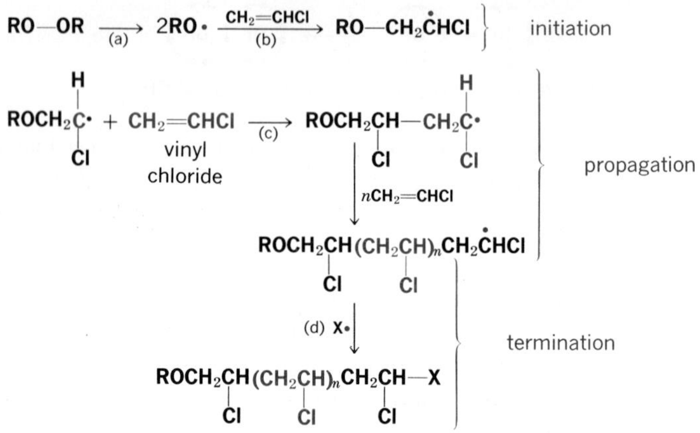

$$\text{RO—OR} \xrightarrow[\text{(a)}]{} 2\text{RO}\cdot \xrightarrow[\text{(b)}]{\text{CH}_2{=}\text{CHCl}} \text{RO—CH}_2\dot{\text{C}}\text{HCl} \quad \text{initiation}$$

$$\text{ROCH}_2\underset{\text{Cl}}{\overset{\text{H}}{|}}\!\text{C}\cdot + \underset{\text{vinyl chloride}}{\text{CH}_2{=}\text{CHCl}} \xrightarrow[\text{(c)}]{} \text{ROCH}_2\underset{\text{Cl}}{\text{CH}}\text{—CH}_2\underset{\text{Cl}}{\overset{\text{H}}{\text{C}}}\cdot$$

$$\downarrow n\text{CH}_2{=}\text{CHCl}$$

$$\text{ROCH}_2\underset{\text{Cl}}{\text{CH}}(\text{CH}_2\underset{\text{Cl}}{\text{CH}})_n\text{CH}_2\dot{\text{C}}\text{HCl} \quad \text{propagation}$$

$$\text{(d) X}\cdot \downarrow$$

$$\text{ROCH}_2\underset{\text{Cl}}{\text{CH}}(\text{CH}_2\underset{\text{Cl}}{\text{CH}})_n\text{CH}_2\underset{\text{Cl}}{\text{CH}}\text{—X} \quad \text{termination}$$

Figure 4-6. *Steps in the free-radical polymerization of vinyl chloride; the same steps are followed in the polymerization of other substituted alkenes.*

even oxygen from the air, because oxygen contains two unpaired electrons. When this happens, termination is said to occur: The chain ceases to grow. Thus the steps in a polymerization reaction as in any other chain reaction are *initiation, propagation,* and *termination* (Fig. 4-6).

Simple alkenes are the source of many common plastics. In Table 4-1 the names, structures, and products formed in some polymerization reactions are listed. Certain well-known polymers like nylon and Dacron are conspicuously missing from this list, however. These fibers are the products of an entirely different chemical process, which we shall discuss in later chapters.

Table 4-1 shows clearly that the properties of a polymer depend largely on the structure of the monomer selected. This is to be expected. A whole new spectrum of polymers is obtained if a mixture of two monomers is allowed

Table 4-1
Some Common Monomers and the Polymers Formed from Them by Free-Radical Polymerization

Monomer	Polymer Name	Trade Name	A Typical Use
$CH_2{=}CH_2$	polyethylene	Polythene	squeeze bottles
$CF_2{=}CF_2$	polytetrafluoroethylene	Teflon	bearings
$CH_2{=}CHCl$	polyvinyl chloride	Koroseal	raincoats
$CH_2{=}CHCN$	polyacrylonitrile	Orlon	fibers
$CH_2{=}C(CH_3)C(=O)OCH_3$	polymethyl methacrylate	Lucite, Plexiglass	safety glasses

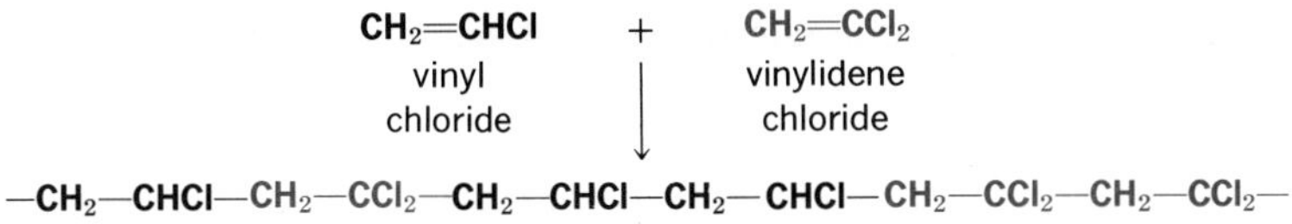

$$\underset{\text{vinyl chloride}}{CH_2{=}CHCl} + \underset{\text{vinylidene chloride}}{CH_2{=}CCl_2}$$
$$\downarrow$$
$$-CH_2-CHCl-CH_2-CCl_2-CH_2-CHCl-CH_2-CHCl-CH_2-CCl_2-CH_2-CCl_2-$$

Figure 4-7. *A portion of the chain of a one-to-one copolymer of vinyl chloride and vinylidene chloride. The distribution of the two monomers is approximately random throughout the chain.*

to polymerize. The resulting polymer, called a *copolymer,* usually contains both monomers, more or less randomly distributed along the chain. The plastic Saran, familiar in kitchen wrapping and auto seat covers, is a copolymer formed by the polymerizing of vinyl chloride and vinylidene chloride (Fig. 4-7). The properties of a copolymer may be altered by changing the proportions of the two monomers employed in its production.

Prob. 4-2. Write equations for the copolymerization of vinyl chloride and vinylidene chloride using a peroxide ROOR as initiator and terminating by reaction with RO·.

Dienes also polymerize under radical conditions. Interest in their reactions was originally stimulated by the observation that natural rubber could be considered, at least on paper, to result from the polymerization of isoprene (Fig. 4-8). A great deal of research has been carried out on the polymerization of isoprene and other dienes, but unfortunately normal radical polymerization of isoprene does not give natural rubber; the product is instead an isomer whose double bonds are mainly in the *trans* rather than the *cis* arrangement. A similar polymer, known as gutta percha, occurs naturally but lacks elastic

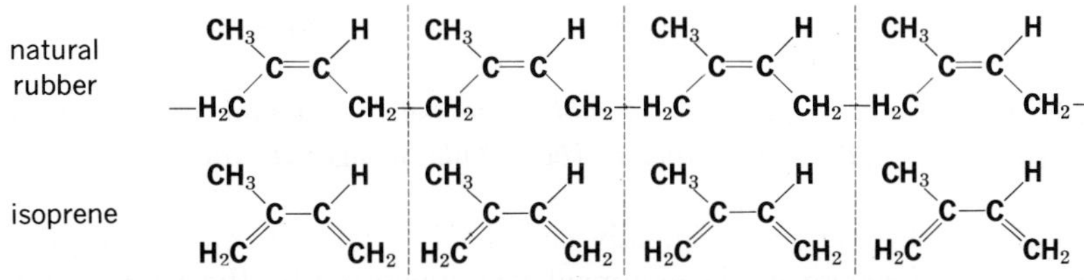

natural rubber:

$$-H_2C-C(CH_3){=}CH-CH_2-CH_2-C(CH_3){=}CH-CH_2-H_2C-C(CH_3){=}CH-CH_2-H_2C-C(CH_3){=}CH-CH_2-$$

isoprene:

$$H_2C{=}C(CH_3)-CH{=}CH_2 \quad H_2C{=}C(CH_3)-CH{=}CH_2 \quad H_2C{=}C(CH_3)-CH{=}CH_2 \quad H_2C{=}C(CH_3)-CH{=}CH_2$$

Figure 4-8. *Natural rubber is a highly unsaturated hydrocarbon which is obtained from the rubber tree. It might be considered to be obtainable from isoprene (2-methylbutadiene) by 1,4-polymerization. Note that all the double bonds are* cis-substituted *in natural rubber.*

properties. Recently, several unusual catalyst systems (mentioned later in this chapter), some of which do not involve radicals, have been developed. These systems convert isoprene to a polymer indistinguishable from natural rubber.

It is amusing that, contrary to the original speculations about natural rubber, the rubber tree clearly does not use isoprene as such in its synthesis of rubber, and chemists have been led to duplicate a natural product by a process entirely different from that used by nature, the chemical and technological importance of industrial rubber syntheses is immense.

When the ordinary polymerization of isoprene failed to produce natural rubber, chemists turned to investigations of the polymerization of other dienes. Some of these polymerizations led to useful synthetic rubbers. One important example is Neoprene, formed from the polymerization of 2-chlorobutadiene (chloroprene, Fig. 4-9). This monomer, in which the methyl group of isoprene

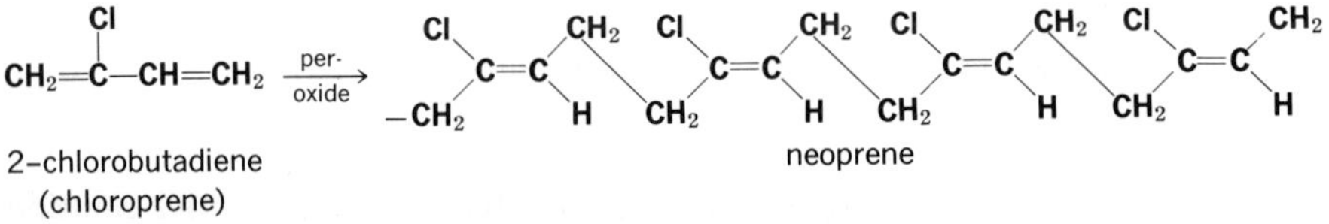

$$CH_2{=}C(Cl){-}CH{=}CH_2 \xrightarrow[\text{oxide}]{\text{per-}} {-}CH_2{-}C(Cl){=}CH{-}CH_2{-}CH_2{-}C(Cl){=}CH{-}CH_2{-}CH_2{-}C(Cl){=}CH{-}CH_2{-}CH_2{-}C(Cl){=}CH{-}CH_2{-}$$

2-chlorobutadiene (chloroprene) — neoprene

Figure 4-9. Chloroprene upon polymerization is converted to Neoprene, an important synthetic rubber. Note the trans *double bonds.*

is replaced by a chlorine atom, is prepared by the addition of hydrogen chloride to vinylacetylene. Neoprene is not affected by many organic solvents that cause natural rubber to swell, and is used in the hoses of gasoline pumps and many similar applications.

The most important synthetic rubber is, however, a copolymer of butadiene and styrene (Fig. 4-10). A large percentage of the tires in use today

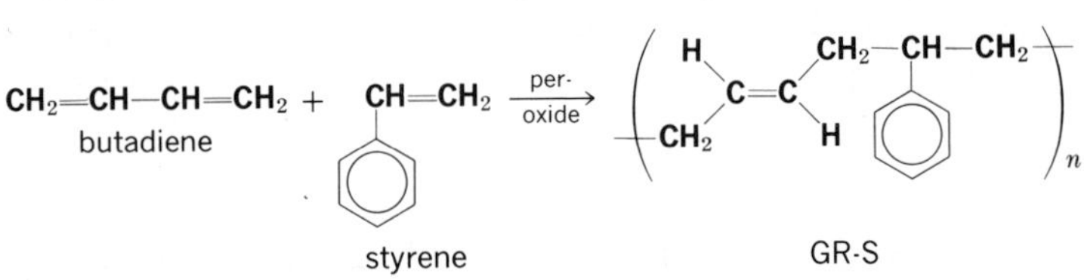

$$CH_2{=}CH{-}CH{=}CH_2 + CH(C_6H_5){=}CH_2 \xrightarrow[\text{oxide}]{\text{per-}} \left({-}CH_2{-}CH{=}CH{-}CH_2{-}CH(C_6H_5){-}CH_2{-} \right)_n$$

butadiene — styrene — GR-S

Figure 4-10. The most widely used rubber, synthetic or natural, is GR-S rubber, a copolymer of butadiene and styrene. The olefinic linkages are trans.

are made of this rubber (originally called GR-S for Government Rubber–Styrene type, because its development was sponsored by the United States Government in a crash program to replace natural rubber during World War II).

An interesting aspect of rubber technology is the necessity to "cure" both synthetic and natural rubbers. This is often accomplished by heating them with sulfur. The sulfur reacts at the double bonds of two polymer chains, linking them together (cross-linking). The need to cure explains why a small amount of isoprene is added to isobutylene before it is polymerized to butyl rubber, since pure polyisobutylene would not have any double bonds available for cross-linkage.

Free Radical Additions to Double Bonds

Free radicals of many types add to carbon–carbon double bonds. Among the most useful reactions of this type are the *free-radical, anti-Markovnikov addition of HBr* to an alkene or acetylene. The Markovnikov and anti-Markovnikov addition of HBr to 1-butene is compared in Fig. 4-11; in the

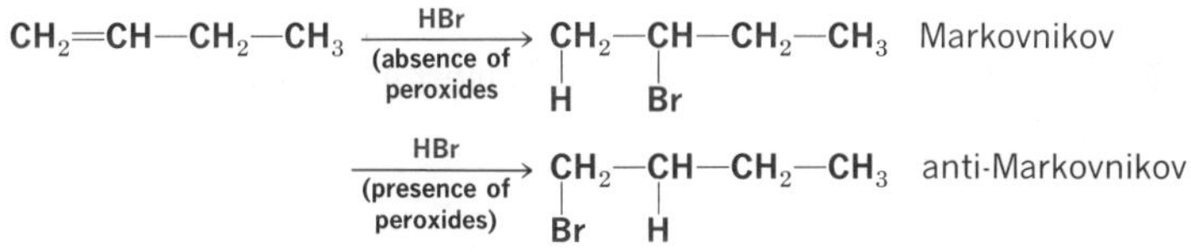

$$CH_2{=}CH{-}CH_2{-}CH_3 \xrightarrow[\text{(absence of peroxides)}]{HBr} \underset{H}{CH_2}{-}\underset{Br}{CH}{-}CH_2{-}CH_3 \quad \text{Markovnikov}$$

$$\xrightarrow[\text{(presence of peroxides)}]{HBr} \underset{Br}{CH_2}{-}\underset{H}{CH}{-}CH_2{-}CH_3 \quad \text{anti-Markovnikov}$$

Figure 4-11. *In the presence of peroxides, HBr adds in an anti-Markovnikov direction to alkenes and alkynes. Neither HCl nor HI undergo this reversal of direction of addition in the presence of peroxides.*

presence of peroxides the direction of addition of HBr is reversed and the bromine atom becomes attached to the carbon with the most hydrogens. The addition is a free-radical reaction, as might be expected by the necessity of using peroxides as initiators. A small amount of peroxide breaks down into free radicals, which then react with HBr to form the bromine atom. These

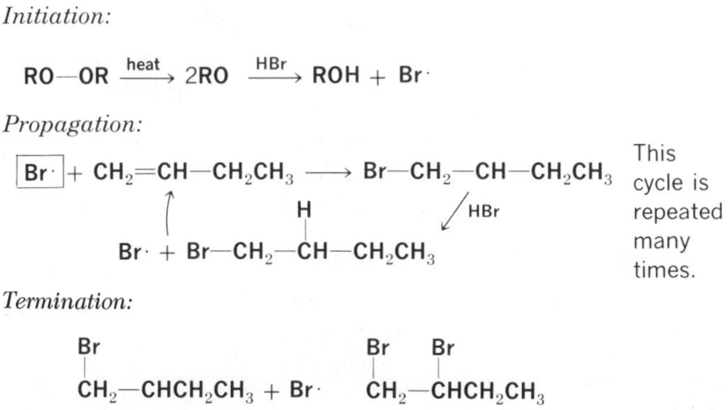

Initiation:

$$RO{-}OR \xrightarrow{heat} 2RO \xrightarrow{HBr} ROH + Br\cdot$$

Propagation:

$$\boxed{Br\cdot} + CH_2{=}CH{-}CH_2CH_3 \longrightarrow Br{-}CH_2{-}\dot{C}H{-}CH_2CH_3$$

$$\xrightarrow{HBr} Br\cdot + Br{-}CH_2{-}\overset{H}{CH}{-}CH_2CH_3$$

This cycle is repeated many times.

Termination:

$$\overset{Br}{CH_2}{-}CHCH_2CH_3 + Br\cdot \quad \overset{Br}{CH_2}{-}\overset{Br}{C}HCH_2CH_3$$

Figure 4-12. *Steps in the peroxide-catalyzed, free-radical, anti-Markovnikov addition of HBr to an alkene. The propagation steps occur thousands of times for each initiation or termination reaction. Other termination steps are possible.*

in turn add to 1-butene; the direction of addition is controlled by the bromine atom rather than by the proton as in normal addition. By bromine atom addition, a carbon free radical is formed, which quickly removes a hydrogen atom from HBr, reforming a bromine atom. The addition is thus a *chain reaction,* analogous to the chain chlorination of methane (Fig. 1-30). The steps are given in Fig. 4-12.

A number of other reagents also add across carbon–carbon double bonds in the presence of peroxide catalysts. Some examples are given in Fig. 4-13.

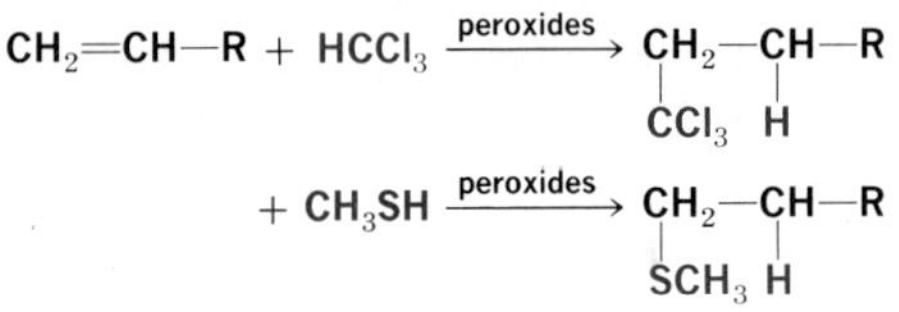

$$CH_2{=}CH{-}R + HCCl_3 \xrightarrow{\text{peroxides}} \underset{CCl_3}{CH_2}{-}\underset{H}{CH}{-}R$$

$$+ CH_3SH \xrightarrow{\text{peroxides}} \underset{SCH_3}{CH_2}{-}\underset{H}{CH}{-}R$$

Figure 4-13. *In the presence of peroxides a number of reagents add across carbon–carbon double bonds.*

Other Types of Polymerization

In recent years the chemical industry has worked to develop new polymerization catalysts that will produce polymeric materials with better properties.

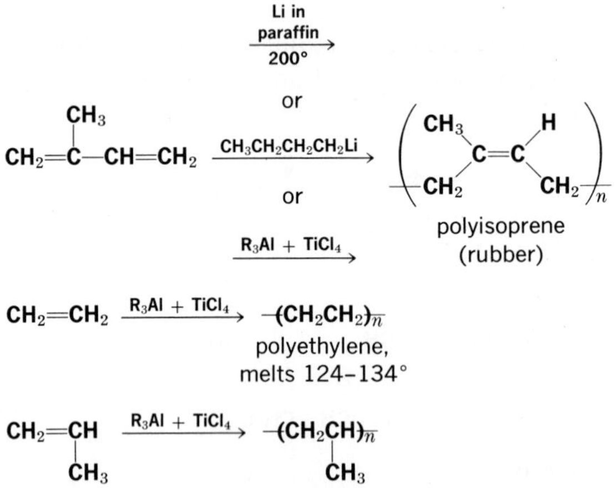

$$CH_2{=}\overset{CH_3}{C}{-}CH{=}CH_2 \xrightarrow[\text{200°}]{\text{Li in paraffin}} \text{ or } \xrightarrow{CH_3CH_2CH_2CH_2Li} \text{ or } \xrightarrow{R_3Al + TiCl_4} \left(\begin{matrix} CH_3 & & H \\ & C{=}C & \\ {-}CH_2 & & CH_2{-} \end{matrix} \right)_n$$

polyisoprene (rubber)

$$CH_2{=}CH_2 \xrightarrow{R_3Al + TiCl_4} {-}(CH_2CH_2)_n{-}$$

polyethylene, melts 124–134°

$$\underset{CH_3}{CH_2{=}CH} \xrightarrow{R_3Al + TiCl_4} {-}(CH_2\underset{CH_3}{CH})_n{-}$$

polypropylene, melts 176°

Figure 4-14. *New catalysts give all* cis-*1,4-polyisopropene, identical with natural rubber. They also allow the polymerization of ethylene and propylene under mild conditions to polymers of high melting points and crystallinity.*

As we have seen, a synthetic material with all the properties of natural rubber has now been prepared from isoprene. Three different catalysts can be employed for this polymerization (Fig. 4-14): One is simply metallic lithium dispersed in paraffin oil; a second is an organometallic reagent, *n*-butyllithium ($CH_3CH_2CH_2CH_2Li$); the third is a complex catalyst composed of another organometallic reagent, a trialkylaluminum, and titanium chloride. Butadiene itself and simple alkenes can be polymerized by the same catalysts. Karl Ziegler (Germany) and Giulio Natta (Italy) were awarded the Nobel Prize for these important developments.

The carbonium-ion polymerization of isobutylene, the free-radical polymerization of alkenes and dienes, and the polymerization of isoprene by means of metals and organometallic compounds are all examples of *addition polymerization.*

PROBLEMS

1. Explain what is meant by the following terms:
   (a) peroxide (b) copolymer
   (c) termination (d) free radical
   (e) anti-Markovnikov addition (f) chain reaction
2. Write equations for the free-radical polymerization of acrylonitrile ($CH_2{=}CHCN$) initiated by peroxide and terminated by reaction with peroxy radicals.
3. A *tetramer* ($C_{12}H_{24}$) resulting from the acid-catalyzed combination of four propylene molecules was produced in large amounts for use in the manufacture of synthetic detergents. Write reactions for this *tetramerization.*
4. Chloroprene (Fig. 4-9) is made industrially by the addition of HCl to vinyl acetylene (1-buten-3-yne). Write the equation for this reaction.
5. The following table compares the electronic structure of various acids, bases, and free radicals. Complete the table, and add one or two examples of your own choosing.

acid	*radical*	*base*
Cl^+	$Cl\cdot$	Cl^-
	$H_3C\cdot$	
H^+		
		HO^-

6. Give the structure of the product you would expect to obtain from

addition of HBr to each of the following alkenes (1) in the absence of peroxides and (2) in the presence of peroxides:
(**a**) isobutylene (**b**) 1-methylcyclohexene
(**c**) 2-butene (**d**) 2-methyl-1-pentene

7. Boron trifluoride (BF_3) is a powerful Lewis acid, despite its neutrality. Draw its electronic structure and explain its acidity.

8. Certain substances are so effective in terminating free-radical polymerizations that they are called *inhibitors* because they prevent any reaction at all. The gas nitric oxide (NO) is one of these. Draw its electronic formula and suggest a reason for its inhibitory action.

9. Polymerization of butadiene or any of its derivatives (chloroprene, isoprene) is complicated by the fact that 1,2-, as well as 1,4-, polymerization can occur (see Fig. 3-27). Draw a segment of polybutadiene three units long which would be formed only by 1,2-polymerization. Draw another segment three units long in which one butadiene molecule has polymerized by 1,2-addition, one by 1,4-*cis* and one by 1,4-*trans* addition.

10. Basing your answer on Fig. 4-12, write equations for the steps in the free-radical additions in Fig. 4-13.

5
aromatic hydrocarbons

The contrast in reactivity between the saturated and the unsaturated hydrocarbons is sharp. As we have seen, the single-bonded alkanes are inert to most reagents. When forced to react, they do so by substitution—with a halogen, for example, replacing a hydrogen atom. The unsaturated double- and triple-bonded alkenes and alkynes, on the other hand, react with acids and oxidizing agents, and undergo addition with halogens and other reagents.

A third class of hydrocarbons does not easily fit this complex of single-, double-, and triple-bonded compounds. This class comprises the *aromatic hydrocarbons*. They were originally given this name because many of them have spicy or otherwise pleasant fragrances, and the name has persisted, although aromatic chemistry now embraces the chemistry of all compounds like benzene that behave in a special manner.

Benzene

The mobile liquid benzene boils at 80°. Once considered atypical in behavior, benzene is now known to be truly representative of the aromatic compounds. Its formula, C_6H_6, suggests unsaturation—a saturated alkane would have the formula C_6H_{14} and a cycloalkane, C_6H_{12}; however, much of the behavior of benzene more closely resembles that of the saturated rather than of the unsaturated hydrocarbons. For example, benzene will not react with either bromine or chlorine under conditions where alkenes and dienes like hexene and hexadiene rapidly add both. Benzene is also inert to hydrogen bromide, hydrogen chloride, and other reagents that readily add to double and triple bonds. When it does react, it usually does not undergo addition, but rather substitution, as in its reaction with bromine in the presence of ferric bromide (Fig. 5-1).

Benzene shows that it is actually unsaturated because it adds hydrogen or chlorine, although only when reacted under very vigorous conditions compared to those required for alkenes and alkynes. When reduced with

$$C_6H_6 + Br_2 \xrightarrow{FeBr_3} C_6H_5Br + HBr$$

$$+ 3H_2 \xrightarrow[\text{heat, press}]{\text{catalyst}} \text{cyclohexane}$$

Figure 5-1. *As compared to most unsaturated compounds, benzene is highly unreactive. Under vigorous reaction conditions it is* substituted *by bromine but* adds *hydrogen.*

hydrogen, it forms cyclohexane, and 1,2,3,4,5,6-hexachlorocyclohexane (the insecticide Lindane) can be prepared from benzene by free-radical addition of chlorine. These reactions suggest that benzene has its six carbon atoms arranged in a ring, and it must have the equivalent of three double bonds, since three molecules of hydrogen or chlorine react if forced to do so.

These addition reactions could be accounted for if benzene were 1,3,5-cyclohexatriene (Fig. 5-2), but such a molecule should be highly reactive

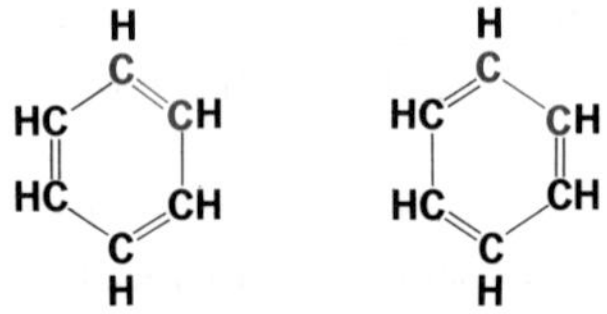

Figure 5-2. *Either of the cyclohexatriene structures shown would account for the addition reactions of benzene, but not for its substitution reactions, and not for its failure to react with hydrogen bromide or bromine.*

in marked contrast to benzene. Although the peculiar nature of benzene has occupied the attention of organic chemists for a century, they have only acquired a general understanding of the compound's structure and its unusual chemical properties since the 1930s. The explanation now accepted has required the best modern physical evidence and mathematical theories.

Some of the evidence involves bond lengths. If the left-hand structure in Fig. 5-2 were the correct one for benzene, the bond between the two carbon atoms in red should be a double bond, with a bond length of 1.34 Å. However, if the correct structure were the one on the right, the bond between those atoms would be single and its length, 1.54 Å. Each of these structures should contain three carbon–carbon single bonds and three carbon–carbon double bonds. Physical measurements of benzene not only confirm that the six carbon atoms are arranged in a hexagon, but also show that *all* of the carbon–carbon bonds are of the same length, 1.39 Å, intermediate between the lengths of double and single bonds.

We discussed the variation of bond length with bond order in Chapter 3 and presented this variation in graphic form in Fig. 3-5. That graph is

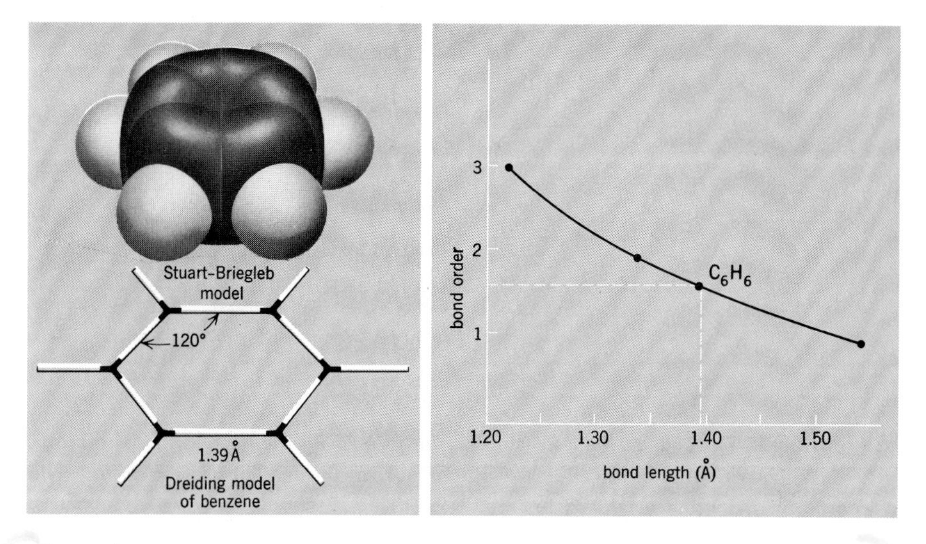

Figure 5-3. *Models of benzene. All the carbon–carbon bonds in benzene are the same length, 1.39 Å. This corresponds to a bond order of 1.5, a one-and-one-half bond.*

reproduced in Fig. 5-3. Note that a bond length of 1.39 Å, as in benzene, corresponds almost exactly to a bond order of 1.5, namely, to one and one-half bonds. How can we account for half bonds?

Orbital Picture of Aromatic Compounds

The simplest pictorial presentation of the equivalence of all C—C bonds in benzene employs the orbital scheme that applies to unsaturated molecules. The internal bond angles in benzene are 120°, exactly the same as for carbon with sp^2 hybridized bonds (see Fig. 3-11). Six such carbon atoms can be joined to form a perfect hexagon. Owing to the symmetry of the molecule, the p orbital on each carbon can overlap equally well with the p orbitals of the two adjacent carbon atoms. This gives rise to a continuous cloud of p electrons on top of, and below, the six-carbon structure (Fig. 5-4). These p electrons are not localized between any pair of carbons, as in a double bond, but are *delocalized* in π molecular orbitals over the entire ring. Each of the C—C bonds is identical, containing a pair of electrons in its σ bond, which unifies the framework, and also a one-sixth share of the six π electrons in the molecule. This makes a total of three electrons per bond, or one and one-half bonds.

Because the bonding in benzene is different from that in alkenes and alkynes, a new symbol is sometimes used to represent the compound. A

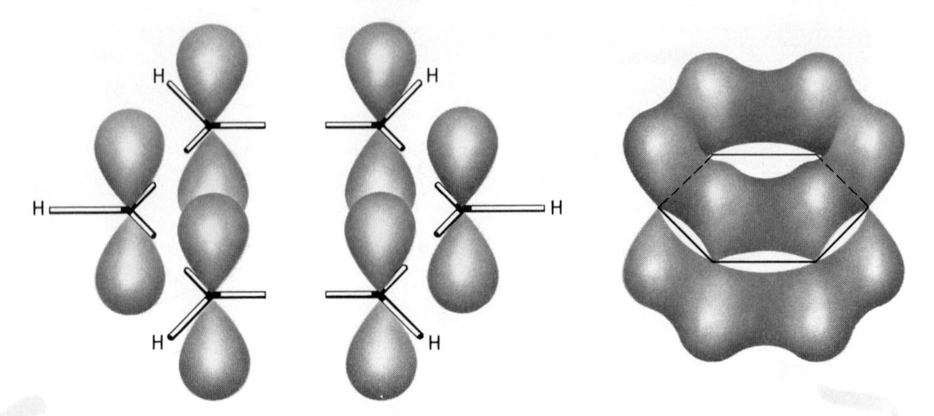

Figure 5-4. *The sigma bonds in benzene are formed by the overlap of* sp^2 *orbitals, while each carbon contributes to the* π *bond system by overlap of its* p *orbital with* p *orbitals of its two neighbors.*

common one is a hexagon with a circle inside to indicate the circular distribution of π electrons. Two or more hexagonal units of carbon atoms can fit together, thus forming molecules that contain more than one benzene ring. The simplest of these is naphthalene, $C_{10}H_8$, which has two six-membered rings "fused" together. Figure 5-5 indicates how the orbital picture of naphthalene accounts for the delocalization of the π electrons over the entire ten-carbon skeleton.

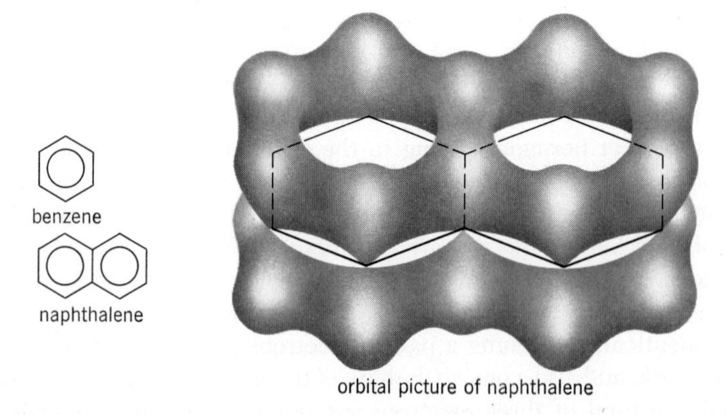

Figure 5-5. *A hexagon surrounding a circle is often used as a symbol for a benzene ring. In the orbital picture for naphthalene, each of the carbons shared by both rings is involved in* π *bonding with three other carbons.*

Resonance Picture of Aromatic Compounds

The molecular orbital picture of bonding in unsaturated molecules is only a relatively recent concept. Long before it was developed, however, chemists made an interesting observation: Whenever an unsaturated molecule displayed the peculiar properties we have termed aromatic (resistance to oxidation, substitution in place of addition, etc.), any attempt to write its structure using only single and double bonds resulted in ambiguity. It is possible to place three double bonds in a six-membered ring in two different ways without disturbing the position of any of the atoms of the molecule. Only the positions of the electrons of the double bond will differ in the two structures.

The reader should notice in Fig. 5-2 that each pair of carbon atoms is joined by a double bond in one structure and by a single bond in the other. Physical measurements have shown that the structure of this aromatic molecule is intermediate between the two normal structures. All the carbon–carbon bonds are the same length: halfway between the length of a double and a single bond. It is possible to represent the structure of an aromatic molecule as the sum of all possible normal structures for the molecule. Though rather cumbersome, this method of representing the molecule is quite useful.

This technique of accounting for aromatic properties was given theoretical justification and made quantitative by Linus Pauling (Nobel Prize in Chemistry in 1954). Pauling's method is now called the *resonance* picture of aromatic molecules (Fig. 5-6). It is important to remember that no single resonance structure or form adequately represents an aromatic molecule. Rather, a weighted average of all the forms is the true state. The molecule itself, called a *resonance hybrid,* partakes of some of the properties of each resonance form, much as a hybrid corn inherits some of the characteristics of each variety from which it springs. This is indicated by the special two-headed arrows in Fig. 5-6. They do not describe an equilibrium but an intermediate state (the resonance hybrid), in which the actual molecule resembles both forms.

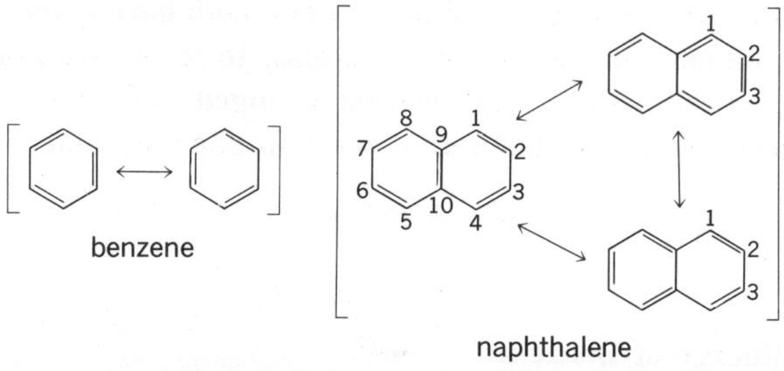

benzene

naphthalene

Figure 5-6. *The* resonance *picture of aromatic molecules starts from the observation that double bonds can be formally arranged in more than one way in standard formulas: two ways in benzene, and three in naphthalene. The actual molecule is considered to be a* hybrid *of these structures, i.e., to have an intermediate structure.*

Bond Lengths in Naphthalene

If we compare the compact pictures of benzene and naphthalene in the orbital representation of Fig. 5-5 with the cumbersome *resonance* representation of Fig. 5-6, we might wonder why the latter picture has survived. The answer is simple: The resonance picture contains a great deal of valuable information which is often obscured in the orbital picture. As an example, suppose we wished to estimate the bond lengths in naphthalene. If we used Fig. 5-5, we might guess that they were all the same length. If, however, we chose the resonance picture of Fig. 5-6 and assumed that all three structures made equal contributions to the hybrid we would see that the bond between C-1 and C-2 was double in two of the structures and single in the third. This bond is a $1\frac{2}{3}$ bond ($\frac{2}{3}$ double and $\frac{1}{3}$ single). By contrast the C-2, C-3 bond is a $1\frac{1}{3}$ bond, $\frac{1}{3}$ double and $\frac{2}{3}$ single. If we refer to the graph of bond order *versus* bond length in Fig. 5-3, we observe that a $1\frac{2}{3}$ bond is predicted to be 1.36 Å long and a $1\frac{1}{3}$ bond 1.42 Å long. The precise values found for these bonds in naphthalene are 1.364 and 1.404 Å. Although agreement is not perfect, we thus arrive by a simple process at reasonable estimates of bond lengths. Other uses of the resonance picture will become obvious as we proceed, and we shall use both resonance and molecular orbital pictures interchangeably, depending upon which one appears to illustrate a point more clearly.

Prob. 5-1. What is the bond order of the 9,10- bond in naphthalene? The 1,9- bond?

Benzene rings may be fused together indefinitely to form larger molecules called *polynuclear aromatic hydrocarbons* (Fig. 5-7). Three benzene rings are linked linearly in anthracene, angularly in phenanthrene.

Prob. 5-2. Anthracene and phenanthrene both have enough π electrons (one from each of the 14 carbon atoms) to form seven double bonds. Show that the double bonds can be arranged in four different ways in anthracene and in five different ways in phenanthrene.

Resonance Energy of Benzene

Since either the orbital or resonance picture of benzene accounts reasonably well for the bond lengths in aromatic molecules, let us now see how to explain other typical aromatic characteristics, such as lowered reactivity relative to

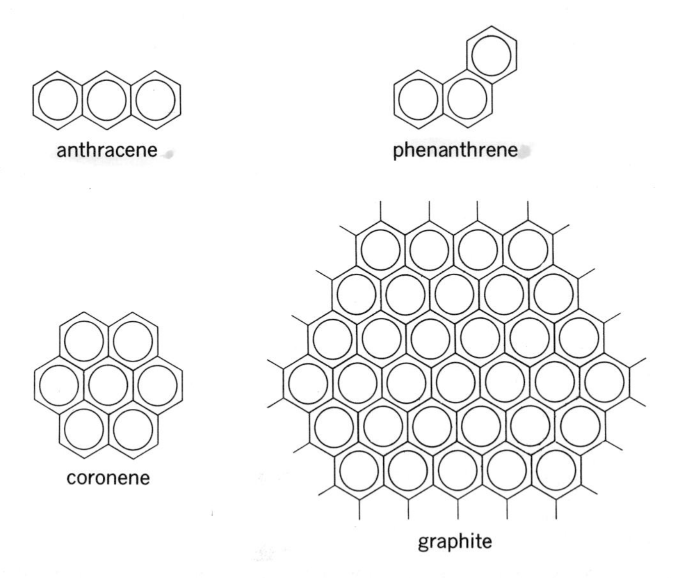

Figure 5-7. *Benzene rings may be fused together to form polynuclear hydrocarbons. Anthracene and phenanthrene are colorless and coronene is yellow; the color deepens as the number of rings increases, and graphite, which is essentially pure carbon, is black.*

alkenes and substitution (rather than addition) with acid reagents. Suppose that benzene were an ordinary cyclic triene with three normal double bonds. Such a compound, 1,3,5-cyclohexatriene, could not actually be prepared, since it would immediately transform itself into benzene. Nevertheless, we can make excellent forecasts of how such a molecule should react. By comparing an actual benzene molecule with this hypothetical model, we can estimate how much benzene will be stabilized by the delocalization of the π electrons. The amount of this stabilization is known as the *resonance energy of benzene.*

When an ordinary double bond is reduced with hydrogen, a large amount of energy is released, since an alkane is more stable than an alkene. For instance, hydrogenation of cyclohexene releases 28,600 calories (28.6 kcal) of heat per mole of alkene. If a molecule contains two normal double bonds, approximately twice as much energy is released on hydrogenation (55,400 calories per mole from cyclohexadiene). One might expect that a molecule such as our hypothetical cyclohexatriene with three normal double bonds would release somewhere in the neighborhood of $3 \times 28{,}600 = 85{,}800$ calories. Hydrogenation of a mole of benzene, however, releases only 49,800 calories. The resonance energy of benzene may then be estimated as $85{,}800 - 49{,}800 = 36{,}000$ calories (36 kcal) of energy (Fig. 5-8).

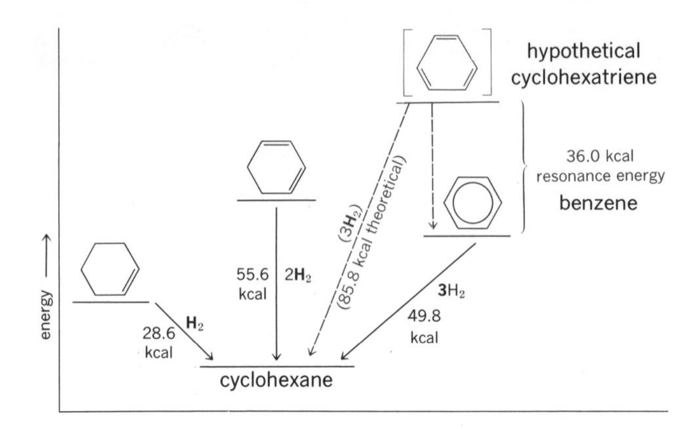

Figure 5-8. *The actual molecule of benzene is 36 kilocalories per mole more stable than the hypothetical cyclohexatriene. In order to observe normal alkene reactions of benzene, we should have to expend this additional amount of energy to localize the* π *electrons. Thus more energy, usually in the form of higher reaction temperatures, is required for reactions of benzene than for similar reactions of nonaromatic alkenes.*

Prob. 5-3. If 82.0 kcal of heat is released in the hydrogenation of naphthalene with 5 moles of hydrogen, calculate the resonance energy of naphthalene. Assume 28.6 kcal for the heat of hydrogenation of a normal double bond.

Substitution in Benzene

The benzene model now accounts satisfactorily for the equality of its carbon–carbon bonds and its lack of reactivity toward addition, as compared with other unsaturated molecules. We must still deal with the problem of why benzene, in contrast to alkenes, undergoes substitution rather than addition with halogens and other reagents.

We have said that benzene does not react with hydrogen chloride or hydrogen bromide and can be recovered unchanged after prolonged boiling with these reagents. In fact, a reaction does occur, but does not lead to any observable change. It is of interest to compare this reaction with the similar reaction of hydrogen chloride with cyclohexene. As we saw in considering the addition of halogen acids to alkenes, the reaction starts with the addition of the electron-deficient proton to the electrons of the unsaturated system. With benzene, too, a proton may attack the ring, attaching itself to one of

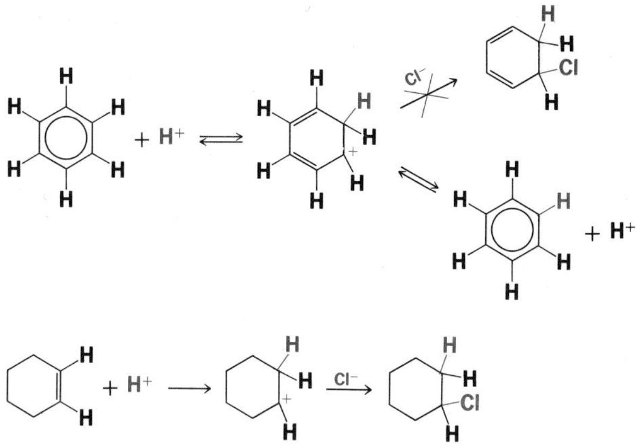

Figure 5-9. *Addition of a proton to benzene is more difficult than addition of one to cyclohexene because additional energy must be supplied to localize the π electrons (i.e., to overcome the resonance energy of benzene). No observable reaction takes place because the benzene ring is formed again by loss of the same or an equivalent proton.*

the carbon atoms (Fig. 5-9). In doing so, however, it at least partially localizes the π electrons, since the completely symmetrical system of overlapping *p* orbitals in benzene is disrupted in the carbonium-ion product. Proton addition will therefore be more difficult with benzene than with cyclohexene.

In the addition to cyclohexene, when the carbonium ion is once formed, it rapidly combines with a chloride ion to produce the addition product, cyclohexyl chloride. Similar combination with the carbonium ion formed by addition of a proton to benzene *does not* occur because the resulting chloride would no longer have an aromatic system: The special stabilizing energy (resonance energy) of benzene would have been lost. Instead, a proton is ejected, and the benzene ring is regenerated. No net change has been accomplished, although one proton may have replaced another.

It is possible to demonstrate that the reaction actually takes place if, instead of hydrogen chloride, *deuterium chloride* is used (Fig. 5-10). Deuterium

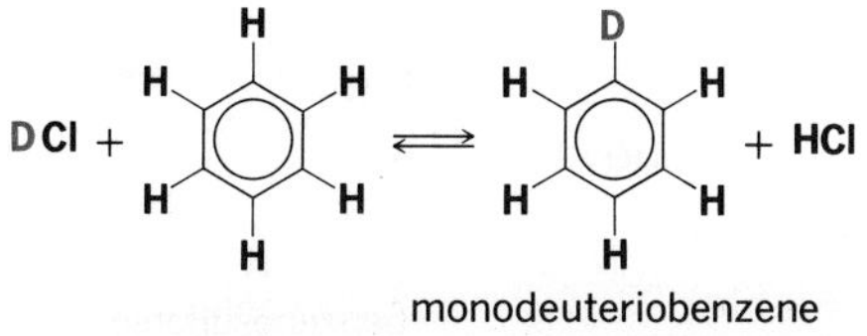

monodeuteriobenzene

Figure 5-10. *Any of the protons in benzene may be replaced with deuterium (an isotope of hydrogen) by heating with deuterium chloride.*

is an isotope of hydrogen, identical in chemical properties but distinguishable because it contains a neutron, as well as a proton, in its nucleus. If benzene is heated with a large excess of deuterium chloride, the recovered benzene has all of its hydrogen atoms replaced by deuterium (C_6D_6). Hexadeuteriobenzene is nearly identical to normal benzene but has slightly different physical properties.

If any acidic reagent other than a proton reacts with benzene, and the conditions used are vigorous enough to overcome the resonance energy of benzene, addition to form a carbonium ion may occur. As with hydrogen chloride, the reaction is not completed by a subsequent addition but by loss of a proton to reform a substituted benzene ring (Fig. 5-11).

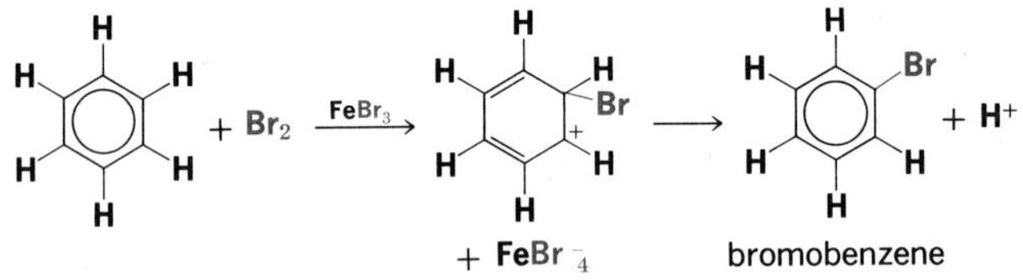

$C_6H_6 + Br_2 \xrightarrow{FeBr_3} C_6H_6Br^+ + FeBr_4^- \longrightarrow C_6H_5Br + H^+$

bromobenzene

Figure 5-11. *Steps in the bromination of benzene. The intermediate carbonium ion regenerates the aromatic ring rather than combining with an anion.*

Benzene undergoes substitution reactions with nitric acid and with sulfuric acid to form nitrobenzene and benzenesulfonic acid, respectively. Analogous reactions with alkenes are not usually observed. In the case of nitration the reactive species is known to be the positively charged group formed by loss of hydroxide from the acid ($HONO_2 + H^+ \longrightarrow H_2O + {}^+NO_2$). (See Fig. 5-12.)

Carbon–carbon bonds may be formed by an aromatic substitution reac-

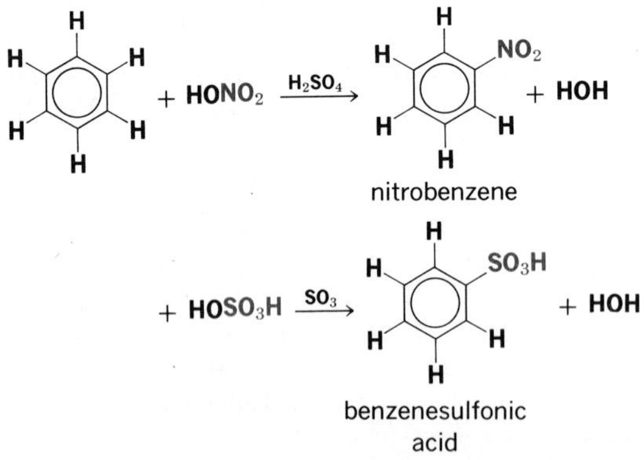

$C_6H_6 + HONO_2 \xrightarrow{H_2SO_4} C_6H_5NO_2 + HOH$

nitrobenzene

$C_6H_6 + HOSO_3H \xrightarrow{SO_3} C_6H_5SO_3H + HOH$

benzenesulfonic acid

Figure 5-12. *Nitration and sulfonation of benzene are important synthetic reactions.*

tion known as the Friedel–Crafts reaction. The attacking reagent is a carbonium ion that may be formed, as happens in the polymerization of isobutylene, by addition of a proton to a double bond or by reaction of an alkyl halide with aluminum chloride (Fig. 5-13). By means of the Friedel–Crafts reaction, alkylated benzenes may be prepared that are useful in the production of synthetic detergents and many other commercial products.

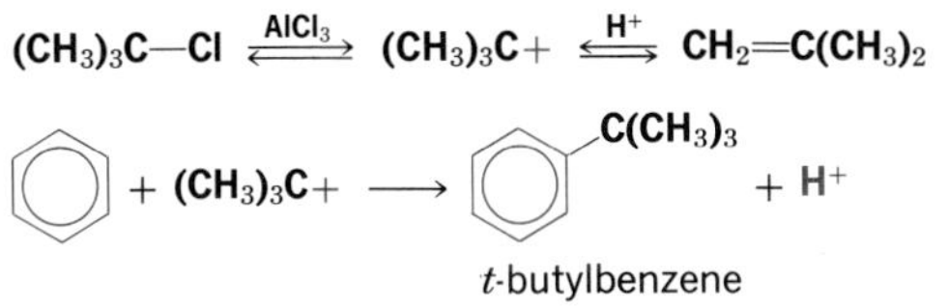

$$(CH_3)_3C{-}Cl \xrightleftharpoons{AlCl_3} (CH_3)_3C^+ \xleftrightharpoons{H^+} CH_2{=}C(CH_3)_2$$

$$C_6H_6 + (CH_3)_3C^+ \longrightarrow C_6H_5C(CH_3)_3 + H^+$$

t-butylbenzene

Figure 5-13. *Benzene undergoes substitution reactions with carbonium ions. The sequence is called Friedel-Crafts alkylation.*

Substituted benzenes may also undergo reaction. This results in di-, tri-, and polysubstituted benzenes. Suppose, for example, that bromobenzene reacts with nitric acid (using sulfuric acid as a catalyst) under nitrating conditions. From reactions at a position adjacent to the bromine atom, those at a position once removed from the bromine atom, and those at a position twice removed from the bromine atom, three isomeric products are possible. These are called the 2- or *ortho,* 3- or *meta,* and 4- or *para* positions, respectively, with reference to the bromine atom (Fig. 5-14). In fact, only two of the three possible products are formed in appreciable amounts, *ortho*-nitrobromobenzene (abbreviated

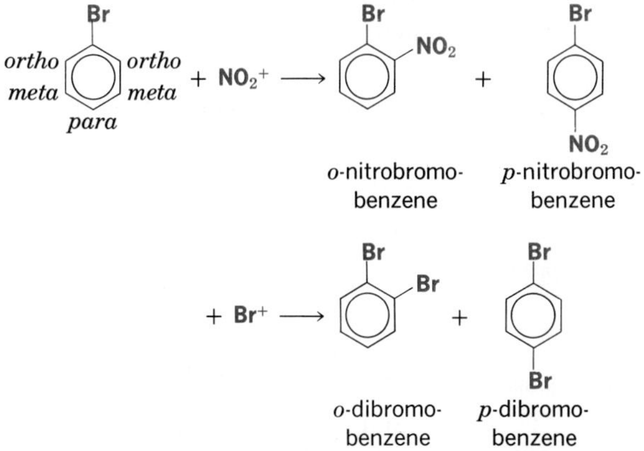

Br; *ortho*, *ortho*; *meta*, *meta*; *para*

$$C_6H_5Br + NO_2^+ \longrightarrow o\text{-}C_6H_4BrNO_2 + p\text{-}C_6H_4BrNO_2$$

o-nitrobromobenzene *p*-nitrobromobenzene

$$C_6H_5Br + Br^+ \longrightarrow o\text{-}C_6H_4Br_2 + p\text{-}C_6H_4Br_2$$

o-dibromobenzene *p*-dibromobenzene

Figure 5-14. *A bromine atom attached to a benzene ring is an* ortho-para *directing group, so that further substitution of bromobenzene leads to a mixture of* ortho- *and* para-*disubstituted benzenes.*

o-nitrobromobenzene) and *para*-nitrobromobenzene (abbreviated *p*-nitrobromobenzene). *The substituent group already in the molecule, in this case the bromine atom, directs the incoming group to these positions.*

Bromine on a benzene ring directs substitution toward both the *ortho* and *para* positions and is therefore said to be an *ortho-para* directing group. Groups that favor *ortho* substitution also favor *para* substitution. The nitro group, on the other hand, is *meta* directing (Fig. 5-15). If nitrobenzene is

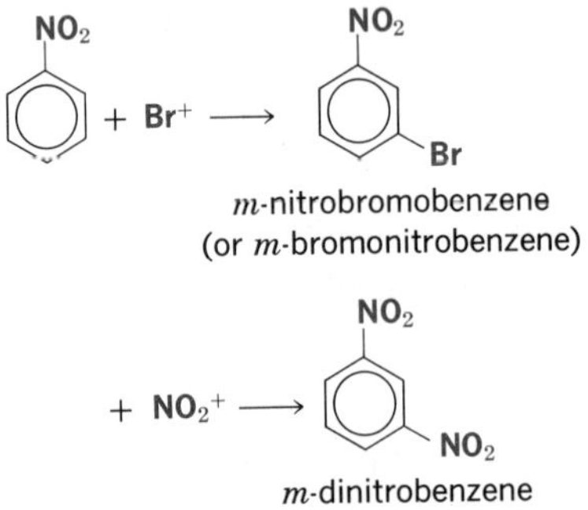

$$C_6H_5NO_2 + Br^+ \longrightarrow m\text{-}BrC_6H_4NO_2$$

m-nitrobromobenzene
(or *m*-bromonitrobenzene)

$$+\ NO_2^+ \longrightarrow m\text{-}C_6H_4(NO_2)_2$$

m-dinitrobenzene

Figure 5-15. *The nitro group is* meta *directing.*

brominated, the main product is *meta*-nitrobromobenzene (abbreviated *m*-nitrobromobenzene). It is easy to see that, in preparing disubstituted benzenes, the order in which the groups are introduced can be extremely important in determining the products formed.

Groups attached to a benzene ring may be classified either as *ortho-para* directing groups or as *meta* directing groups toward further substitution. Some of the more important members of each group are listed in Fig. 5-16. The reader must remind himself that *the substituent already present on the ring,*

Mark. ⊕ ← *ortho-para* directing
- $—NH_2$
- $—OH$, OCH_3
- $—CH_3$, C_2H_5, and other alkyl groups
- —Br, Cl, I

anti Mark ⊕ ← *meta* directing
- $—C{\equiv}N$, $—\overset{O}{\overset{\|}{C}}—$,
- $—SO_3H$, $—NO_2$, $—\overset{+}{N}(CH_3)_3$

Figure 5-16. *Some important* ortho-para *and* meta *directing groups. When two groups are present in the same molecule, the influence of the group higher on the list takes precedence over that below.*

not the incoming group, *decides where further substitution occurs.* Within the table the groups are ranked in order of decreasing directing power. A hydroxyl (OH) group, for example, is a much more powerful director than a halogen atom. If both groups were present in the same molecule, the hydroxyl group would control the position of further substitution.

Prob. 5-4. Predict what products would be formed in the mononitration of methylbenzene (toluene) and of cyanobenzene, and in the monobromination of toluene, chlorobenzene, and *p*-hydroxybromobenzene.

Origin of Directive Effect

The action of *ortho-para* and *meta* directing groups may be explained by applying the same concepts used in the interpretation of Markovnikov's rule. Directive effects in aromatic substitution are really little more than an extension of that rule. Let us consider the nitration of toluene (methylbenzene) as an example (Fig. 5-17).

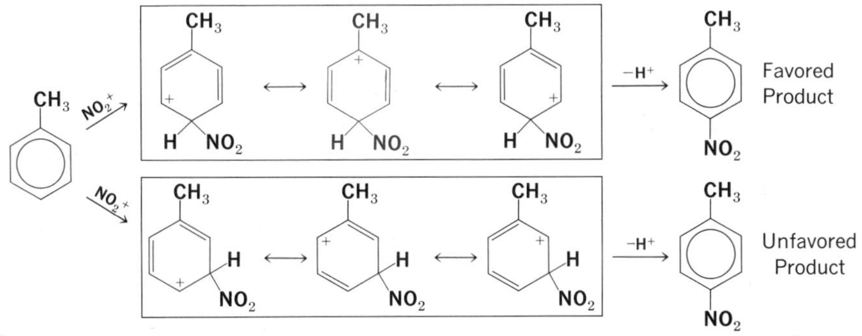

Figure 5-17. *Addition of* $^{+}NO_2$ *to either the* para *or* meta *position of toluene gives an intermediate carbonium ion in which the charge is shared by three carbons. In* para *attack, one of these carbons bears the methyl group that can help stabilize the adjacent charge. (The structure in red is favorable.) In* meta *attack, the methyl group is never attached to a positively charged carbon.*

A Lewis acid like $^{+}NO_2$ might, in theory, attack a toluene molecule in any of the three positions: *ortho, meta,* or *para.* We shall consider why it is easier to attack *para* than *meta,* leaving the case for *ortho* attack to be worked out in Prob. 5-5. When the aromatic ring is attacked, a carbonium ion is formed, and this ion is adjacent to a double bond. As a consequence the charge is delocalized, i.e., it can be shared by other carbon atoms in the molecule (compare the discussion for addition reactions to butadiene, Fig. 3-28). Note in Fig. 5-17 that, for both para and meta attack, the positive charge is shared by three ring carbons, as shown by the three resonance structures that may be written for these charged species.

We know from our discussion of the origin of Markovnikov's rule that addition to an unsaturated system will be easiest in that direction which produces the more stable carbonium ion. The carbonium ion produced by *para* attack on toluene is a hybrid of two secondary carbonium ions (the first and third ions in Fig. 5-17) and one tertiary carbonium ion (the second ion). From *meta* attack we also get a hybrid, but now a hybrid of three secondary cations. *In para attack the methyl substituent is able to help stabilize the cation because the adjacent ring carbon shares the positive charge.* In meta attack the carbon adjacent to the methyl is never positively charged. The methyl group makes attack at the *para* position easier than attack at the *meta* position.

Prob. 5-5. Draw structures for the carbonium ion formed by *ortho* attack of NO_2^+ on toluene and show that here, too, the methyl group is able to stabilize the positive charge.

Consider next aromatic substitution in a benzene ring substituted with a *meta* directing group. We shall use the $—N(CH_3)_3$ group as an example, since we have already considered its effect on Markovnikov's rule (Fig. 3-25). Again we can write resonance structures for the cation formed by attack by the $^+NO_2$ (Fig. 5-18). Now, however, attack at the *para* position results in

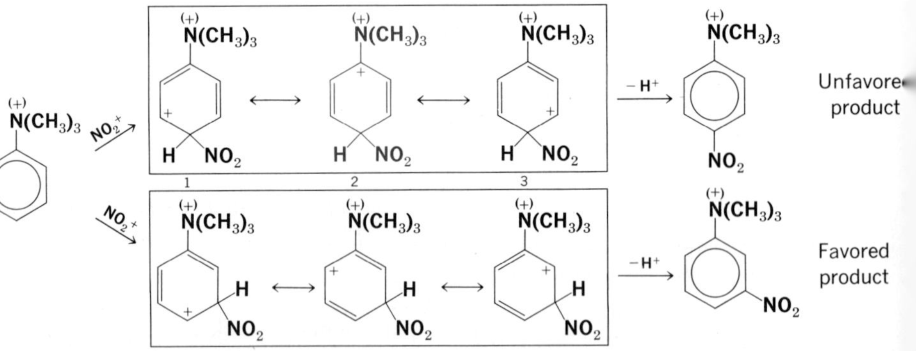

Figure 5-18. *If the benzene ring is substituted with a* meta *directing group, attack at the* para *position is unfavorable because these groups repel a positive charge on an adjacent carbon. (The structure in red is unfavorable.) Attack occurs* meta, *so as to keep the two charges as far away from one another as possible.*

part of the charge being placed on the carbon adjacent to the already positively charged nitrogen atom (the second structure of Fig. 5-18). As we have seen previously, this is very unfavorable, since the two positive charges repel one another. *Meta* attack, where the charges are always farther apart, is more favorable than *para* (or *ortho*) attack. So it is with all the other *meta* directors—they have either a permanent positive charge, like the nitro group (—N(=O)O), or a partial positive charge resulting from charge separation (>C=O $\longleftrightarrow$ >C$^+$—O$^-$). All of these *meta* directing groups oppose introduction of a second positive charge on an adjacent carbon atom and direct incoming groups to the *meta* position.

The key to understanding directive effects in aromatic substitution lies in structures like those in red in Figs. 5-17 and 5-18. If the substituent stabilizes an adjacent positively charged carbon cation, it will be *ortho,para* directing. For the same reason, if this substituent is attached to ethylene, it will promote Markovnikov addition to the double bond. On the other hand, if the substituent repels an adjacent positive charge, it will be *meta* directing and will also cause anti-Markovnikov addition if attached to ethylene. These generalizations are summarized in Fig. 5-19.

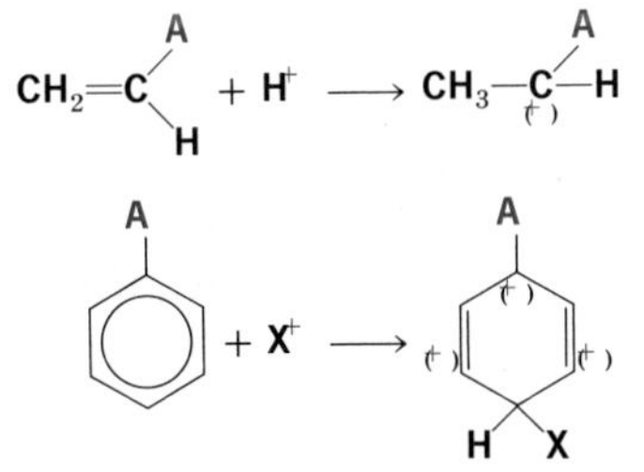

$$CH_2{=}C(A)H + H^+ \longrightarrow CH_3{-}\overset{(+)}{C}(A){-}H$$

$$C_6H_5A + X^+ \longrightarrow [C_6H_5A(H)(X)]^{(+)}$$

Figure 5-19. *If a substituent A makes an adjacent positive charge more stable, it will be an* ortho,para *director in aromatic substitution and will cause Markovnikov addition to a double bond. If A repels an adjacent positive charge, it will be* meta *directing and cause anti-Markovnikov addition.*

We see then how closely related Markovnikov addition and aromatic substitution are, and how we can use the results from one system to make predictions in the other. For instance, we see from Fig. 5-16 that the methoxyl (—OCH_3) group is *o,p* directing. We can then predict with confidence that the compound in Fig. 5-20 will add HX in a Markovnikov fashion, as shown. A cyano group (—CN), which is *meta* directing, will cause anti-Markovnikov addition. It should also be remembered that the *o,p* directing groups of Fig. 5-16 are arranged in order of their directing power; groups at the top of the list (—NH_2, OH) are better at stabilizing an adjacent charge than those at the bottom (alkyl, halogen).

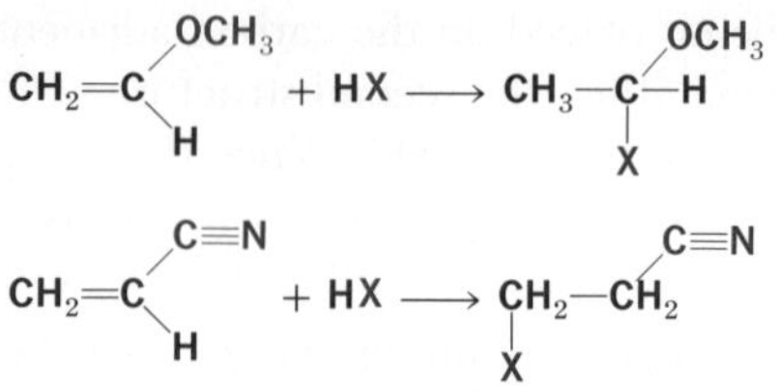

$$CH_2{=}C(OCH_3)H + HX \longrightarrow CH_3{-}C(OCH_3)(X){-}H$$

$$CH_2{=}C(C{\equiv}N)H + HX \longrightarrow CH_2(X){-}CH_2{-}C{\equiv}N$$

Figure 5-20. *Since —OCH_3 is* o,p. *directing, it will cause Markovnikov addition to ethylene, while the* meta *directing —C≡N group causes anti-Markovnikov addition.*

Prob. 5-6. What product would you expect if you added HCl to $CH_2{=}CH{-}NO_2$? To $CH_2{=}CHBr$?

Reactivity of Substituted Benzenes

There is an interesting corollary to this explanation of the directing effects. Since *ortho-para* directing groups achieve their effect by stabilizing the positive charge formed, and *meta* directing groups achieve theirs by destabilizing it, we might expect benzene derivatives containing *ortho-para* directing groups to be more reactive than unsubstituted benzene, and those containing *meta* directing groups to be less reactive than benzene itself. That this is usually the case can be seen from Fig. 5-21, where conditions required to nitrate a variety of substituted benzenes are compared with the conditions required for benzene itself. Further substitution on nitrobenzene is quite difficult, and it is nearly impossible to nitrate dinitrobenzene further to trinitrobenzene. Toluene, on the other hand, can be trinitrated, yielding trinitrotoluene, the explosive TNT, because the methyl group activates the ring. The halogens are the main exception. These *ortho-para* directing atoms also make the ring less reactive.

Nomenclature

We have already seen a number of examples of the naming of aromatic compounds. Common names are largely employed for the unsubstituted hydrocarbons, such as benzene, naphthalene, and anthracene, and sometimes for simple substituted compounds, like toluene, xylene (dimethylbenzene), and

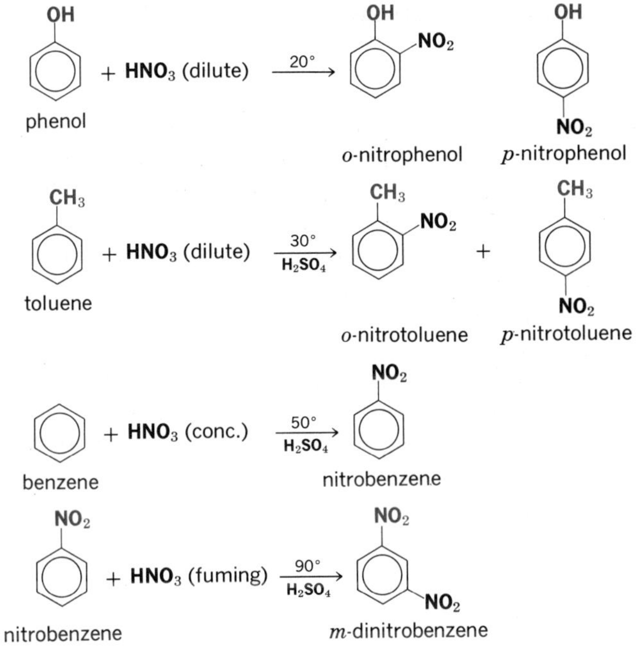

OH
+ HNO3 (dilute)
20°
OH
NO2
OH
NO2
phenol
o-nitrophenol
p-nitrophenol
CH3
+ HNO3 (dilute)
30°
H2SO4
CH3
NO2
+
CH3
NO2
toluene
o-nitrotoluene
p-nitrotoluene
+ HNO3 (conc.)
50°
H2SO4
NO2
benzene
nitrobenzene
NO2
+ HNO3 (fuming)
90°
H2SO4
NO2
NO2
nitrobenzene
m-dinitrobenzene

Figure 5-21. *Further substitution of a benzene ring already containing an* ortho-para *directing group (OH, CH_3) is usually easier than substitution of benzene itself. Further substitution of a benzene ring containing a* meta *directing group (NO_2) is more difficult.*

phenol. The *o*, *m*, *p* nomenclature in Fig. 5-21 is common for aromatic compounds containing disubstituted rings, but for rings with more than two substituents numbers are usually employed (Fig. 5-22). When a benzene ring is named as a substituent, its group name is *phenyl.* The name *benzyl* for the group $C_6H_5CH_2$— is also commonly used in nomenclature.

CH3
Cl
Cl
2,4-dichlorotoluene
CH2CH3
phenylethane
(ethylbenzene)
CH2—Cl
benzyl
chloride

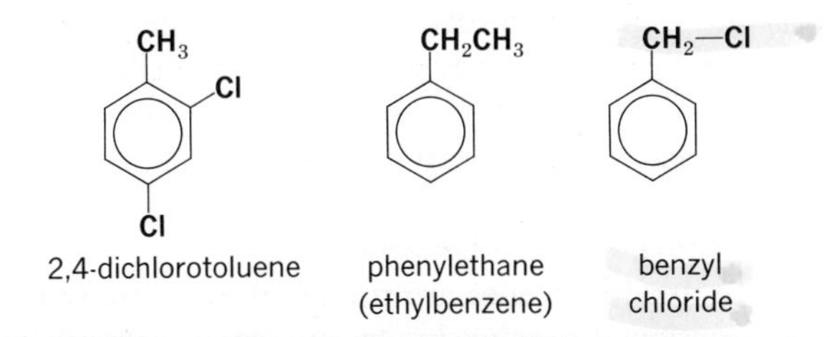

Figure 5-22. *Examples of the use of numbers and of the* phenyl (C_6H_5—) *and* benzyl ($C_6H_5CH_2$—) *groups in the naming of aromatic compounds.*

Sources of Aromatic Hydrocarbons

Aromatic hydrocarbons were originally obtained mainly from coal, and aliphatic hydrocarbons mainly from petroleum. Heating coal in the absence of air leads to the distillation of *coal tar,* a complex mixture of compounds from which benzene, alkyl benzenes, naphthalene, anthracene, and more complex aromatic compounds may be isolated. The residue from this distillation is *coke,* which is mainly carbon. *Coal gas,* a mixture consisting mostly of hydrogen and methane, is also produced. A variety of compounds containing oxygen and nitrogen is also obtained from coal tar.

These aromatic compounds do not all exist in coal itself, but some form during the strong heating involved in the distillation process. A number of the polycyclic hydrocarbons found in coal tar have the ability to cause cancer when applied to the skin. They are called *carcinogens.* For example, 3,4-benzpyrene is quite carcinogenic (Fig. 5-23) and occurs to the extent of

$$\text{coal} \xrightarrow[\text{(no } O_2)]{\text{heat}} \text{coal gas} + \text{coal tar} + \text{coke}$$

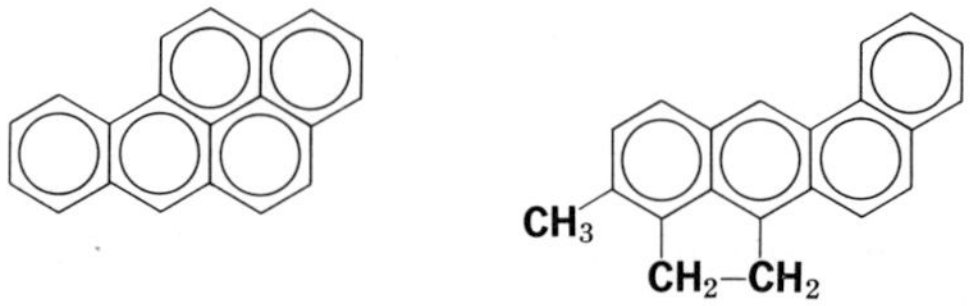

CH_3

CH_2-CH_2

3,4-benzpyrene 3-methylcholanthrene

Figure 5-23. *Distillation of coal is an important source of many aromatic hydrocarbons, including benzene and naphthalene. The two polynuclear hydrocarbons shown are* carcinogens. *The process by which they cause cancer is not known.*

1.5% in coal tar. Before this was recognized, skin cancer was a common occupational hazard among people like chimney sweeps who came into frequent contact with coal tar. Polynuclear hydrocarbons are almost always formed to some extent whenever organic molecules are heated to high temperatures, and it is presumed that carcinogenic hydrocarbons formed in the burning of cigarettes are a main cause of lung cancer.

Coal tar has not proved, however, to be an adequate source for aromatic hydrocarbons; therefore, methods for converting aliphatic hydrocarbons from petroleum into aromatics have been developed. Passing *n*-heptane over a catalyst at 500° converts it in high yield into toluene (Fig. 5-24); similarly, acetylene can be converted into benzene.

A number of alkyl benzenes are prepared industrially by the Friedel–Crafts reaction mentioned previously. An important example is the reaction of benzene with tetrapropylene, whose product is then sulfonated to furnish a synthetic detergent. A simpler example is the formation of ethylbenzene from benzene and ethylene (Fig. 5-25). Dehydrogenation gives the extremely

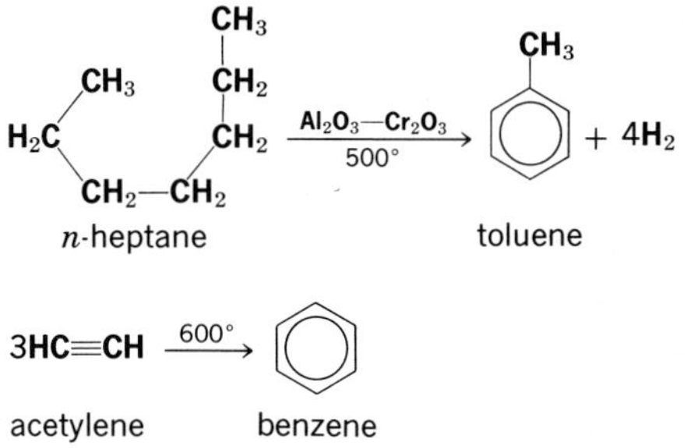

n-heptane $\xrightarrow[500^\circ]{Al_2O_3-Cr_2O_3}$ toluene $+ 4H_2$

$3HC{\equiv}CH \xrightarrow{600^\circ}$ benzene

acetylene, benzene

Figure 5-24. *Toluene and other aromatic hydrocarbons may be formed by* dehydrogenation *of aliphatic hydrocarbons, and benzene by the trimerization of acetylene.*

important compound *styrene,* a useful monomer for the formation of a variety of plastics. Polystyrene is widely used in such items as toothbrush handles and plastic toys, and a copolymer of styrene and butadiene is the synthetic rubber GR-S (see Fig. 4-10).

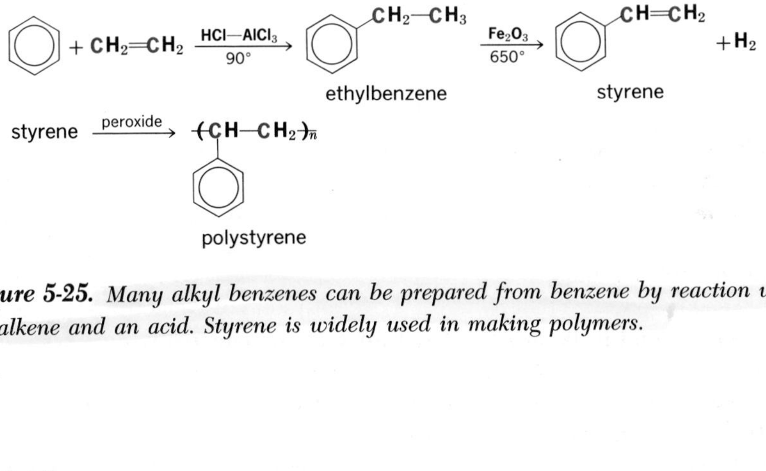

benzene $+ CH_2{=}CH_2 \xrightarrow[90^\circ]{HCl-AlCl_3}$ ethylbenzene ($C_6H_5CH_2-CH_3$) $\xrightarrow[650^\circ]{Fe_2O_3}$ styrene ($C_6H_5CH{=}CH_2$) $+ H_2$

styrene $\xrightarrow{\text{peroxide}}$ $(CH(C_6H_5)-CH_2)_n$ polystyrene

Figure 5-25. *Many alkyl benzenes can be prepared from benzene by reaction with an alkene and an acid. Styrene is widely used in making polymers.*

Aryl Alkanes

The alkyl benzenes contain both aromatic and aliphatic parts; in their reactions they provide an interesting comparison between the two classes of compounds. Such molecules are often called *aryl alkanes.* The great stability of the aromatic ring system is shown in the oxidation of these compounds. The aliphatic portion of the molecule is attacked and the side chain, no matter how long, is removed (Fig. 5-26). Again, while ionic halogenation with acids as catalysts causes the benzene ring to react, halogenation with free radicals in the presence of light induces substitution in the aliphatic portion of the

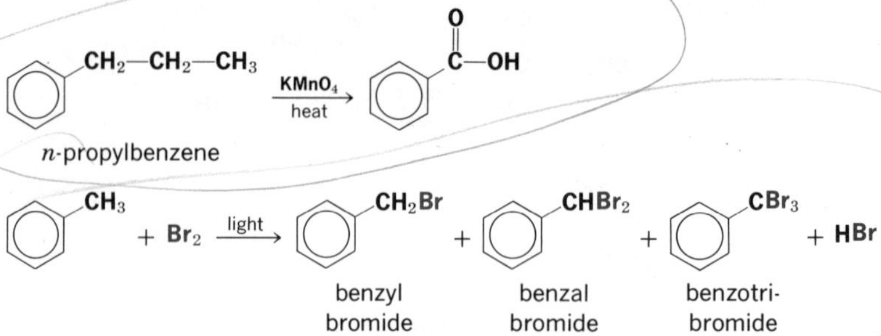

$$C_6H_5{-}CH_2{-}CH_2{-}CH_3 \xrightarrow[\text{heat}]{KMnO_4} C_6H_5{-}\overset{O}{\overset{\|}{C}}{-}OH$$

n-propylbenzene

$$C_6H_5{-}CH_3 + Br_2 \xrightarrow{\text{light}} C_6H_5{-}CH_2Br + C_6H_5{-}CHBr_2 + C_6H_5{-}CBr_3 + HBr$$

benzyl bromide, benzal bromide, benzotribromide

Figure 5-26. *In some reactions the aliphatic portion of an alkyl benzene is attacked more readily than the aromatic portion.*

molecule, most readily at the carbon adjacent to the benzene ring, the *benzylic* carbon.

Nonbenzenoid Aromatic Hydrocarbons

One might well inquire whether it is possible to have an aromatic compound that does not contain a benzene ring. A four-membered ring containing two double bonds (cyclobutadiene) or an eight-membered ring with four double bonds (cyclooctatetraene) might be expected to display delocalization of electrons and other aromatic properties. Yet these two compounds do not. Cyclooctatetraene is a well-known chemical whose double bonds are normally reactive; the molecule contains alternate double and single bonds. Cyclobutadiene has proved too reactive to be isolated, although it has been formed and observed near absolute zero.

Theory predicts that, whereas a cyclic system of π electrons is necessary for a compound to be aromatic, not all such systems really are. In fact, we predict that only those systems with 2, 6, 10, 14, 18, . . . π electrons in the ring will be aromatic. In these systems the number of π electrons is given by the series $(4n + 2)$, where n is zero or an integer 1, 2, 3, etc. Those cyclic systems with only $4n$ π electrons, like cyclobutadiene with 4 π electrons or cyclooctatetraene with 8 π electrons, will not be aromatic (Fig. 5-27). In this respect, there are closed shells of aromatic stability analogous to the closed shells of 2, 8, 18, . . . electrons observed in atomic structure.

One would predict, then, that a 10-membered ring with 5 double bonds (a total of 10 π electrons) would be an aromatic system. A derivative of such a system has recently been prepared and does indeed display aromatic characteristics, despite the inability of 10 atoms with 120° bond angles to fit comfortably into a ring. An 18-membered ring containing 9 double bonds has also been prepared and shows definite aromatic properties.

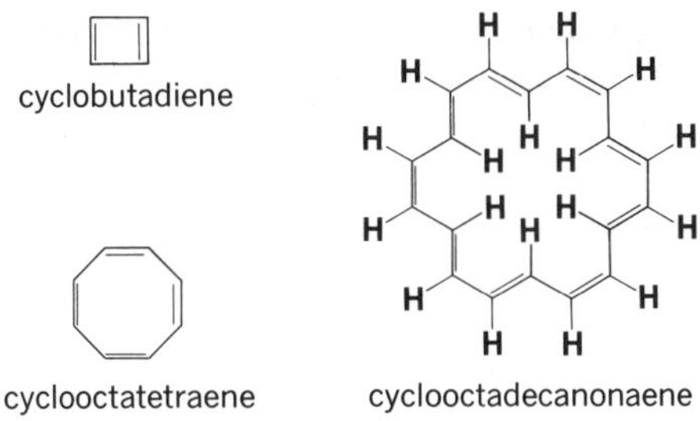

cyclobutadiene

cyclooctatetraene

cyclooctadecanonaene

Number of π electrons necessary for aromatic stability = $(4n + 2)$, $n = 0, 1, 2, 3, \ldots.$

Figure 5-27. *Cyclobutadiene and cyclooctatetraene, with 4 and 8 π electrons respectively, do not meet the* 4n + 2 *requirement for aromatic stability. Cyclooctadecanonaene, with 18 π electrons, fits the rule* (n = 4) *and has some aromatic characteristics.*

Prob. 5-7. Which of the following systems would you expect to be "aromatic" and which "nonaromatic," according to the $4n + 2$ rule, assuming they could be made without undue bond strain: (a) a 20-membered ring with 10 double bonds and (b) a 30-membered ring with 15 double bonds?

The aromatic character shows up in an even more striking way in certain organic ions. The carbonium ions discussed up to this point have been highly reactive and short-lived. Nevertheless, 1-bromocycloheptatriene exists as an ionic salt that is stable, high-melting, and soluble in water; notice in Fig. 5-28 that its seven-membered ring contains three double bonds, has a positive

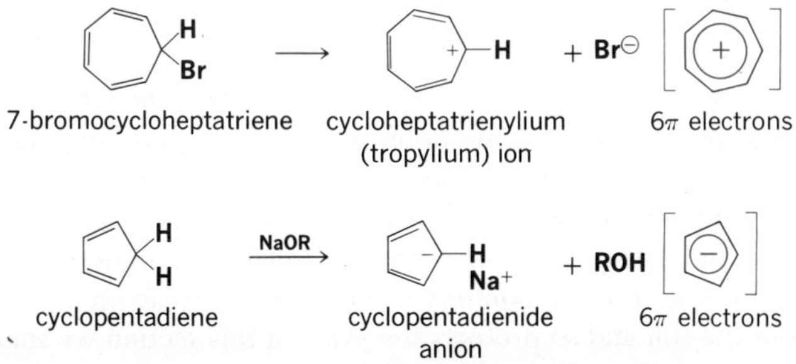

7-bromocycloheptatriene → cycloheptatrienylium (tropylium) ion + $Br^{\ominus}$ — 6π electrons

cyclopentadiene —NaOR→ cyclopentadienide anion (Na^+) + ROH — 6π electrons

Figure 5-28. *Certain organic ions are especially stable because they have aromatic sextets of π electrons.*

charge and *six* π electrons, and is therefore aromatic. Analogously, cyclopentadiene, in contrast to nearly all other hydrocarbons, is as acidic as some alcohols, and is readily converted into its sodium salt; notice in the figure that its five-membered ring contains two double bonds and a pair of unbonded electrons and thus has an aromatic sextet of electrons.

Absorption of Light by Organic Molecules

The electromagnetic spectrum is a vast continuum stretching from the very short wave cosmic, gamma, and X rays through the ultraviolet (uv), visible, and infrared to the long wavelength radio waves (Fig. 5-29). When electro-

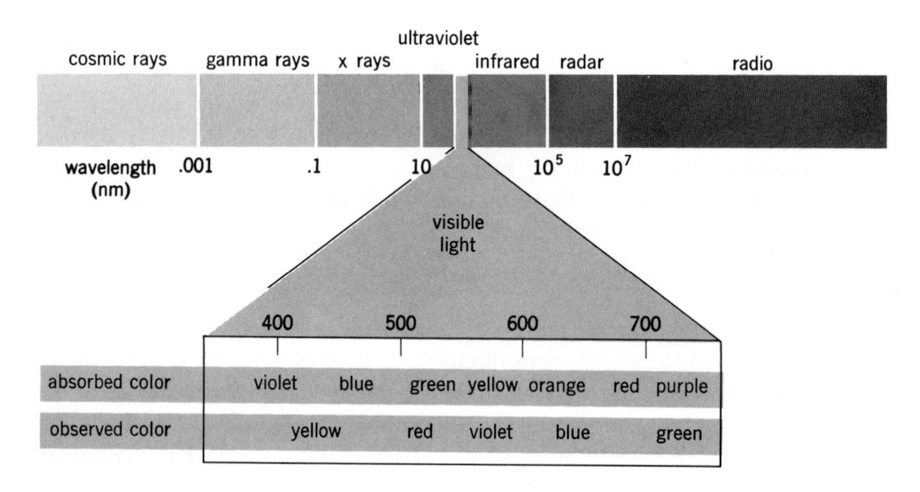

Figure 5-29. *Visible light comprises only a tiny fraction of the spectrum of electromagnetic radiation.*

magnetic radiation falls on a molecule, whether organic or inorganic, it may or may not be absorbed, depending upon the wavelength of the radiation and the structure of the molecule. If the molecule absorbs light in the visible region, it will appear to be colored; if not, the molecule will be colorless if it transmits the light (like water) or white if it reflects the light. Even colorless compounds will absorb *ultraviolet light;* for example, clear suntan oil absorbs some uv from the sun and so protects the skin. In this section we shall discuss the absorption of visible and ultraviolet light by organic compounds, and show how this absorption may be used to identify and characterize the compounds.

When ultraviolet or visible light is absorbed by an atom or molecule, an electron is *excited,* i.e., it is raised to an orbital of higher energy than the one it usually occupies. In an atom, for example, an electron might move from an *s* to a *p* orbital upon absorption of light; in a molecule an electron moves from one molecular orbital to another, more energetic one. This excitation is possible because light in the visible and ultraviolet regions of the spectrum contains a great deal of energy—of the order of 50 to 200 kcal/mole, or even more at very short wavelengths. Since all matter contains electrons, there will be some portions of the visible-ultraviolet spectrum in which light absorption will occur for any molecule.

For practical reasons the ultraviolet-visible region of the spectrum is divided into three parts. Light whose wavelength is shorter than 200 nm (2000 Å) is referred to as the *vacuum ultraviolet;* here air and most solvents absorb, and measurements must be carried out, with great difficulty, in a vacuum. Every organic molecule absorbs light very strongly in the vacuum ultraviolet part of the spectrum. The region from 200 nm to 360 nm is called the *near ultraviolet* (or simply the *ultraviolet*), and from 360 to about 700 nm the *visible.* Light in these regions has less energy and is only able to excite relatively loosely bound electrons, like those in π bonds. Note the absorption spectrum of benzene, phenanthrene and naphthacene in Fig. 5-30. Colorless

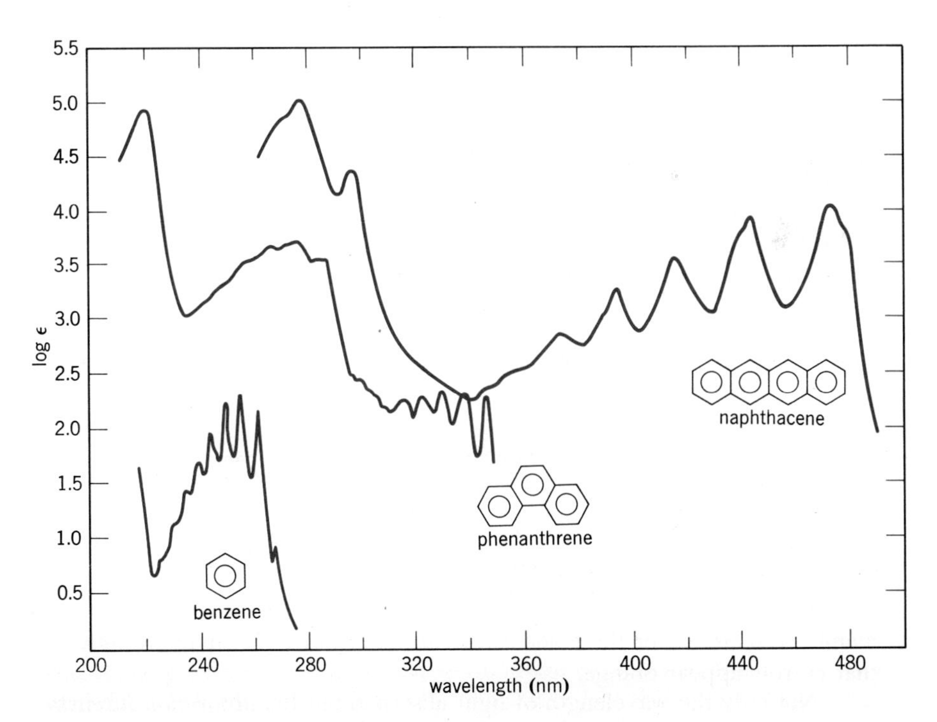

Figure 5-30. *The absorption spectrum of benzene, phenanthrene, and napthacene in the ultraviolet and visible regions of the spectrum.*

benzene does not absorb in the visible part of the spectrum, while the yellow naphthacene absorbs in the violet and blue part of the visible spectrum. The light we see when we look at a sample of naphthacene is white light from which the violet and blue portions have been removed by absorption; i.e., yellow light is reflected. It is important to remember this complementary nature of color: The color observed on reflection is always white light from which the absorbed color has been removed. An object that appears green actually absorbs purple light, for example, while a compound that absorbs green light will appear red. The relationship between absorbed and observed color is shown in Fig. 5-29.

As a general rule, molecules containing conjugated systems of π electrons absorb light closer to the visible region than saturated molecules or those with isolated double or triple bonds. The longer the conjugated system, the longer the wavelength of the light absorbed.

Ethylene absorbs light of wavelength in the far ultraviolet part of the spectrum, at 180 nm. Butadiene, with two conjugated double bonds, absorbs at 217 nm, a wavelength closer to the visible zone than that absorbed by ethylene. 1,3,5-Hexatriene absorbs still closer to the visible region, at 258 nm. All three compounds are colorless; however, as the number of conjugated double bonds increases, the position of absorption falls nearer the visible region, and with enough conjugation, the molecules are colored. The pigment β-carotene (Fig. 5-31) contains 11 conjugated double bonds and is the dye

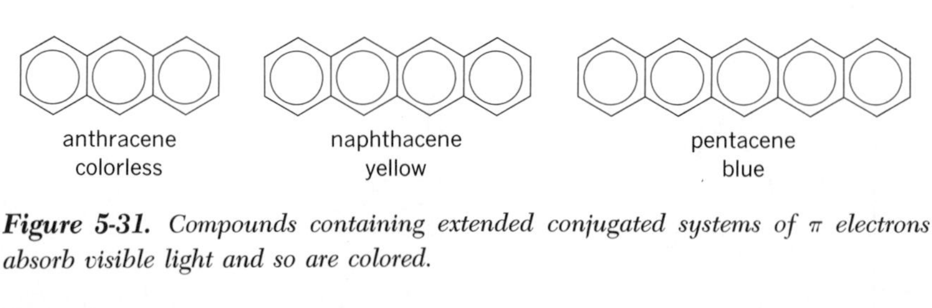

Figure 5-31. *Compounds containing extended conjugated systems of π electrons absorb visible light and so are colored.*

mainly responsible for the color of carrots. It absorbs blue light strongly, so that carrots appear orange.

Not only the wavelength of light absorbed but the *absorption intensity* is characteristic of a given molecule. Note that both benzene and naphthacene absorb light in the near ultraviolet but that the latter does so much more intensely. A solution of naphthacene will absorb almost 100 times as much light at 250 nm as a solution of benzene of the same molar concentration. The extinction coefficient ε is defined as the ratio of the light intensity (I_0) that enters a 1 *M* solution of the compound in a cell 1 cm long to the intensity of the light that emerges after absorption (I).[1] Since this ratio can vary from 1 (no absorption) to 10,000 or more, it is usually plotted as log ε, so that both large and small values can be easily seen on a small graph (see Fig. 5-30). For molecules that absorb strongly (large ε), very dilute samples have to be employed so that I_0/I can be measured accurately.

Prob. 5-8. A 0.01 *M* solution of a compound in a 1-cm cell absorbed 91% of the light at 400 nm. What is its ε value at this wavelength? What is its log ε?

Ultraviolet-visible light may sometimes be used to identify a molecule by using its absorption spectrum as a "fingerprint" and comparing it with the spectrum of an authentic sample of the material. Even if an authentic spectrum is not available for comparison, clues to the structure of an unknown compound can often be gained from its uv spectrum. A hydrocarbon that does not absorb in the uv, for example, is either saturated or contains only nonconjugated multiple bonds.

Visual Chemical Tests for Aromatic Hydrocarbons

Since aromatic hydrocarbons do not react with bromine in carbon tetrachloride, this test can be used to distinguish them from other unsaturated molecules. Aromatic hydrocarbons are relatively soluble in concentrated sulfuric

[1] The compound is dissolved in some solvent that does not absorb in the near uv or visible. Water, alcohol, and hexane are typical solvents used.

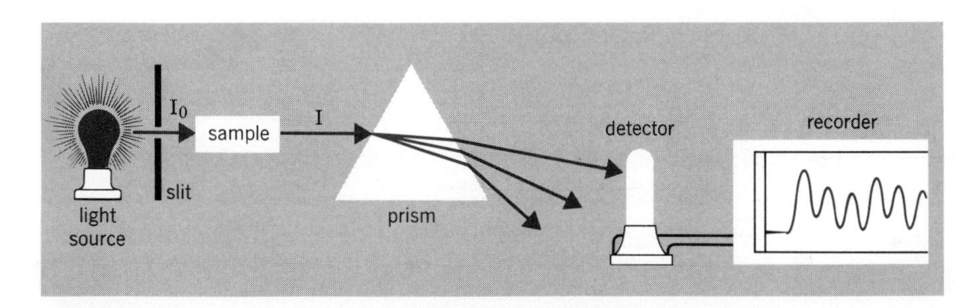

Figure 5-32. *A schematic diagram of an apparatus for measuring the uv-visible absorption spectrum. The prism is used to select a particular wavelength; the absorption coefficient (ε) at that wavelength is given by* I_0/I *for a 1* M *solution 1 cm long.*

acid, while alkanes are not. If a few crystals of naphthalene, for example, are shaken with H_2SO_4, they will quickly and obviously go into solution. Almost any organic molecule except an alkane will also dissolve in concentrated H_2SO_4, so other tests (e.g., Br_2 in CCl_4) must be used first to distinguish aromatic hydrocarbons from compounds that contain functional groups (such as double and triple bonds).

PROBLEMS

1. Draw structures for the following compounds:

(a) *p*-dichlorobenzene
(b) 1,3,5-trinitrobenzene
(c) *m*-bromochlorobenzene
(d) *p*-nitrobenzenesulfonic acid
(e) isopropylbenzene
(f) toluene
(g) *p*-chlorostyrene
(h) 1-phenyl-2-butene
(i) benzyl bromide
(j) *o*-chlorotoluene
(k) 1-bromo-3-chloro-4-nitrobenzene
(l) *p*-di-*t*-butylbenzene

2. Name the following compounds:

(a) Br— —Cl
(b) —NO_2, NO_2
(c) CH_3— —I
(d)
(e) —CH_2—CH_3
(f) —CH_2CH_2Cl
(g) Cl— —NO_2, Cl
(h) —CH_3, NO_2
(i)

3. Complete the following reactions:

(a) + $(CH_3)_2CHCl$ $\xrightarrow{AlCl_3}$

(b) CH_3—CH=CHNO_2 + HCl $\longrightarrow$

(c) + $14H_2$ $\xrightarrow[\text{heat, pressure}]{Pt}$

(d)

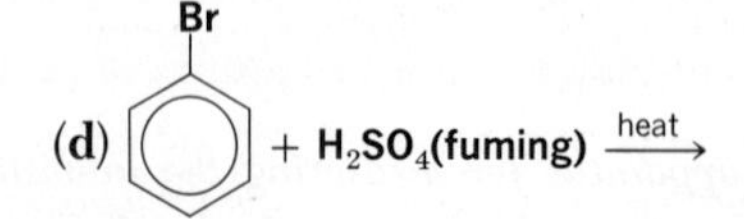

(e) NO_2 $+ Cl_2 \xrightarrow{FeCl_3}$

(f) OCH_3 $+ HNO_3 \longrightarrow$

(g) CH_3, CH_3 $+ KMnO_4(\text{excess}) \xrightarrow{\text{heat}}$

(h) CH_3, OCH_3 $+ Br_2 \xrightarrow{FeBr_3}$

(i) $CH_3OCH{=}CHCH_3 + HCl \longrightarrow$

(j) $CH_2{-}CH_3$ $+ \quad Br_2 \xrightarrow{\text{light}}$
(one equivalent)

4. What would be the expected monochlorinated products if each of the following compounds were reacted with chlorine and ferric chloride:
   (a) $C_6H_5CH_3$ (b) $C_6H_5NO_2$
   (c) p-$O_2NC_6H_4OH$ (d) C_6H_5CN
   (e) m-$O_2NC_6H_4CH_3$ (f) p-$ClC_6H_4OCH_3$
5. How could you prepare each of the following compounds in the laboratory starting with either benzene or toluene? Assume an *o*-, *p*-mixture can be separated into its constituents.
   (a) bromobenzene (b) ethylbenzene
   (c) *p*-nitrochlorobenzene (d) *m*-nitrochlorobenzene
   (e) *p*-methylbenzenesulfonic acid (f) *p*-dimethylbenzene(*p*-xylene)
6. *p*-Cymene occurs in eucalyptus oil and oil of thyme and has been shown to be *p*-isopropyltoluene. How might this compound be prepared from toluene?
7. Natural musk (muscone) is obtained from a small male musk deer. Artificial musk is 3-*tert*-butyl-2,4,6-trinitrotoluene. Draw its structure.
8. The three isomeric dimethylbenzenes are known as xylenes and are obtained from coal oil or, by a special cracking process, from petroleum. The three have very similar boiling points and so are difficult to separate by distillation. A process for separating them has been devised; it depends for its success on the fact that *m*-xylene undergoes aromatic substitution reactions more rapidly than the other two isomers. Draw the structures

for the carbonium ions that would be formed upon sulfonation of the three isomers, and show how the relative stabilities of the carbonium ions can be used to explain why the *m*-isomer is sulfonated the most rapidly of the three.

9. Azulene ($C_{10}H_8$) is a blue, aromatic hydrocarbon isomeric with naphthalene, but it contains a five and a seven-membered ring fused together. Draw the structure of azulene and show that there are two different ways the five double bonds can be arranged. If the heat of hydrogenation of azulene with five molecules of H_2 is 113 kcal/mole, what is its resonance energy? Assume a value of 28.6 kcal/mole for the heat of hydrogenation of a double bond.

10. A synthetic detergent can be synthesized by the following series of reactions. An alkene is formed by the acid-catalyzed tetramerization of propylene (Prob. 3, Chapt. 4). In the presence of acid this alkene can be used to carry out a Friedel–Crafts reaction on benzene. The resultant product is sulfonated by H_2SO_4; after neutralization with NaOH a sodium alkyl benzene sulfonate is formed. Write reactions for this synthesis. Would you expect the product to be a single isomer?

6 alcohols, phenols, ethers

In introducing the basic ideas of organic structure and reactivity we have, to this point, emphasized compounds of carbon and hydrogen only, the hydrocarbons. Most organic molecules also contain oxygen, and our subsequent discussions will involve the chemistry of functional groups containing oxygen and, to a lesser extent, those containing nitrogen and sulfur. In this chapter we shall consider the organic derivatives of water, namely the alcohols, phenols, and ethers.

The simplest alcohol, methyl alcohol, may be obtained by the dry heating of wood in the absence of air, hence the common name "wood alcohol." It was obtained in relatively pure form by Robert Boyle as early as 1661, but its structure and the recognition of its relationship to ethyl alcohol dates from the 1830s, when, the word "methyl," derived from the Greek words for wine and wood, was coined. Ethyl alcohol has, of course, been known and used for thousands of years, during which time its relaxing and inebriating properties have both blessed and cursed society. Many other alcohols, both simple and complex, occur naturally.

Alcohols contain the hydroxyl group (—OH), and phenols are compounds in which the hydroxyl group is directly attached to an aromatic ring. The presence of hydroxyl groups gives alcohols many of the reactions of water. Since they are in part hydrocarbons, alcohols also bear some physical and chemical resemblance to alkanes; ethyl alcohol, for example, is soluble both in water and in saturated hydrocarbons, and it burns easily. Its boiling point (78°) is lower than that of water, although its molecular weight is much higher.

Physical Properties of Alcohols

The chemical behavior of alcohols is almost entirely determined by the hydroxyl group, since the alkyl portion of the molecule is relatively unreactive.

Nevertheless, the latter does affect the physical properties of an alcohol, especially such properties as boiling point and solubility in water. The solubility of an alcohol in water decreases as the size of its alkyl group increases (Table 6-1); within the homologous series of *normal* alcohols, the boiling point increases with greater molecular weight.

Table 6-1
Correlation of the Size of the Alkyl Group in Alcohols with their Solubility in Water and Boiling Point

Common Name	Alcohol	Solubility in Water, g/100 ml	B.P.,°
methyl alcohol	CH_3OH	∞	64.9
ethyl alcohol	CH_3CH_2OH	∞	78.5
n-propyl alcohol	$CH_3CH_2CH_2OH$	∞	97.2
n-butyl alcohol	$CH_3(CH_2)_2CH_2OH$	9	117.5
n-pentyl alcohol	$CH_3(CH_2)_3CH_2OH$	2.7	138.0
n-hexyl alcohol	$CH_3(CH_2)_4CH_2OH$	0.6	158.0

Just as they do in water, the hydroxyl groups in alcohols form *hydrogen bonds* with one another—that is, they form loose bonds between the hydrogen atom of one hydroxyl group and the oxygen atom of another. This hydrogen bonding accounts for the high boiling point of ethanol (78°) compared to that of propane (−42°), even though both have about the same molecular weight. Hydrogen bonding is not as extensive in ethanol as in water, and so its boiling point is lower, despite its greater molecular weight. Ethanol is completely soluble in water, but when a liter of ethanol is added to a liter of water at 20°, only 1.93 liters of the mixture is produced. Since the hydrocarbon-like ethyl groups of the alcohol cannot hydrogen bond, they slip into holes in the water structure and do not increase the total volume quite proportionally.

When two or more hydroxyl groups are present in an organic molecule, hydrogen bonding increases and the boiling point rises sharply. Common examples are ethylene glycol ($HOCH_2CH_2OH$, b.p. 197°) and glycerol, also called glycerine ($HOCH_2\underset{\displaystyle OH}{\underset{|}{C}}HCH_2OH$, b.p. 290°).

IUPAC Nomenclature of Alcohols

In the systematic nomenclature of alcohols the presence of the hydroxyl group is indicated by the ending *-ol*. Methyl alcohol then is called *methanol* and ethyl alcohol *ethanol.* The parent hydrocarbon chain is numbered so as to give the carbon bearing the hydroxyl group the smallest possible number. If a molecule contains both a hydroxyl group and a double or triple bond, both endings are used but the hydroxyl group is given the lower number and its

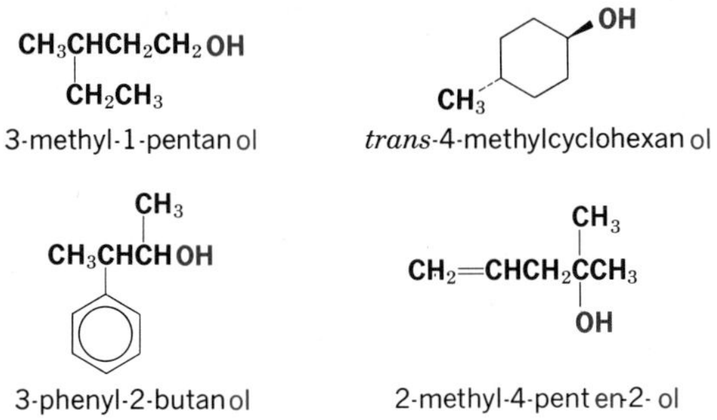

$CH_3CHCH_2CH_2OH$ with CH_2CH_3 on C-3
3-methyl-1-pentanol

trans-4-methylcyclohexanol (CH_3, OH)

$CH_3CH(C_6H_5)CH(CH_3)OH$
3-phenyl-2-butanol

$CH_2{=}CHCH_2C(CH_3)_2OH$ (OH on C bearing CH_3)
2-methyl-4-penten-2-ol

Figure 6-1. *Application of the IUPAC system of nomenclature to alcohols is illustrated by the compounds shown.*

ending is placed last. Some examples of the IUPAC nomenclature of alcohols are given in Fig. 6-1 and in the problems at the end of the chapter. Common names are also used for alcohols, based on the structure of the alkyl group they contain, e.g., *t*-butyl alcohol, benzyl alcohol.

Alcohols as Acids

Alcohols mainly undergo two types of reactions: those that break the O—H bond and those that break the C—O bond (Fig. 6-2). Reactions of the former

$$2CH_3CH_2{-}O{\wr}H + 2\,Na \longrightarrow 2CH_3CH_2O^-Na^+ + H_2\uparrow$$

ethyl alcohol → sodium ethoxide

$$CH_3CH_2CH_2{\wr}O{-}H + HBr \longrightarrow CH_3CH_2CH_2Br + HOH$$

n-propyl alcohol → *n*-propyl bromide

Figure 6-2. *Alcohols may undergo reactions which involve the breaking of either the C—O or O—H bond. The evolution of bubbles of hydrogen gas when a piece of sodium metal is added provides a good visual test for the presence of an O—H group in an organic molecule.*

type find their parallels in the chemistry of water. As we shall see, reactions of the C—O bond involve intermediates related to those we have met in the chemistry of alkenes.

The sodium salt of an alcohol, conveniently formed by the reaction of an alcohol with metallic sodium, is the organic analog of sodium hydroxide, and in alcoholic solution is often used in organic reactions in which a water-free basic solution is needed. It can be used in the same way as sodium hydroxide to neutralize an acid (Fig. 6-3).

$$RO^- Na^+ + H^+X^- \longrightarrow ROH + NaX$$

Figure 6-3. *Sodium alkoxides are strong bases and react with acids.*

In discussing the chemistry of alcohols, it is convenient to classify them as primary (1°) alcohols if the hydroxyl group is attached to a primary carbon, secondary (2°) if it is attached to a secondary carbon, and tertiary (3°) if it is attached to a tertiary carbon. As acids, primary alcohols are about as weak as water, while secondary and tertiary alcohols are weaker. Conversely, the salt of a tertiary alcohol (for instance, potassium *tert*-butoxide) is a considerably stronger base than potassium hydroxide. Some alcohols and their corresponding alkoxides are compared as acids and bases in Fig. 6-4.

Acidic strength increases (↑)	Alcohols	Alkoxides	Basic strength increases (↓)
	HOH	HO^-	
	CH_3OH	CH_3O^-	
	CH_3CH_2OH	$CH_3CH_2O^-$	
	$(CH_3)_2CHOH$	$(CH_3)_2CHO^-$	
	$(CH_3)_3COH$	$(CH_3)_3CO^-$	

Figure 6-4. *In solution the salts of tertiary alcohols are stronger bases than those of secondary alcohols which in turn are stronger bases than the salts of primary alcohols.*

Prob. 6-1. Classify the alcohols of Fig. 6-1 as primary, secondary, or tertiary.

Esters

By elimination of a molecule of water, primary and secondary alcohols react with cold sulfuric acid to form alkyl hydrogen sulfates. Molecules of this type are known as *esters*. Analogous esters may be formed from nitric and nitrous acid, and from organic acids (Chapter 11). In the formation of an ester the acid loses a hydroxyl group, and the alcohol a proton. Thus only the O—H bond of the alcohol is broken. The reactions are reversible; if an ester is treated with water under the proper conditions, it can usually be hydrolyzed back to the alcohol and the acid.

Among the esters in Fig. 6-5 is nitroglycerine, a pale yellow, oily liquid first discovered in 1846. Nitroglycerine is sensitive to shock and detonates on impact with the ground when a container encasing the compound is dropped. It found little use until Alfred Nobel of Sweden discovered, in 1866, that a mixture of the liquid with diatomaceous earth, sawdust, or other materials

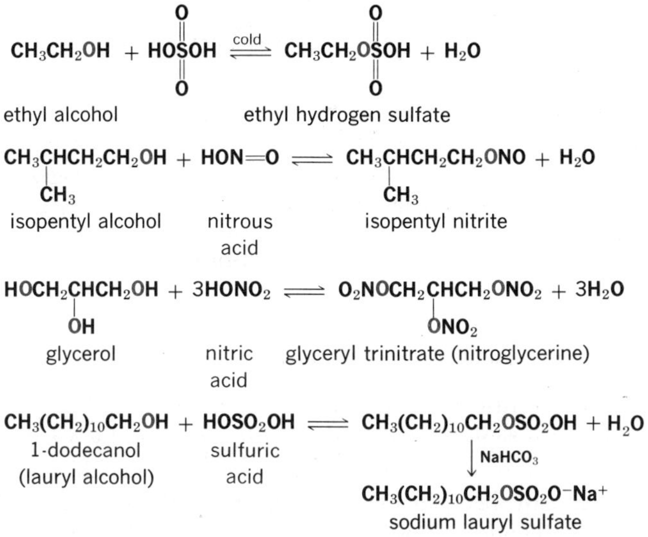

$$CH_3CH_2OH + HOSO_2OH \xrightleftharpoons{\text{cold}} CH_3CH_2OSO_2OH + H_2O$$

ethyl alcohol — ethyl hydrogen sulfate

$$CH_3CH(CH_3)CH_2CH_2OH + HON{=}O \rightleftharpoons CH_3CH(CH_3)CH_2CH_2ONO + H_2O$$

isopentyl alcohol — nitrous acid — isopentyl nitrite

$$HOCH_2CH(OH)CH_2OH + 3HONO_2 \rightleftharpoons O_2NOCH_2CH(ONO_2)CH_2ONO_2 + 3H_2O$$

glycerol — nitric acid — glyceryl trinitrate (nitroglycerine)

$$CH_3(CH_2)_{10}CH_2OH + HOSO_2OH \rightleftharpoons CH_3(CH_2)_{10}CH_2OSO_2OH + H_2O$$

1-dodecanol (lauryl alcohol) — sulfuric acid

$$CH_3(CH_2)_{10}CH_2OSO_2OH \xrightarrow{NaHCO_3} CH_3(CH_2)_{10}CH_2OSO_2O^-Na^+$$

sodium lauryl sulfate

Figure 6-5. *Alcohols react with acids to form esters. Among useful esters of inorganic acids, isopentyl nitrite is used for the relief of pain in certain heart conditions; glyceryl trinitrate is the correct name for nitroglycerine; and sodium lauryl sulfate is a detergent used in toothpaste.*

left its explosive properties intact but removed its sensitivity to shock. In this form, known as dynamite, it even can be burned without exploding. A small amount of a shock-sensitive explosive (called a detonator) will cause dynamite to explode (Fig. 6-6). Nobel amassed a great fortune and left the bulk of it for the establishment of the Nobel prizes. Income from the investment of this fortune still provides the money for these prizes.

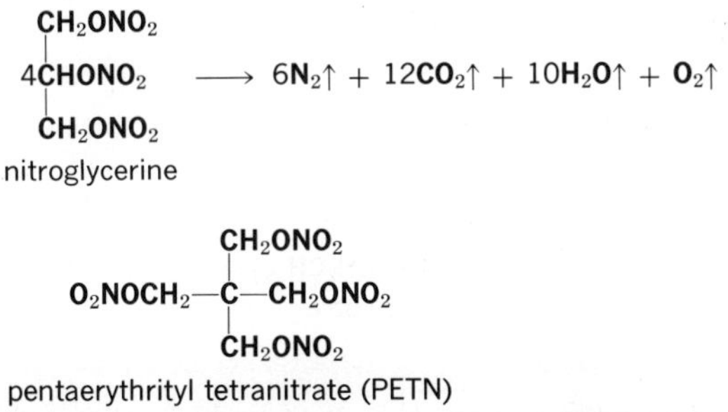

$$4\,CH_2ONO_2{-}CHONO_2{-}CH_2ONO_2 \longrightarrow 6N_2\uparrow + 12CO_2\uparrow + 10H_2O\uparrow + O_2\uparrow$$

nitroglycerine

$$C(CH_2ONO_2)_4$$

($O_2NOCH_2{-}C(CH_2ONO_2)_3$)

pentaerythrityl tetranitrate (PETN)

Figure 6-6. *Nitroglycerine explodes by a process of internal oxidation-reduction, liberating large amounts of gas and heat. Nitrate esters of a number of other polyhydroxy compounds, including PETN, are also used as explosives.*

Prob. 6-2. Esters of phosphoric acid (H_3PO_4) play important roles in biological systems. Genes, for instance, which carry hereditary information, have phosphate esters as an integral part of their structure. It is possible to replace all three hydrogen atoms in phosphoric acid by alkyl groups. Write equations for the formation of ethyl dihydrogen phosphate, diethyl hydrogen phosphate, and triethyl phosphate from ethyl alcohol and phosphoric acid.

Alkyl Halides and Alkenes from Alcohols

One of the characteristic and important reactions of the carbon–oxygen bond of an alcohol is the replacement of the hydroxyl group by a halogen atom. The formation of *n*-propyl bromide from *n*-propyl alcohol and hydrogen bromide has already been mentioned (Fig. 6-2). Primary, secondary, and tertiary alcohols all undergo this reaction, but the conditions required to bring about conversion are quite different for each. These differences form the basis of the *Lucas test*, which is used to differentiate the three types of alcohols. When a tertiary alcohol is dissolved in a solution of zinc chloride in strong hydrochloric acid, an immediate reaction occurs and a distinct layer of the tertiary chloride separates. With the same reagents secondary alcohols usually react within 5 minutes, while primary alcohols require a much longer time for any evidence of reaction (Fig. 6-7).

Heating an alcohol with H_2SO_4 results in the loss of water and the formation of an alkene (Fig. 6-8). In many cases the dehydration may proceed

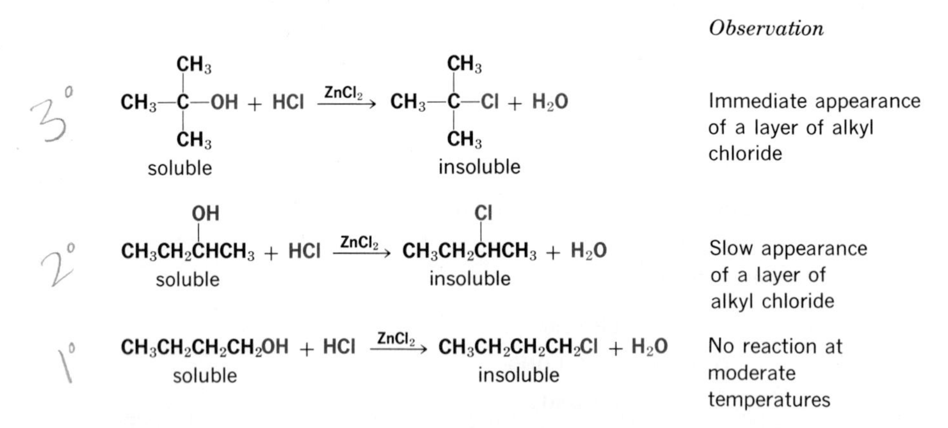

Figure 6-7. *The Lucas test is a visual method to differentiate primary, secondary, and tertiary alcohols by their varying ease of conversion to chlorides. A separate layer of alkyl chloride is observed in a positive test.*

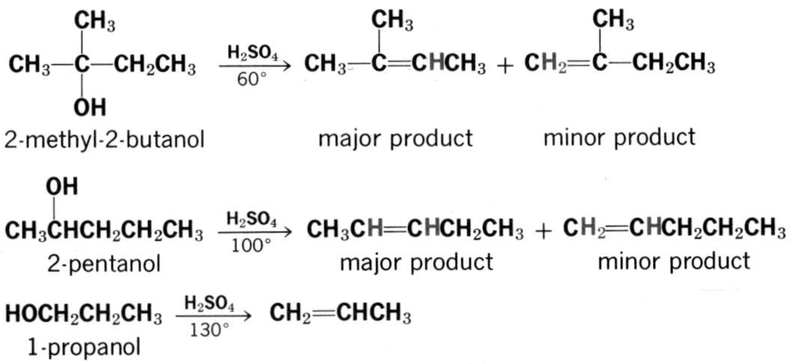

$$CH_3-\underset{OH}{\overset{CH_3}{C}}-CH_2CH_3 \xrightarrow[60°]{H_2SO_4} CH_3-\overset{CH_3}{C}=CHCH_3 + CH_2=\overset{CH_3}{C}-CH_2CH_3$$

2-methyl-2-butanol — major product — minor product

$$CH_3\overset{OH}{C}HCH_2CH_2CH_3 \xrightarrow[100°]{H_2SO_4} CH_3CH=CHCH_2CH_3 + CH_2=CHCH_2CH_2CH_3$$

2-pentanol — major product — minor product

$$HOCH_2CH_2CH_3 \xrightarrow[130°]{H_2SO_4} CH_2=CHCH_3$$

1-propanol

Figure 6-8. *Alcohols are dehydrated when heated with sulfuric acid or other dehydrating agents. A greater amount of the alkene with the fewer hydrogens on the double bond is formed. A higher temperature is required to dehydrate a primary alcohol than a secondary alcohol, and a higher temperature for a secondary than a tertiary alcohol.*

in more than one direction. The major product may then be predicted by *Saytzeff's rule,* which says that *the alkene with the fewest hydrogen atoms attached to the double bond is formed in greatest amount* (Fig. 6-8). In this reaction, too, 3° alcohols react quickly, 2° alcohols slowly and 1° alcohols only with very strong heating.

The close relationship among *tert*-butyl alcohol, isobutylene, and *tert*-butyl chloride is summarized in Fig. 6-9. In the formation of all three, the

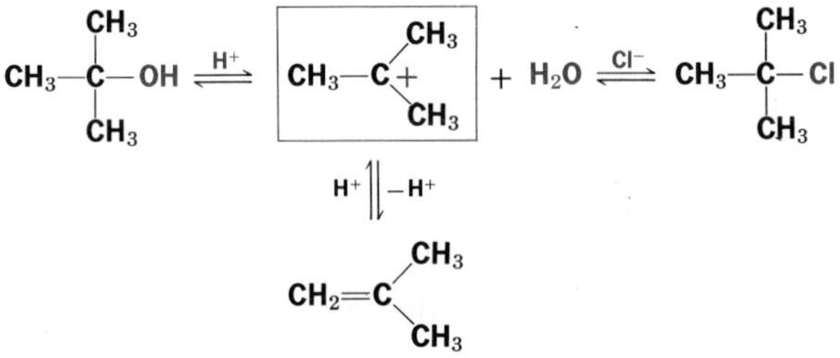

$$CH_3-\underset{CH_3}{\overset{CH_3}{C}}-OH \overset{H^+}{\rightleftharpoons} \boxed{CH_3-\overset{CH_3}{\underset{CH_3}{C^+}}} + H_2O \overset{Cl^-}{\rightleftharpoons} CH_3-\underset{CH_3}{\overset{CH_3}{C}}-Cl$$

$$H^+ \Big\uparrow\Big\downarrow -H^+$$

$$CH_2=C(CH_3)_2$$

Figure 6-9. *Carbonium ions are intermediates in many acid-catalyzed reactions of alcohols, alkenes, and halides, and explain why 3° alcohols are more reactive than 2° alcohols, which in turn are more reactive than 1° alcohols in conversions to alkenes and alkyl halides.*

t-butyl cation plays a central role. A positive charge can be placed with relative ease on a tertiary carbon compared to a secondary or primary carbon, since alkyl groups have more electrons than hydrogen atoms with which to stabilize a positive charge. It is therefore easier to form *t*-butyl chloride or isobutylene from *t*-butyl alcohol by way of the *t*-butyl cation than it is, for example, to form 2-chlorobutane or butenes from 2-butanol by way of the *sec*-butyl cation. The varying reactivities of alcohols are thus seen to have the same underlying explanation as Markovnikov's rule.

Oxidation of Alcohols

Primary and secondary alcohols are readily attacked by oxidizing agents. Tertiary alcohols do not react, since the point of attack is usually the hydrogen atom attached to the carbon bearing the hydroxyl group (Fig. 6-10).

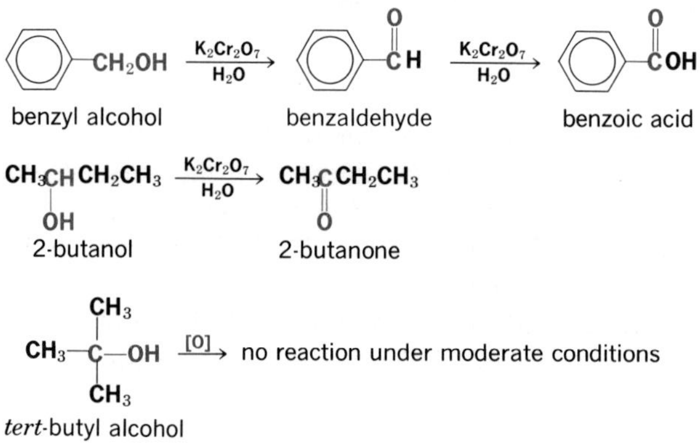

$$C_6H_5CH_2OH \xrightarrow[H_2O]{K_2Cr_2O_7} C_6H_5\overset{O}{\overset{\|}{C}}H \xrightarrow[H_2O]{K_2Cr_2O_7} C_6H_5\overset{O}{\overset{\|}{C}}OH$$

benzyl alcohol — benzaldehyde — benzoic acid

$$CH_3\underset{OH}{\underset{|}{C}H}CH_2CH_3 \xrightarrow[H_2O]{K_2Cr_2O_7} CH_3\underset{O}{\underset{\|}{C}}CH_2CH_3$$

2-butanol — 2-butanone

$$CH_3-\underset{CH_3}{\underset{|}{\overset{CH_3}{\overset{|}{C}}}}-OH \xrightarrow{[O]} \text{no reaction under moderate conditions}$$

tert-butyl alcohol

Figure 6-10. *Primary and secondary (but not tertiary) alcohols are readily oxidized by dichromate or permanganate.*

These oxidations serve as practical methods for the synthesis of aldehydes, ketones and acids and will be discussed further in later chapters.

Syntheses of Alcohols

Addition of the elements of water across a carbon–carbon double bond is a general laboratory and industrial method for the synthesis of alcohols. The reaction is catalyzed by acid, and the steps are exactly the reverse of those by which an alkene is formed from an alcohol (Fig. 6-9). Some examples are given in Fig. 6-11. Note that Markovnikov's rule is followed in these reactions.

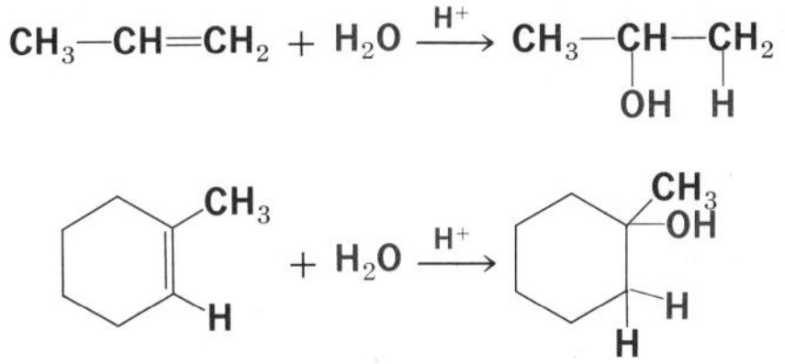

$$CH_3{-}CH{=}CH_2 + H_2O \xrightarrow{H^+} CH_3{-}\underset{OH}{CH}{-}\underset{H}{CH_2}$$

$$\text{(1-methylcyclohexene)} + H_2O \xrightarrow{H^+} \text{(1-methylcyclohexanol: }CH_3\text{, OH, H, H)}$$

Figure 6-11. *Acid-catalyzed addition of water to a double bond is a general synthetic method for the preparation of alcohols.*

Prob. 6-3. Making use of the reactions of Figs. 6-8 and 6-11, show how 1-butanol could be converted to 2-butanol using only sulfuric acid and water.

A second general laboratory method for the synthesis of alcohols is important because it leads to anti-Markovnikov addition of the elements of water. In this method *diborane,* B_2H_6, is allowed to react with an alkene in an inert solvent like ether. Diborane is in ready equilibrium with the Lewis acid *borane,* BH_3, which adds to the alkene. Here the electron-seeking (acidic) part of the reagent is boron, and addition of BH_3 proceeds according to Markovnikov's rule to give an intermediate organoboron compound (Fig. 6-12). Oxidation of this intermediate with basic hydrogen peroxide converts it to an alcohol.

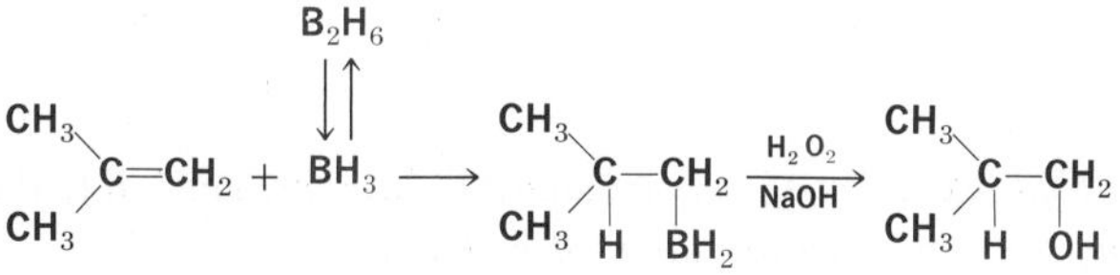

$$(CH_3)_2C{=}CH_2 + BH_3 \;(\rightleftarrows B_2H_6) \longrightarrow (CH_3)_2\underset{H}{C}{-}\underset{BH_2}{CH_2} \xrightarrow[NaOH]{H_2O_2} (CH_3)_2\underset{H}{C}{-}\underset{OH}{CH_2}$$

Figure 6-12. *Addition of borane (from diborane) across the carbon–carbon double bond, followed by oxidation, results overall in the anti-Markovnikov addition of water to an alkene.*

Prob. 6-4. Making use of the reactions of Figs. 6-8 and 6-12, show how 2-propanol could be converted to 1-propanol.

Borane has a second important advantage in the synthesis of alcohols: Both the addition of BH_3 and the oxidation to an alcohol are *stereospecific;* i.e., they occur so as to form only a single stereoisomer when more than one might have been expected. Addition of hydrogen and BH_2 occur from the same side of the double bond, and the hydroxyl group replaces the boron without any change in stereochemistry. The result is particularly obvious in cyclic systems, as shown in Fig. 6-13. Acid catalyzed addition of water across a double bond is usually not stereospecific.

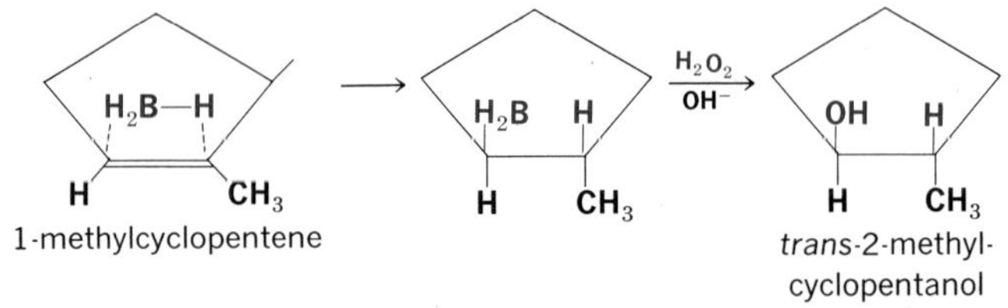

H_2B—H, H, CH_3, 1-methylcyclopentene → H_2B, H, H, CH_3 $\xrightarrow[OH^-]{H_2O_2}$ OH, H, H, CH_3, *trans*-2-methyl-cyclopentanol

Figure 6-13. *Addition of water to a double bond by the two-step diborane method gives* stereospecifically *the alcohol with the hydrogen atom and hydroxyl group on the same side of the ring.*

Other Alcohols

Methanol is now produced industrially by special processes, one of which is the hydrogenation of carbon monoxide (Fig. 6-14). It still finds some use as

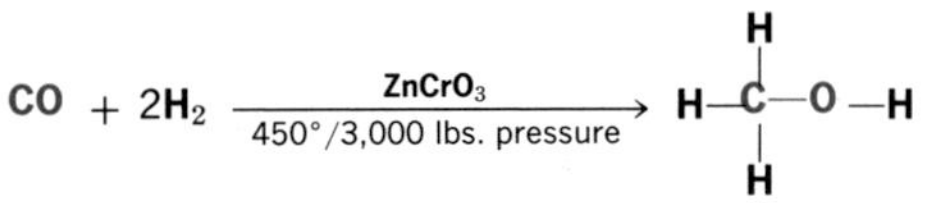

$$CO + 2H_2 \xrightarrow[450°/3,000\text{ lbs. pressure}]{ZnCrO_3} H{-}\overset{H}{\underset{H}{C}}{-}O{-}H$$

Figure 6-14. *Hydrogenation of carbon monoxide gives methanol. It is also synthesized industrially by the controlled oxidation of methane.*

a nonpermanent antifreeze, and is converted industrially into formaldehyde and a variety of other molecules. It has also been suggested as a synthetic replacement for hydrocarbons in gasoline, since it burns cleanly and has a high octane number. Like ethanol, methanol is inebriating. However, it is much more poisonous, probably because it is oxidized in the body to formaldehyde and formic acid. For this reason methanol is sometimes added to ethanol to denature it, i.e., to make ethanol unfit for drinking. Death from methanol poisoning occurs from respiratory paralysis preceded by visual disturbances; blindness can result in nonfatal cases. Both may sometimes be prevented by administration of sodium bicarbonate or other bases.

A few other alcohols besides ethanol derive from agricultural sources. Among these are *n*-pentyl and isopentyl alcohols (3-methyl-1-butanol). These are the principal constituents of fusel oil, a fermentation product that flavors many liquors but leads to hangovers.

Isopropyl alcohol (b.p. 82.4°) is formed by the hydration of propylene with sulfuric acid (Fig. 6-11); propylene, in turn, is cheap because it is available from petroleum cracking. The good solvent properties of isopropyl alcohol and its ability to absorb water make it useful as an industrial solvent and also as a deicer for airplane wings. Rubbing alcohol is isopropyl alcohol. *n*-Propyl alcohol is synthesized industrially by the controlled oxidation of propane.

Ethylene glycol, used as permanent antifreeze, is made from ethylene by addition of hypochlorous acid, followed by replacement of the chlorine atom with base (Fig. 6-15). An alternative preparation is shown in Fig. 6-27. Large amounts of ethylene glycol are used in the manufacture of Dacron.

$$CH_2{=}CH_2 + HOCl \longrightarrow \underset{OH}{CH_2}{-}\underset{Cl}{CH_2} \xrightarrow{OH^-} HOCH_2CH_2OH$$

Figure 6-15. *Ethylene glycol is formed from ethylene by addition of hypochlorous acid followed by basic hydrolysis.*

Phenols

Phenols are a class of compounds containing the hydroxyl group directly attached to an aromatic ring (Fig. 6-16). Phenols are closely related to alcohols, but the aromatic nucleus exerts a strong modifying effect on the properties of the hydroxyl group. For example, phenols are much more acidic than alcohols, as attested by the name carbolic acid for phenol itself.

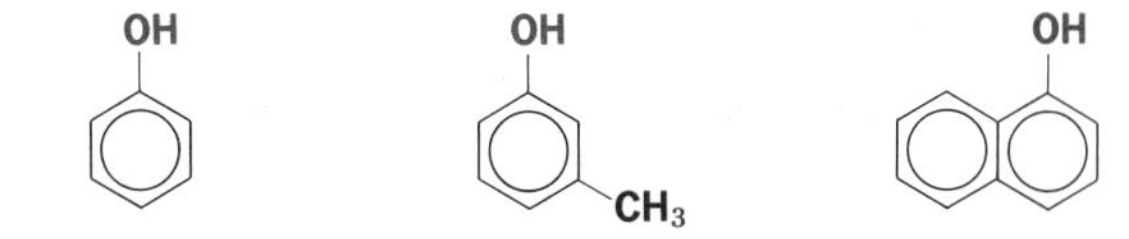

OH | OH, CH_3 | OH

phenol (carbolic acid) | *m*-cresol (*m*-cresylic acid) | α-naphthol

Figure 6-16. *Some common phenols. The characteristic functional group is a hydroxyl group attached to an aromatic ring. The development of a deep color in the presence of a solution of ferric chloride is a visual chemical test for phenols.*

The most striking modification the benzene ring achieves on the properties of the hydroxyl group is shown by comparing the acidity of phenol with that of aliphatic alcohols. Phenol is approximately 10^6 (1 million) times more acidic than ethanol and about the same amount less acidic than acetic acid. As a consequence of this acidity, phenol is converted completely to its salt, sodium phenoxide, when treated with sodium hydroxide (Fig. 6-17). Because of this salt formation, phenols are soluble in aqueous sodium hydroxide solution, even if they do not dissolve in water.

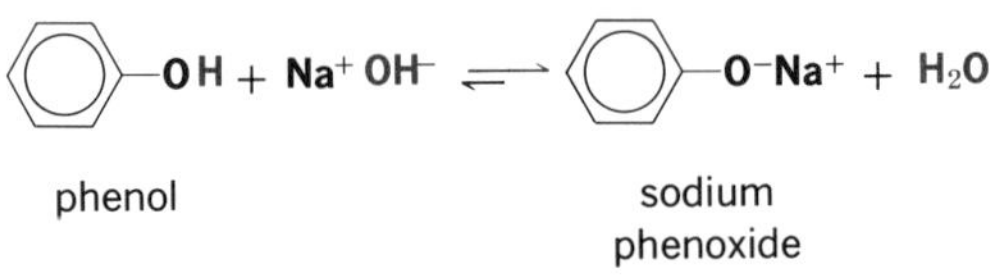

$$C_6H_5{-}OH + Na^+\,OH^- \rightleftharpoons C_6H_5{-}O^-Na^+ + H_2O$$

phenol — sodium phenoxide

Figure 6-17. *Attaching a hydroxyl group directly to a benzene ring, as in phenol, increases its acidity nearly 1 million times compared to an alcohol. As a consequence, phenol is converted entirely to its salt by sodium hydroxide.*

Resonance in the Phenoxide Ion

Phenols are much more acidic than alcohols because the negative charge in the phenoxide ion is not localized on the oxygen atom, as it is in an alkoxide ion, but is delocalized, i.e., shared by a number of carbon atoms in the benzene ring. Just as a positive charge can be delocalized if it resides on a carbon adjacent to a double bond or other unsaturated center (Fig. 3-28), so too a negative charge can disperse, i.e., spread out and so achieve a lower energy if it is next to a π electron system. The possibility for charge delocalization in the phenoxide ion can be recognized by our ability to write resonance structures for the anion (compare resonance structures for cations in Fig. 5-17). Note in Fig. 6-18 that in addition to residing on the oxygen atom, the charge may also reside on three ring carbons (two *ortho*, one *para*). Because the charge

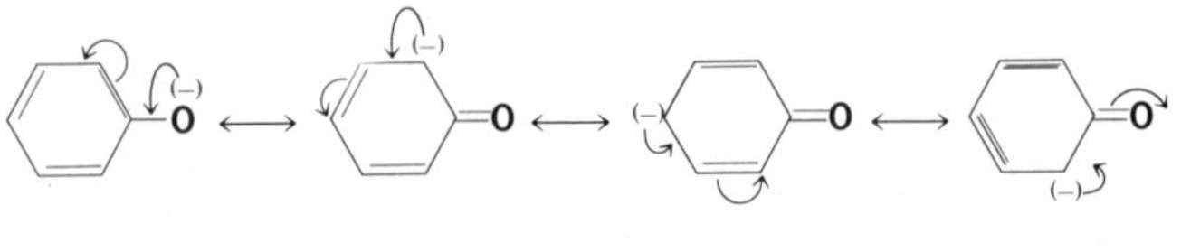

Figure 6-18. *In the phenoxide ion the negative charge is delocalized, i.e., shared by the ortho and para carbons of the ring. As a result the phenoxide ion has less tendency than an alkoxide ion to combine with a proton.*

is less concentrated on oxygen, the anion will have a smaller tendency to combine with a proton; therefore, the phenoxide ion is a weaker base than an alkoxide ion (and, conversely, phenol is a stronger acid than an alcohol).

If groups that stabilize negative charge are attached to the benzene ring, the charge may be further delocalized, and the phenol becomes even more acidic. *p*-Nitrophenol, for example, is 500 times more acidic than phenol itself because the negative charge in the phenoxide ion may be delocalized into the nitro group (Fig. 6-19); while 2, 4, 6-trinitrophenol (picric acid), in which all three nitro groups can help stabilize the negative charge, is nearly as strong as hydrochloric acid. Groups that are *meta* directing in aromatic substitution (because they repel an adjacent positive charge) will *increase* the acidity of phenol because they stabilize an adjacent negative charge.

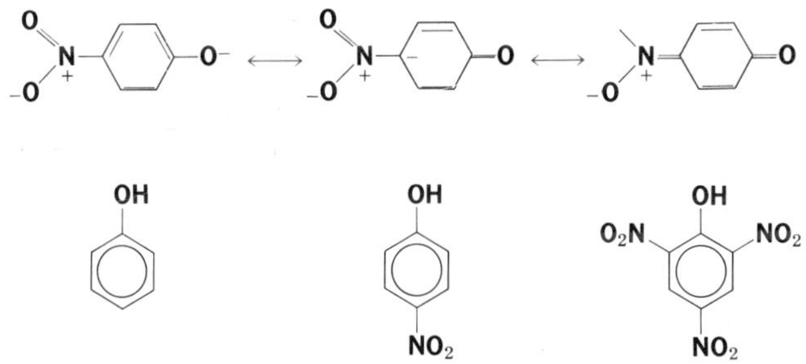

phenol $K_a = 1.3 \times 10^{-10}$ | *p*-nitrophenol 6.9×10^{-8} | picric acid 4.2×10^{-1}

Figure 6-19. *Attachment of a nitro group to phenol increases its acidity. In the* p-*nitrophenoxide ion the charge can be shared by the oxygens of the nitro group as well as by the ring carbons. Picric acid is a strong acid, almost as strong as hydrochloric acid.*

Reactions of Phenols

Just as the benzene ring affects the acidity of a hydroxyl group attached to it, so does it modify other reactions of the functional group, sometimes drastically and sometimes subtly. In particular, the hydroxyl group is much more firmly attached to the benzene ring than to an alkyl group, so that replacement reactions of the entire —OH group by cleavage of the C—O bond do not occur with phenols. For example, phenol cannot be converted to bromobenzene with hydrogen bromide.

In reactions that involve the O—H bond, phenols resemble alcohols. They liberate hydrogen when treated with sodium and they also form esters of many acids (Fig. 6-20). The interaction of a hydroxyl and a phenyl group

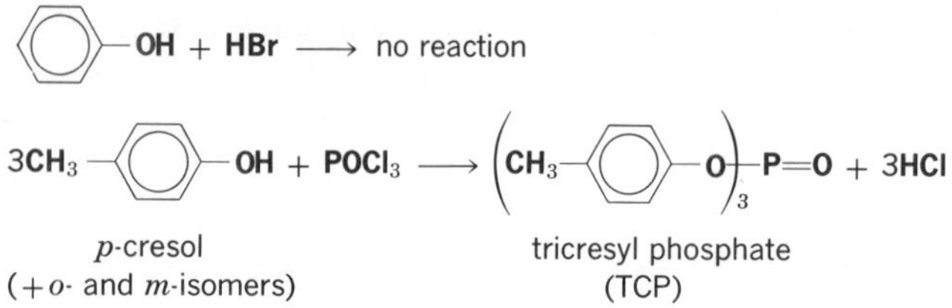

$$C_6H_5OH + HBr \longrightarrow \text{no reaction}$$

$$3CH_3C_6H_4OH + POCl_3 \longrightarrow (CH_3C_6H_4O)_3P{=}O + 3HCl$$

p-cresol (+*o*- and *m*-isomers) → tricresyl phosphate (TCP)

Figure 6-20. *Phenols do not undergo the reactions of alcohols which involve breaking of the C—O bond, but do undergo many of those which involve the O—H bond, as illustrated by preparation of the gasoline additive TCP.*

is, of course, a mutual one, and the benzene ring in phenol differs in reactivity from benzene itself. In particular, phenol is much more reactive than benzene toward substitution reactions. Even with dilute nitric acid, phenol is nitrated at 25°. (Remember that benzene requires concentrated nitric acid, added sulfuric acid as catalyst, and 50°.) As another example, bromine (without catalyst) rapidly trisubstitutes the ring.

Synthesis of Phenols

The hydroxyl group cannot be introduced directly into a benzene ring, so that phenol synthesis always require at least two steps (Fig. 6-21). The most common laboratory preparation of a phenol involves, first, sulfonation of the ring and then fusion (melting together) of the sulfonate with sodium hydroxide. High temperatures are required but, since both reactants and products are high-boiling salts, no special apparatus is needed. A great deal of phenol is made industrially from chlorobenzene and sodium hydroxide. The reaction must be carried out under pressure, or else the chlorobenzene would boil away.

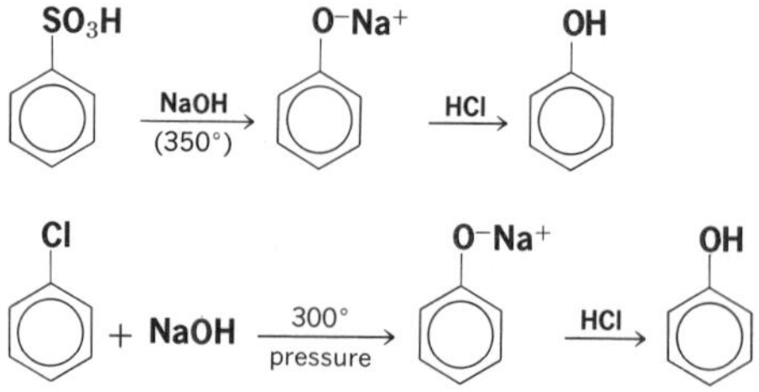

$$C_6H_5SO_3H \xrightarrow[(350^\circ)]{NaOH} C_6H_5O^-Na^+ \xrightarrow{HCl} C_6H_5OH$$

$$C_6H_5Cl + NaOH \xrightarrow[\text{pressure}]{300^\circ} C_6H_5O^-Na^+ \xrightarrow{HCl} C_6H_5OH$$

Figure 6-21. *Phenols can be prepared by the high-temperature reaction of benzenesulfonic acids or halobenzenes with sodium hydroxide.*

In order to make the process continuous, chlorobenzene and aqueous sodium hydroxide are fed together into a mile-long tube heated at 300° and maintained under intense pressure. Sodium phenolate forms and is then converted into phenol by acidification.

A number of phenols are present in coal tar, from which they may be separated by extraction with alkali. Acidification releases the phenols. Phenol itself, and *o*-, *m*-, and *p*-methylphenols (*o*-, *m*-, and *p*-cresols) may be obtained in this way.

Prob. 6-5. Suggest a method for making resorcinol (*m*-dihydroxybenzene) from benzene, sulfuric acid, and sodium hydroxide.

Oxidation of Phenols

A simple phenol cannot be oxidized in quite the same way as a 1° or 2° alcohol, since the hydroxyl group is not on a carbon atom bearing a hydrogen atom. Phenols are easily oxidized, but their products are often complex. This oxidation may occur with air alone (autooxidation) or with other oxidizing agents. The reaction of phenols with oxygen in the air is exploited industrially by the use of phenols as antioxidants in gasoline, rubber, and other products. The phenols react with oxygen more readily than most other organic compounds and protect them from oxidation.

In the special case in which two hydroxyl groups are in the *para* position relative to each other, the compound (*p*-hydroquinone) may be easily, selectively, and reversibly oxidized to a diketone, *p*-benzoquinone, representative of a class of compounds called quinones (Fig. 6-22); *p*-benzoquinone itself is yellow, and substituted quinones are often the coloring material in naturally occurring systems. Hydroquinone is also used as a photographic developer, since it has the ability to reduce exposed grains of AgBr in a photographic emulsion faster than unexposed grains.

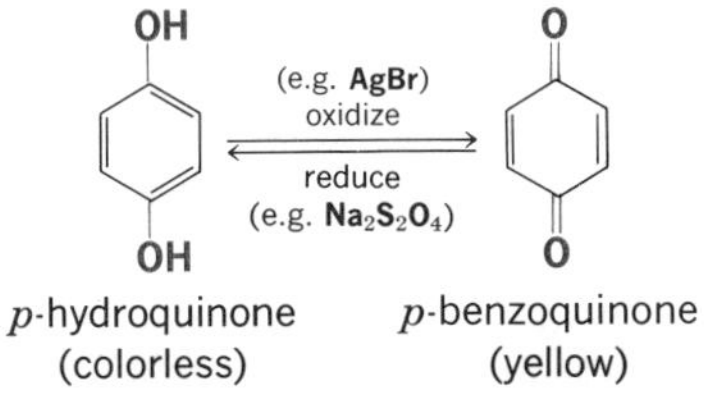

OH, OH — (e.g. **AgBr**) oxidize ⇌ reduce (e.g. $Na_2S_2O_4$) — O, O

p-hydroquinone (colorless)   *p*-benzoquinone (yellow)

Figure 6-22. *The quinone-hydroquinone oxidation-reduction system is common in biological systems.*

The yellow-brown pigment juglone has been extracted from walnut shells; it also occurs in many other colored natural systems. Historically, the beautiful red quinone dye *alizarin* (Fig. 6-23) is of particular interest. This

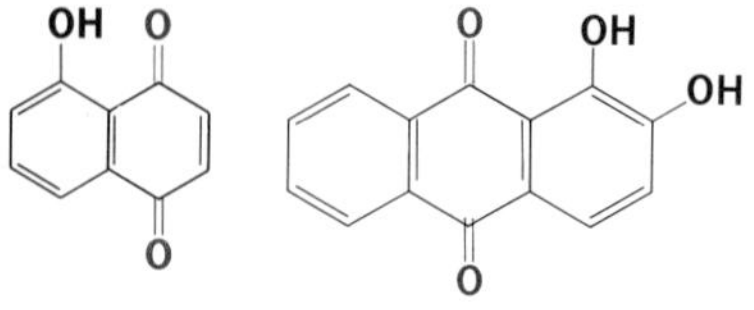

juglone   alizarin

Figure 6-23. *Quinones often occur as natural coloring matters.*

compound, extracted from the roots of the madder plant since the very earliest times, was the most common red dye in Europe until the discovery of dyes from the New World. Alexander the Great probably used alizarin to simulate blood stains on the clothes of his army in a famous battle with the Persians. The Persian soldiers, seeing the Greeks stagger out to battle in their supposedly bloodstained clothes, became overconfident and were overwhelmingly defeated. Pliny tells of the cultivation of madder near Rome in the early years of the Christian era, and similar references can be found throughout ancient manuscripts.

Ethers

If both hydrogen atoms of the water molecule are replaced by alkyl or aryl groups, compounds known as *ethers* are formed. The commonest ether is diethyl ether, usually called simply "ether," and used as an anesthetic and a solvent (Fig. 6-24). Notice the system of nomenclature, in which the groups retain their radical names. If both groups are the same, as they are in *diethyl ether*, the ether is *symmetrical* and is commonly produced by reaction of the

$CH_3CH_2{-}O{-}CH_2CH_3$

diethyl ether, b.p. 35°
("ether")

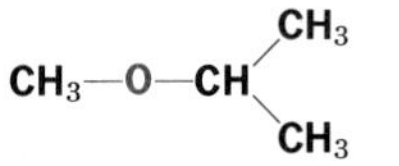

$CH_3{-}O{-}CH(CH_3)_2$

methyl isopropyl ether

$CH_3(CH_2)_4CH_2{-}O{-}CH_2(CH_2)_4CH_3$

di-*n*-hexyl ether

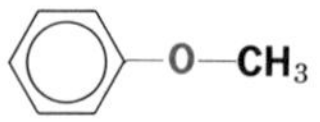

$C_6H_5{-}O{-}CH_3$

methyl phenyl ether
(anisole)

Figure 6-24. *Ethers are usually inert substances with good solvent properties for organic molecules. Since they cannot form hydrogen bonds with each other, their boiling points are close to those of hydrocarbons of the same molecular weight.*

corresponding alcohol with sulfuric acid. Mixed ethers are prepared by the *Williamson ether synthesis,* in which the salt of an alcohol reacts with an alkyl halide (Fig. 6-25). This reaction will be discussed in more detail in the next chapter.

$$2CH_3CH_2OH \xrightarrow[140°]{H_2SO_4} CH_3CH_2—O—CH_2CH_3 + H_2O$$

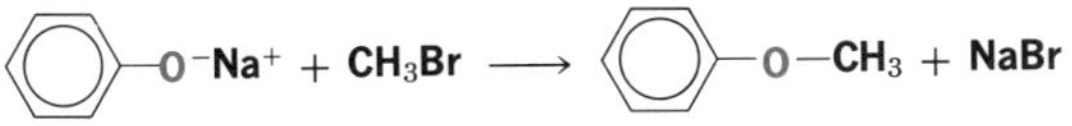

$$C_6H_5—O^-Na^+ + CH_3Br \longrightarrow C_6H_5—O—CH_3 + NaBr$$

Figure 6-25. *Symmetrical ethers are prepared by reactions of alcohols with sulfuric acid at high temperatures. Mixed ethers are prepared by the Williamson ether synthesis.*

Prob. 6-6. Why can mixed ethers not be prepared satisfactorily from reaction of a mixture of alcohols with sulfuric acid? Answer by showing what products would be expected if a mixture of 1-propanol and ethanol were converted to ethers with sulfuric acid.

For an ether to undergo reaction, a C—O or a C—H bond must be broken; however, both of these processes are ordinarily difficult. The strong heating of diethyl ether with concentrated hydriodic acid converts it into ethyl iodide, and ether that is exposed to oxygen and light undergoes air oxidation to become a *peroxide* (Fig. 6-26). These peroxides are highly explosive: Great

$$CH_3CH_2—O—CH_2CH_3 + HI \xrightarrow{120°} \underset{\text{ethyl iodide}}{CH_3CH_2I} + CH_3CH_2OH$$

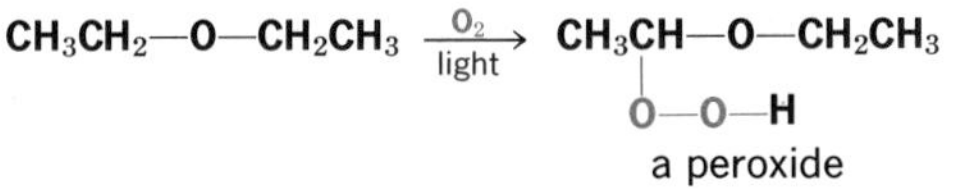

$$CH_3CH_2—O—CH_2CH_3 \xrightarrow[\text{light}]{O_2} \underset{\text{a peroxide}}{CH_3\underset{|}{C}H(O—O—H)—O—CH_2CH_3}$$

Figure 6-26. *Ethers are inert to most chemical reactions. However, heating with concentrated acids eventually cleaves ethers, and exposure to air and light generates small amounts of highly explosive peroxides.*

care must be taken in using ether that has been standing exposed to air for some time.

Because it is quite unreactive, diethyl ether makes an excellent solvent for many reactions. Many more organic compounds will dissolve in ether than in hydrocarbons, because the oxygen atom in ethers imparts some polar character to the ether molecule. Gaseous hydrogen chloride, for example, is rather soluble in ether because the proton of hydrogen chloride forms a highly polar hydrogen bond with the oxygen atom of ether, $R_2O\cdots HCl$, helping to keep the acid in solution.

One important ether is, however, highly reactive. In *ethylene oxide*[1] the ether oxygen is part of a three-membered ring (see Fig. 6-27); the great angle strain in such a small ring (60° bond angles rather than the preferred 109°28′) makes it anxious to enter into reactions that will open the ring. Reaction with water, for example, gives *ethylene glycol.* Since ethylene oxide can be made from ethylene and air, this method is the cheapest for the production of permanent antifreeze. (See also Fig. 6-15.)

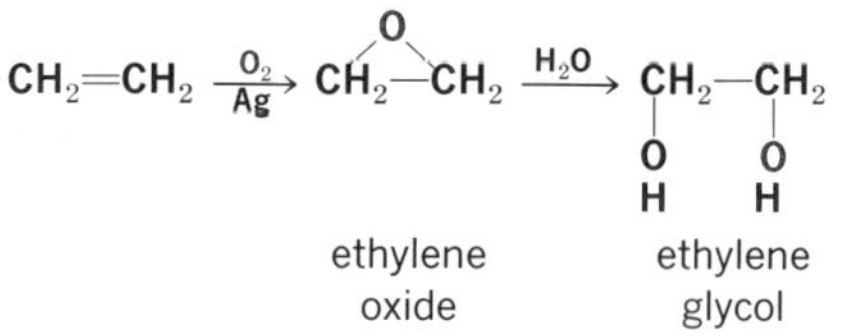

$$CH_2{=}CH_2 \xrightarrow[Ag]{O_2} \underset{\text{ethylene oxide}}{\overset{O}{CH_2{-}CH_2}} \xrightarrow{H_2O} \underset{\text{ethylene glycol}}{\underset{OH\quad OH}{CH_2{-}CH_2}}$$

Figure 6-27. *Permanent antifreeze (ethylene glycol) may also be made by reaction of the highly reactive ether ethylene oxide with water.*

Infrared Absorption Spectra

As a glance at the chart of electromagnetic radiation (Fig. 5-29) will show, the infrared (ir) portion of the spectrum lies to slightly longer wavelengths than the visible. All organic molecules absorb strongly in this part of the spectrum. An apparatus for measuring absorption in this region (an infrared spectrophotometer) does not differ in principle from that used for visible and ultraviolet absorption spectra (Fig. 5-32). The source of radiation is a hot wire, which gives out heat; this infrared radiation is broken up into its constituent wavelengths by a prism. Since glass absorbs infrared radiation, the prisms, cells, and mirrors must be made of salt (cut from large crystals of sodium chloride). The detector is a thermocouple, essentially a thermometer that converts heat into an electrical current. By convention infrared spectra are plotted with the absorption peaks coming down from the top instead of up from the bottom as they were for uv spectra.

[1] The common names ethylene oxide and ethylene glycol are unfortunate, since these compounds have no double bond. The names actually stand for the *oxide of ethylene* and the *glycol of ethylene,* respectively. As a further example of this nomenclature compare dichloroethylene (1,2-dichloroethene) and ethylene dichloride (1,2-dichloroethane).

The ir spectrum of 2-propanol in the most conveniently measured portion of the infrared region (2.5 to 16 microns) is shown in Fig. 6-28, see p. 128. Note the relatively large number of sharp absorptions; because there are usually more peaks in an ir than a uv spectrum, the former serves as a better and sharper "fingerprint" of a molecule. Large collections of ir spectra have been published; often an unknown can be identified with great confidence by comparing it with the published spectrum.

The infrared spectrum of a molecule can reveal a great deal of information about its structure even if its "fingerprint," i.e., its authentic spectrum, is not on file, because different functional groups absorb consistently in different parts of the spectrum. Notice the peak in red in the spectrum of 2-propanol; absorption in this region is characteristic of the presence of an O—H bond in the molecule. An alcohol will always have an absorption just there. Note that benzyl alcohol has some absorption peaks in common with both toluene and 2-propanol.

Absorption of infrared radiation causes an increase in *vibration frequencies* of certain bonds within the molecule. Absorption of radiation of about 3 microns causes the O—H bond to vibrate more rapidly, absorption of 3.5 micron radiation increases the vibration of C—H bonds, and of 4.5 micron radiation the C≡C bond. Other wavelengths will increase the vibrations of C—C bonds, still others double bonds, and so on through the various types of bonds found in molecules. It is often possible for a person skilled in the art of interpreting ir spectra to identify several different functional groups in a complex molecule, to tell if an alkene is *cis*- or *trans*-substituted, or to decide if a disubstituted benzene is *ortho, meta,* or *para.* Only a drop or less of a compound is needed for determining an ir spectrum, and it is possible to collect each component after it passes through a gas chromatograph and to identify the compound by determining its infrared spectrum.

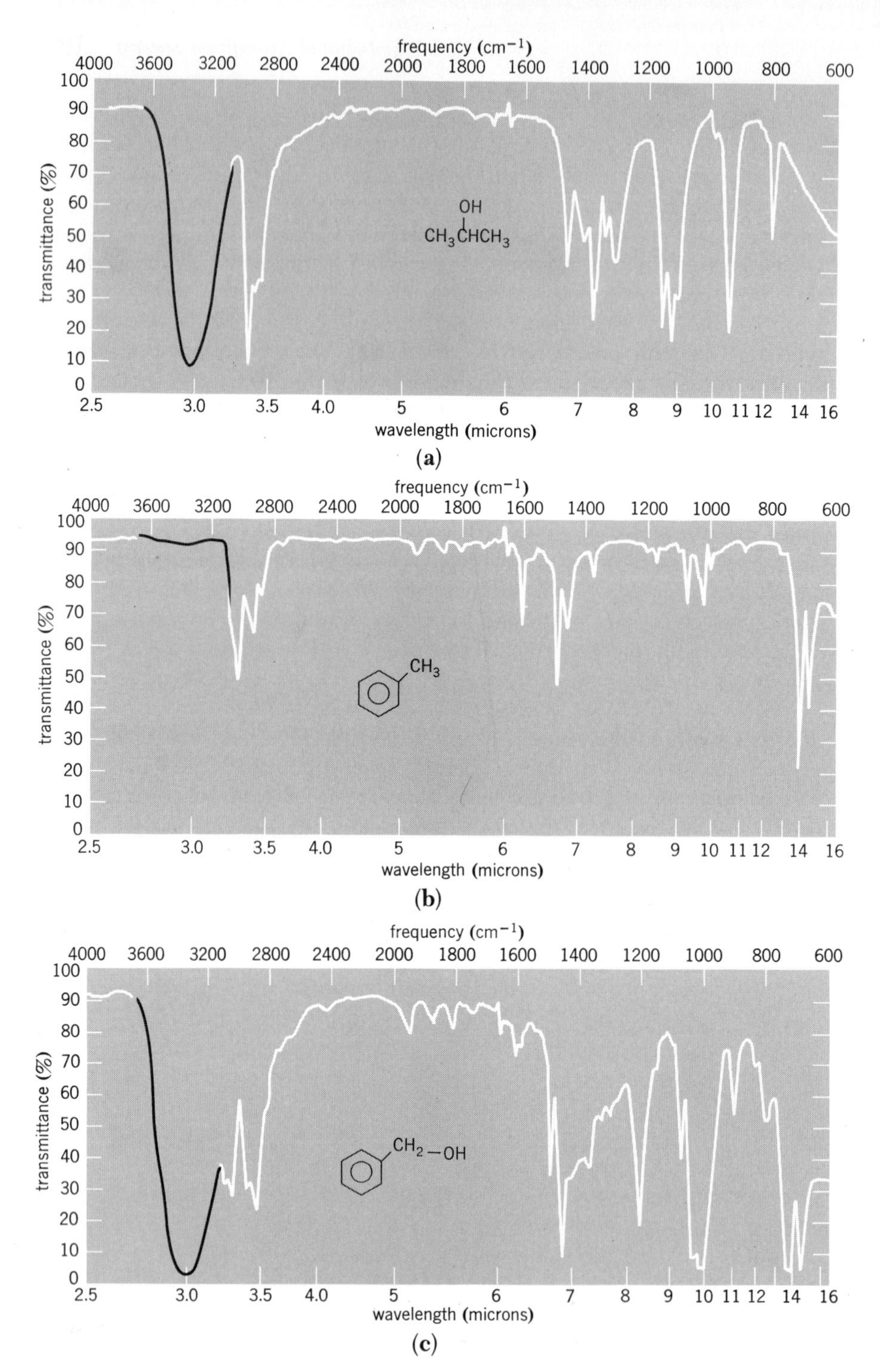

Figure 6-28. *The infrared spectra of three simple organic compounds. Note the characteristic absorption of the hydroxyl group in both 2-propanol and benzyl alcohol.*

PROBLEMS

1. The IUPAC systematic nomenclature ending for an alcohol is *ol*, with numbering from the end of the carbon chain to give the carbon bearing the hydroxyl group its smaller number. Give the structure of:

(a) 1-butanol (b) 2-methyl-2-propanol
(c) cyclopentanol (d) *trans*-2-methylcyclobutanol
(e) 3-phenyl-2-butanol (f) 1-(*p*-chlorophenyl)ethanol
(g) 5-isopropyl-3-octanol (h) *cis*-4-*t*-butylcyclohexanol

2. Give systematic names for the following:

(a) $CH_3CHOHCH_3$ (b) $(CH_3)_2CHCH_2OH$

(c) $CH_3CH_2CHCH_2CH_3$ with CH_2OH on C-3 (d) C_6H_5—CH_2CH_2OH

(e) 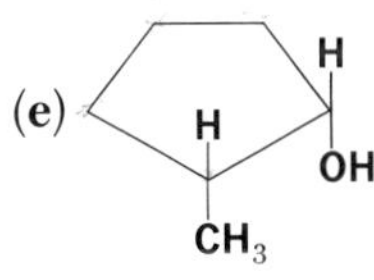

(f) benzene ring (with NO_2 in the meta position) —$CH(CH_3)$—CH_2OH

3. Give the structure of:

(a) methyl ethyl ether (b) *t*-butyl alcohol
(c) *p*-nitrophenol (d) propylene oxide
(e) *p*-cresol (f) 3-methyl-3-penten-2-ol
(g) 2-methyl-4-pentyn-2-ol (h) propylene glycol

4. Complete the following reactions:

(a) $(CH_3)_2CHCH_2CH_2OH + HCl \xrightarrow{\text{heat}}$

(b) $CH_3CH(OH)CH_3 + H_2SO_4 \xrightarrow{\text{heat}}$

(c) $(CH_3)_2C{=}CH_2 \xrightarrow{B_2H_6} \quad \xrightarrow[NaOH]{H_2O_2}$

(d) $C_6H_5SO_3Na + NaOH \xrightarrow{\text{fuse}}$

(e) 1,4-naphthalenediol (OH at C-1 and C-4) $\xrightarrow{\text{mildly oxidize}}$

(f) $CH_3CH_2OH + Na \longrightarrow \quad \xrightarrow{CH_3I}$

(g) 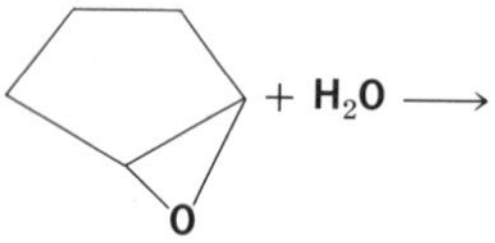 [cycloalkene epoxide] $+ H_2O \longrightarrow$

(h) [1-methylcyclobutene] $\xrightarrow{B_2H_6}$ $\xrightarrow[NaOH]{H_2O_2}$

(include stereochemistry)

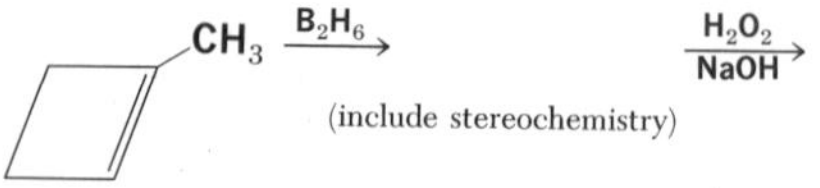

(i) 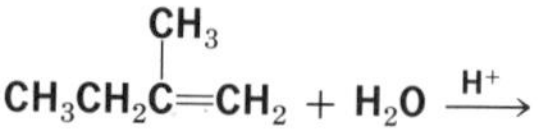 $CH_3CH_2C(CH_3){=}CH_2 + H_2O \xrightarrow{H^+}$

(j) 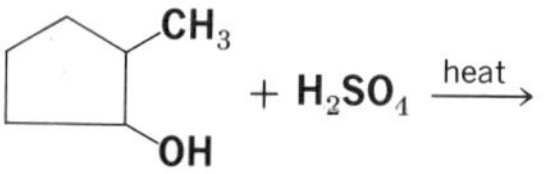 [2-methylcyclopentanol] $+ H_2SO_4 \xrightarrow{heat}$

5. Show all the reagents and conditions that would be necessary to make the following compounds from 1-butene:
   (a) 2-butanol (b) 1-butanol
   (c) di-*n*-butyl ether (d) 2-chlorobutane
   (e) 1-chlorobutane (f) 1-butyl hydrogen sulfate
6. Show all the reagents and conditions that would be necessary to make the following compounds from phenol:
   (a) 2,4,6-tribromophenol
   (b) *o*-nitrophenol
   (c) sodium phenoxide
   (d) anisole (methyl phenyl ether)
7. When the secondary alcohol 3-butene-2-ol ($CH_3CH(OH){-}CH{=}CH_2$) was treated under the conditions of the Lucas reaction, two isomeric unsaturated chlorides (C_4H_7Cl) were obtained. Explain. *Hint:* Remember that a carbonium ion is formed as an intermediate in this reaction. What is its structure?
8. Suppose you had a mixture of 4-*t*-butylcyclohexanol and *p*-*t*-butylphenol. Explain how you could separate the two compounds, using only ether (in which both are soluble), water (in which both are insoluble), sodium hydroxide, and hydrochloric acid. *Hint:* Remember that salts are soluble in water and insoluble in ether.
9. Describe simple, visual chemical tests that would allow you to distinguish among the following:
   (a) cyclohexene, cyclohexanol, 1-methylcyclohexanol
   (b) picric acid, cyclohexanol, anisole
   (c) ethylene glycol, 2-hexene, 2-propen-1-ol
10. Using the Williamson ether synthesis, what two organic molecules would you combine to form the herbicide 2,4-D?

Cl—(2-Cl-C$_6$H$_3$)—O—CH_2COOH

2,4-D

11. Arrange the following phenols in the order of increasing acid strength.

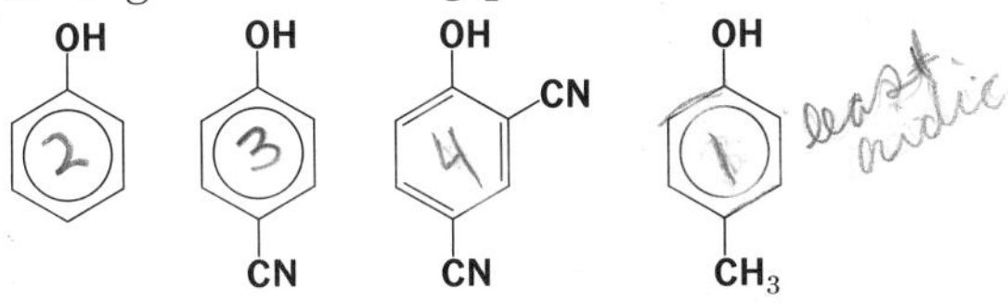

12. Describe simple visual chemical tests that would allow you to distinguish among the following compounds:

(a)

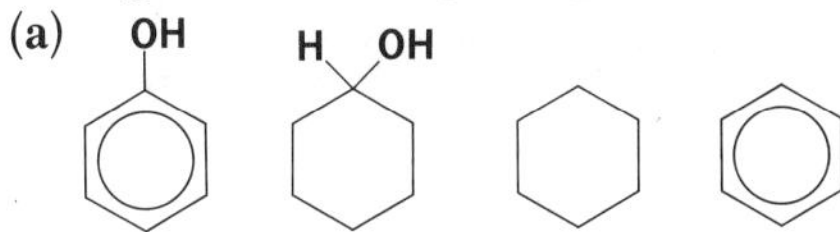

(b)

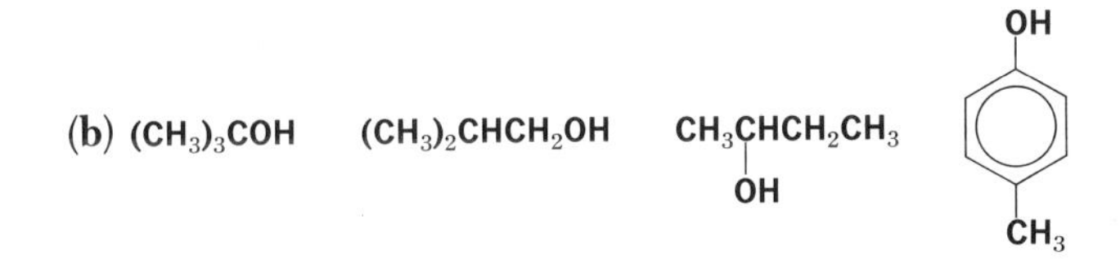

7
organic halides and configurational isomerism

To understand how organic halides react, we must first introduce a new kind of isomerism: *configurational isomerism*. We have already seen how pervasive isomerism is in organic chemistry. At one extreme we have structural isomers such as ethyl alcohol and dimethyl ether or *n*-propyl and isopropyl alcohols, in which the different order of bonding often produces dramatic differences in chemical and physical properties in the isomeric molecules. Geometrical isomerism is subtler. In *cis*- and *trans*-2-butenes, for example, the order of bonding is the same but the distances between some of the atoms differ in the two isomers. Geometrical isomers usually differ only slightly in physical and chemical properties (Fig. 7-1).

Some molecules exist which differ from one another only in the fact that one is the mirror image of the other; these molecules are examples of *configurational isomers*. Configurational isomers are identical in most of their physical and chemical properties. Under certain conditions, however, and particularly in biological systems, they can differ profoundly. Moreover, the mere fact of their identical reactivities combined with their differences in structure can give us a new tool for probing into pathways of organic reactions. In this chapter we shall explore this new type of isomerism and show how it can help us understand the chemistry of halogen-containing organic molecules.

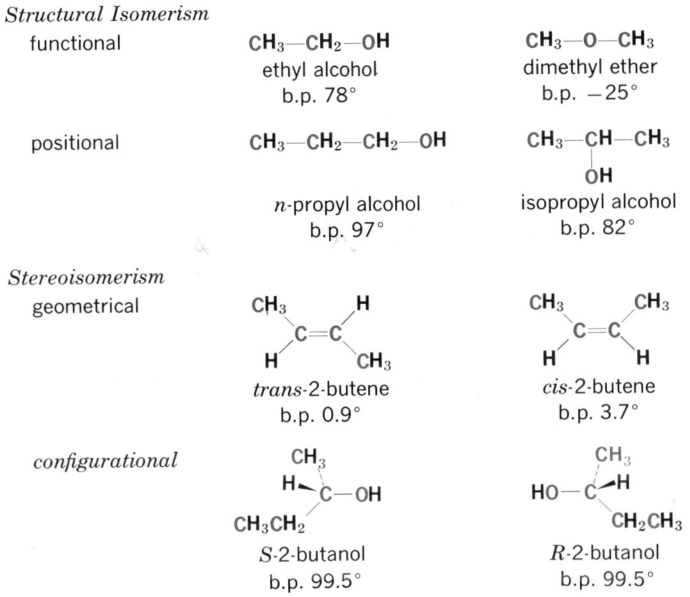

Structural Isomerism		
functional	$CH_3—CH_2—OH$ ethyl alcohol b.p. 78°	$CH_3—O—CH_3$ dimethyl ether b.p. −25°
positional	$CH_3—CH_2—CH_2—OH$ *n*-propyl alcohol b.p. 97°	$CH_3—CH(OH)—CH_3$ isopropyl alcohol b.p. 82°
Stereoisomerism		
geometrical	$CH_3(H)C{=}C(H)CH_3$ *trans*-2-butene b.p. 0.9°	$CH_3(H)C{=}C(H)CH_3$ *cis*-2-butene b.p. 3.7°
configurational	CH_3, H, OH, CH_3CH_2 on C *S*-2-butanol b.p. 99.5°	CH_3, HO, H, CH_2CH_3 on C *R*-2-butanol b.p. 99.5°

Figure 7-1. *Isomerism among organic compounds may be arranged in order of subtlety. Structural isomers are usually quite different, stereoisomers quite similar.*

Nature of Configurational Isomerism

Two molecules that differ from one another only in that they are mirror images are called *enantiomers* (from the Greek word meaning "opposite"). The relationship between enantiomers is the same as that of the left hand to the right (Fig. 7-2); your two hands are enantiomers. Through this analogy one isomer is called the *R*-isomer (from the Latin *rectus,* right) and the other, the S-isomer (*sinister,* left). 2-Butanol is a simple compound that exhibits configurational isomerism; there are two enantiomeric 2-butanol molecules, *R*-2-butanol and S-2-butanol. Their relationship is shown in Fig. 7-2. Converting an object into its mirror image should have no effect on most of its physical properties; *R*- and S-2-butanols do have exactly the same melting point, solubility, density, etc. Similarly, the chemical properties of *R*- and S- isomers are indistinguishable through ordinary reagents; both react in exactly the same way with hydrogen bromide, sodium, and other standard chemicals.

Prob. 7-1. Which of these things are capable of existing in *R*- and S-forms: a sphere, a cube, a foot, a spearfish, a spiral staircase (called a helix), and a tree? Can you make any generalization about the types of

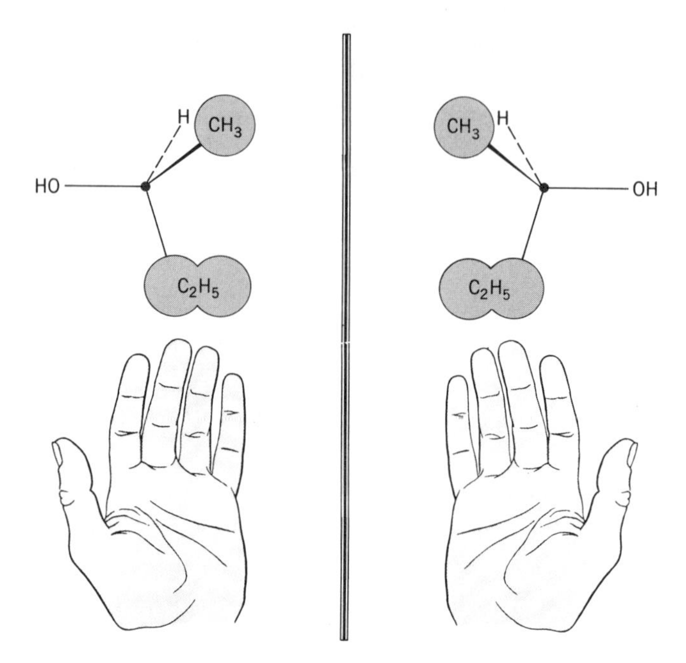

Figure 7-2. Enantiomers *bear the same mirror-image relationship to one another as do the left and right hands. Shown are the two* enantiomers *of 2-butanol.*

objects that are likely to be identical to their mirror image and thus incapable of existing in *R*- and *S*-isomers?

Rotation of the Plane of Polarized Light

Enantiomers rotate the plane of polarized light in opposite directions; this is the most important method of telling them apart. Most people will be somewhat familiar with the properties of polarized light through their experience with a sheet of polarizing material. An ordinary light beam consists of a group of electromagnetic waves that vibrate in all directions, as in Fig. 7-3(a). When such a beam strikes a polarizing film or a Nicol prism,[1] only those waves vibrating in a specific plane with respect to the axis of the film or prism may pass through; all others are blocked out. Upon emergence the light beam is *polarized* as in Fig. 7-3(b); all of its waves vibrate in a single plane (or, more precisely, in parallel planes).

[1] Made from a crystal of calcium carbonate.

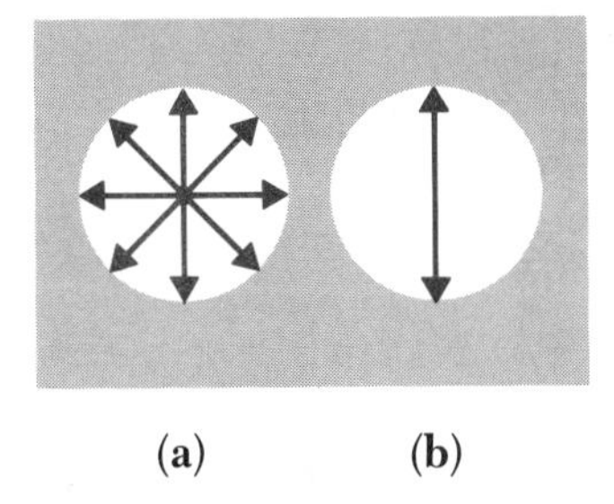

Figure 7-3. *An ordinary beam of light (a), pictured here as emerging from the paper, consists of waves of light vibrating in all planes perpendicular to the direction of propagation of the light. Passage of this beam through a sheet of polarizing material or a Nicol prism results in polarized light (b) in which only those waves vibrating in one plane remain.*

If a beam of polarized light passes through a solution of a single enantiomer of a molecule, the beam will emerge with the plane of its polarization rotated. For example, if a beam of light from a sodium lamp is polarized and then passed through a solution of S-2-butanol, its axis of polarization is rotated in the (+) direction, as shown in Fig. 7-4. The other enantiomer, *R*-2-butanol, rotates the plane in the opposite (−) direction.[2] The amount

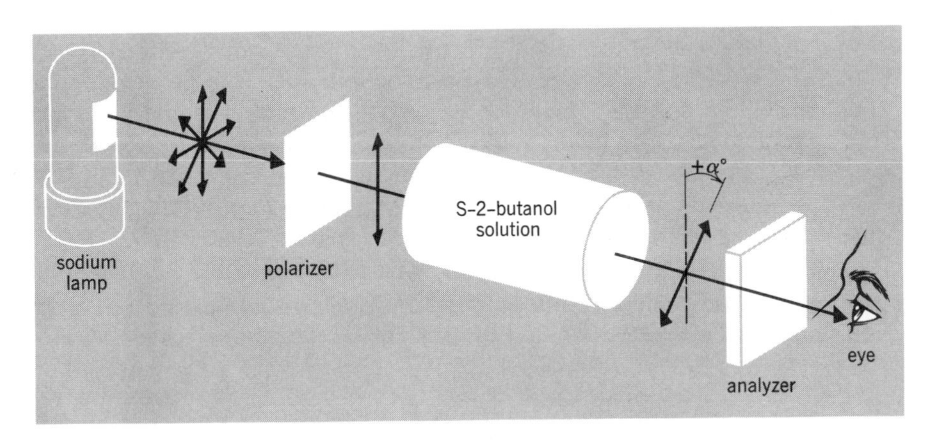

Figure 7-4. *If polarized light from a sodium lamp is passed through a solution of* S*-2-butanol, its plane of polarization is rotated in the* (+) *direction.* R*-2-butanol rotates the plane in the* (−) *direction.*

[2] Sometimes, especially in older books and articles, the letter *d* (from the Latin *dexter*, to the right) is used to indicate a molecule which rotates light in the (+) direction (to the right as one faces the oncoming beam), and such a molecule is called *dextrorotary*. A molecule rotating light in the (−) direction is indicated by *l* (Latin: *laevus*, to the left), and is said to be *levorotatory* according to this terminology.

by which the plane is rotated (i.e., the magnitude in degrees of the angle α) depends upon the number of 2-butanol molecules that the light encounters as it passes through the solution, i.e., upon the length of the tube and the concentration of alcohol within the tube. All we can say for the general case is that, for identical experimental conditions, *enantiomers rotate the plane by equal amounts in opposite directions.* The direction of rotation, and the magnitude of this rotation under a given set of conditions, is a constant for a given molecule, and, like its boiling point, may serve to identify it. Like the boiling point, which changes with pressure, the magnitude and even the direction of rotation may change with the experimental conditions. This is especially true if we change the wavelength of the polarized light we use to measure the rotation; unless otherwise specified the so-called D-line (589.3 nm) from a sodium vapor lamp is understood to be the light used. For purposes of standardization the *specific rotation* $[\alpha]$ of a compound is defined to be that of a solution containing 1 g/ml in a cell 1 decimeter (10 cm) long. Because of their effect on the plane of polarized light, enantiomers are sometimes referred to as *optical isomers.*

Chirality

For a molecule to rotate the plane of polarized light, it must differ from its mirror image. We say such a molecule is *chiral* (pronounced *ky-ral* and derived from the Greek word for "hand"). If a molecule is identical with its mirror image, it is said to be *achiral.* How can we tell whether a given molecule will be chiral or achiral? The safest way is to construct models, either in fact or in your mind's eye, of the molecule and of its mirror image. Now pick up one of the models and try to *superimpose* it upon the other, i.e., try to lay one molecule on top of the other so that the positions of all the atoms correspond in the two structures. If the molecules are chiral, it will not be possible to do this; we say *a chiral molecule is nonsuperimposable on its mirror image.* An *achiral* molecule is *superimposable* on its mirror image. As is shown in Fig. 7-5, the two enantiomers of 2-butanol are nonsuperimposable.

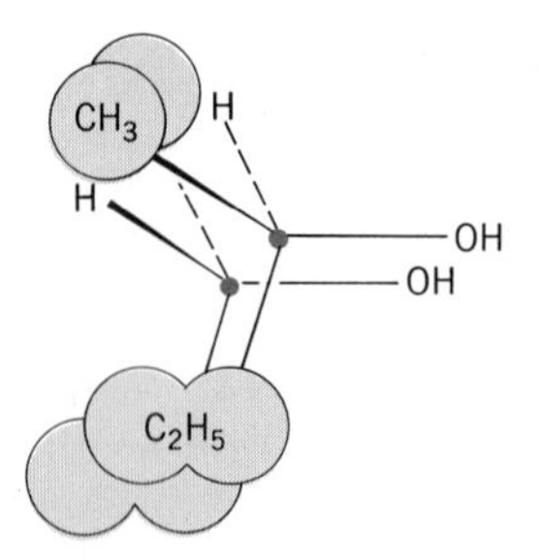

Figure 7-5. *If we lay the two enantiomers of 2-butanol (see Fig. 7-2) on top of one another, we see they are* nonsuperimposable, *and so 2-butanol is a* chiral molecule.

All of the chiral molecules we shall discuss in this book will contain one or more *chiral carbon atoms,* a carbon atom bonded to four different groups. The C-2 of 2-butanol is a chiral carbon, since it is bonded to a hydrogen atom, a hydroxyl group, a methyl group, and an ethyl group. None of the other atoms in 2-butanol is chiral. On C-3, for example, two of the groups are the same (the hydrogen atoms). Some other examples of molecules containing chiral carbon atoms are given in Fig. 7-6. If a molecule contains a single chiral carbon atom, it must differ from its mirror image. If it contains two or more chiral centers, it almost always differs from its mirror image, but there are certain special cases in which it does not. We shall consider molecules with multiple chiral centers in our discussion of the chemistry of carbohydrates in Chapt. 9.

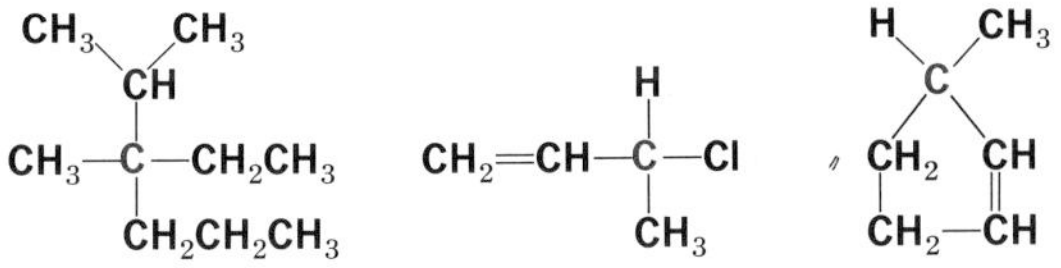

CH_3 CH_3
CH
CH_3—C—CH_2CH_3
$CH_2CH_2CH_3$

H
CH_2=CH—C—Cl
CH_3

H CH_3
C
CH_2 CH
CH_2—CH

Figure 7-6. *Three molecules, each containing a single chiral carbon atom (in red), are shown.*

Racemic Mixtures

If a sample of a compound contains equal amounts of two enantiomers, we say that it is a *racemic mixture,* and designate it *RS*, (±) or *dl*. For example if we purchased a bottle of 2-butanol it would ordinarily be (±)-2-butanol, a racemic mixture. A racemic mixture does not rotate the plane of polarized light, since a beam of polarized light passing through such a solution will meet an equal number of *R*- and *S*-molecules. The physical properties of a racemic mixture need not be identical with those of either pure enantiomer; in particular its melting point will usually differ significantly. This is true because a mixture of left- and right-handed objects ordinarily packs together differently in a crystal than either all right-handed or all left-handed objects.

Nomenclature of Enantiomers

In Fig. 7-1 we labeled one particular three-dimensional structure of 2-butanol *R* and the other *S*. What is the basis of this nomenclature, and how can it be applied to other molecules? Two steps are involved. First, the four groups attached to the chiral carbon atom are ranked 1 (highest priority) through 4 (lowest priority) following a set of *sequence rules,* which are given in the next section. For 2-butanol the hydroxyl group is given the highest priority, 1, the ethyl group, 2, the methyl, 3, and the hydrogen atom has the lowest priority, 4.

Once the group sequence has been determined, the molecule is *examined from the side opposite to the group ranked 4.* Viewing the molecule from this side, one now traces out a circle from the group ranked 1, to that ranked 2, to that ranked 3. If in making this circle, 1 → 2 → 3, you must proceed in a *clockwise* direction, the molecule is designated *R*; if in a *counterclockwise* direction, it is designated S. This is illustrated in Fig. 7-7 for the two enantiomers of 2-butanol.

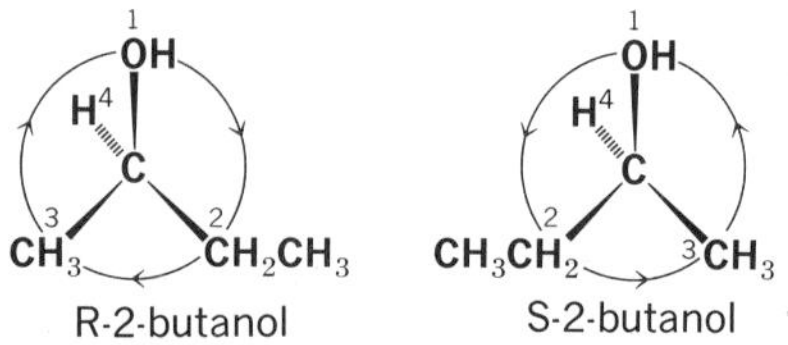

R-2-butanol S-2-butanol

Figure 7-7. *In the* R*-isomer of a molecule the priority sequence 1, 2, 3 is traced out in a clockwise direction when the chiral carbon is viewed from the side opposite to 4. In the* S*-isomer the sequence is counterclockwise.*

The Sequence Rules

In sequencing the groups attached to a chiral carbon, we look first at the *atoms* that are directly bonded to this carbon, and we rank them as far as we can in the order of atomic number, with the atom of higher priority being ranked higher. Suppose, for example, that all the atoms directly bonded to a chiral carbon are different, as in 1-chloro-1-methoxyethane (Fig. 7-8). Of

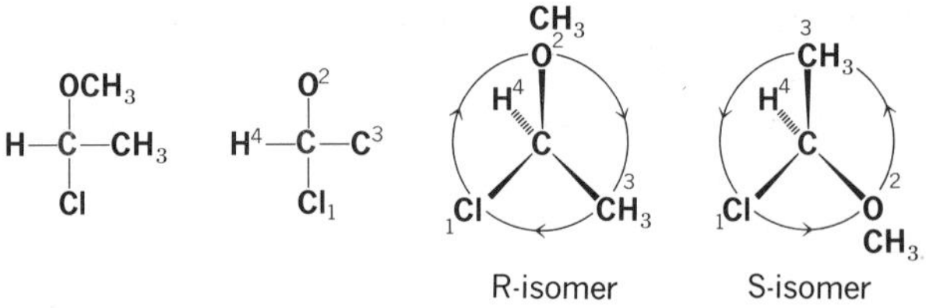

R-isomer S-isomer

Figure 7-8. *In establishing sequence priorities, the* atoms *directly attached to the chiral center are considered first, with the higher priority being given to the atom with higher atomic number.*

the four atoms directly attached to the chiral carbon, chlorine (atomic number 17) is ranked 1, oxygen (atom. no. 8) is 2, carbon (atom. no. 6) is 3, and hydrogen (atom. no. 1) is 4. To draw the *R*-isomer, we view the chiral carbon from the side opposite to 4, the hydrogen atom, and draw the other three groups so that we must proceed in a clockwise direction as we go from chlorine to methoxyl to methyl. In the S-isomer these same three groups are arranged counterclockwise.

Often, two or more of the atoms directly attached to the chiral center will be the same. For example, in 2-butanol two of these are carbon (Fig. 7-9). We make what priority assignments we can at this stage (O = 1, H = 4),

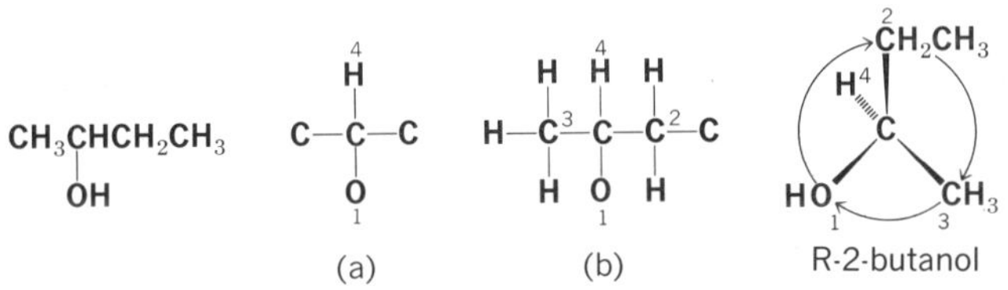

$CH_3CHCH_2CH_3$ (OH on C-2)

(a)

(b)

R-2-butanol

Figure 7-9. *In 2-butanol priorities are made, where possible, on the atoms directly attached to the chiral carbon (a). If two or more of the atoms are the same, priority assignment is made on the basis of the atomic numbers of the next atoms out from the chiral center (b).*

but to assign priority numbers to the carbons we proceed away from the chiral center to the atoms attached to these carbons. One of the carbons (that of the methyl group) has three hydrogens attached, while the other (that of the ethyl group), has two hydrogens and a carbon. The ethyl group will have the higher priority because it has an atom of higher atomic number (C) than any attached to the carbon of the methyl group. The priority sequence is then hydroxyl(1), ethyl(2), methyl(3), and hydrogen(4).

All of the compounds whose *R*, S designations are needed for this book can be handled by a few simple extensions of these rules. First, it should be emphasized that one atom of a higher atomic number outweighs any number of atoms of lower atomic number. So, for instance, a —CH_2OH group is assigned a higher priority than an isopropyl or a *t*-butyl group. Second, if a double bond is encountered on one of the groups, the atoms joined through the double bond are doubled (and tripled for a triple bond). These points are illustrated in Fig. 7-10.

Absolute Configuration

There are two enantiomers of 2-butanol, one that rotates the plane of polarized light in the (+) direction, the other in the (−) direction. We have two

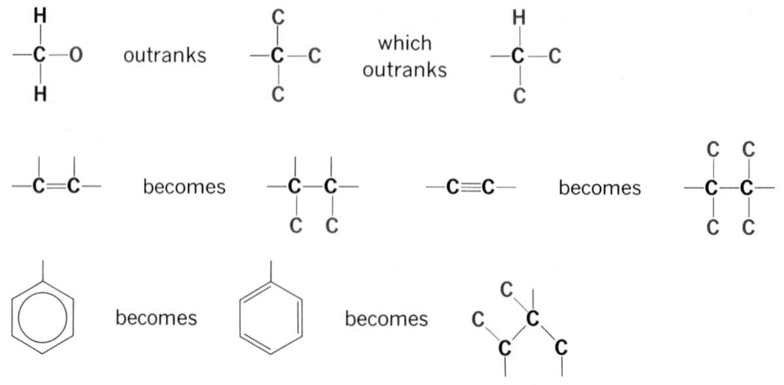

—CH_2—O outranks —C(C)(C)—C which outranks —CH(C)—C

—C=C— becomes —C(C)—C(C)— —C≡C— becomes —C(C)(C)—C(C)(C)—

(benzene ring) becomes (Kekulé structure) becomes (carbons with duplicated atoms)

Figure 7-10. *Some additional rules for determining group priorities. A single atom of higher atomic number is more important than two or three atoms of lower atomic number. Double and triple bonds are rewritten with each atom doubled or tripled before assigning priority.*

three-dimensional formulas, *R*- and S-2-butanol. How do we tell which formula corresponds to the isomer that rotates (+) and which to the (−) isomer? Once we have established this relationship, we can say we know the *absolute configuration* of each isomer of 2-butanol, the actual arrangement in three dimensions of the four groups around the chiral center. Ultimately, the assignment of absolute configuration rests on our ability to measure the position of atoms in certain molecules by the use of X rays. This method was first used for a salt of tartaric acid (Fig. 7-11). Once the absolute configuration of (+)-tartaric acid was known, the absolute configuration of (+)-2-butanol could be determined chemically by synthesis from (+)-tartaric acid.

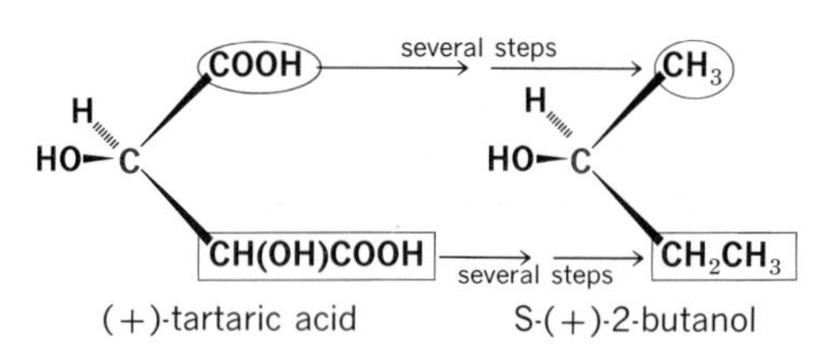

Figure 7-11. *The actual three-dimensional arrangement of the groups around the chiral carbon of (+)-tartaric acid (its* absolute configuration*) has been determined by X-ray analysis. If we start with this isomer of known structure and chemically convert the COOH group to methyl and the CH(OH)COOH group to ethyl, we obtain (+)-2-butanol, which then must have the absolute configuration shown. According to our rules of nomenclature this is the S isomer. Starting from (−)-tartaric acid, we obtain* R-(−)-2-*butanol.*

Separation of Enantiomers

Because enantiomers have identical physical properties, they can be extremely difficult to separate from one another. The simplest way to obtain one pure enantiomer is to isolate it from some natural source, if such a source exists, because nearly all chiral molecules found in living systems occur as single enantiomers. In the absence of a natural source, enantiomers can be obtained by *resolution* of racemic mixtures. There are several possible methods of resolution. The first resolution was accomplished in 1848 by the great French chemist Louis Pasteur. Pasteur was studying crystals of a salt of racemic tartaric acid,[3] when he noticed that some of the crystals were mirror images of others (Fig. 7-12). Using a magnifying glass and a pair of tweezers, he

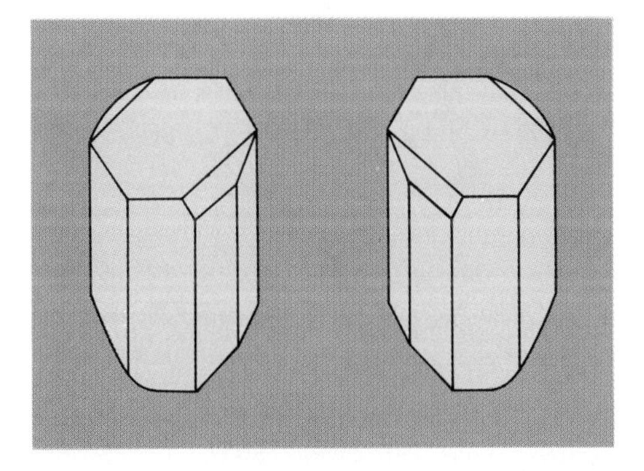

Figure 7-12. *Pasteur separated crystals of racemic tartartic acid that were mirror images of one another. One crystal gave (+)-tartaric acid, the other (−)-tartaric acid.*

separated the two types of crystals; when dissolved in water, one type rotated the plane of polarized light in the (+) direction, the other type in the (−) direction. He had carried out a resolution by mechanical means. Unfortunately, very few enantiomers crystallize in separate mirror image crystals, so that this method is of more historical than practical importance.

It is also possible to resolve enantiomers by a rather straightforward but tedious chemical method, which will be discussed in Chapter 9.

Synthesis of Organic Halides

Two common ways of synthesizing an organic halide in a pure form are reaction of hydrogen halide with an alkene or an alcohol. Both of these

[3] Since the tartaric acids occur in grape juice, and hence in wine, they were well known to French chemists.

reactions occur by way of an intermediate carbonium ion. Suppose that we were to prepare 2-bromobutane by these methods. Addition of a proton to 1-butene forms the 2-butyl cation (Fig. 7-13). Since the cation is planar and has

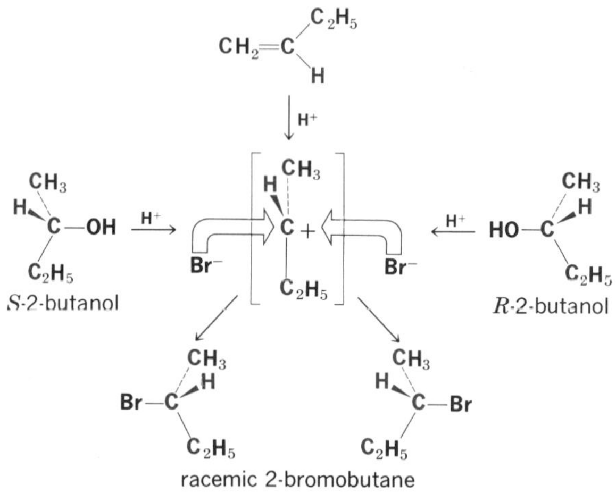

Figure 7-13. *When either 1-butene, pure* R-, *or pure* S-2-*butanol reacts with hydrogen bromide, it is converted into racemic 2-bromobutane, by way of the symmetrical carbonium-ion intermediate.*

only three groups around it, it is achiral; models will show that it is superimposable on its mirror image. This cation can react with bromide ion equally well from either face, giving on the one hand *R*-2-bromobutane and on the other S-2-bromobutane. Simply on the basis of probability, racemic 2-bromobutane will inevitably be formed. In fact, this result illustrates a general rule; *starting with achiral reagents* (and neither 1-butene nor HBr contains a chiral center), *we must always generate a racemic mixture when we create a chiral center.*

What is the result if we produce 2-bromobutane from a chiral molecule, say *R*-2-butanol or S-2-butanol? In this case the result is the same, for again the achiral cation is an intermediate and it leads to racemic-2-bromobutane. This result is some of the strongest evidence we have the cation is an intermediate in the conversion of alcohols to halides (Fig. 7-13).

Several methods for the synthesis of organic halides have been given in earlier chapters. These include halogenation of saturated alkanes (Fig. 1-2), addition of hydrohalic acids and halogens to alkenes and alkynes (Figs. 3-16 and 3-19), the aromatic substitution by halogen (Fig. 5-1) and the replacement of hydroxyl groups by halogen (Fig. 6-2).

Prob. 7-2. If you prepared 2-bromobutane by HBr addition to 2-butene, would you expect to get a racemic product? Suppose that you converted

***R*-2-methyl-1-butanol to its bromide. What would be the stereochemistry of the product, *R*, *S*, or *R,S*?**

Displacement Reactions of Halides

When a primary or secondary alkyl halide is allowed to react with nearly any anion Y^-, the halide is replaced by the anion in what is known as a *displacement reaction*. Even some molecules like ammonia, which are neutral but basic, are effective in this reaction. The displacement reaction is one of the most useful in all of organic chemistry because a very large number of functionally substituted organic molecules can be formed in this way. Some examples are given in Fig. 7-14.

$$CH_3CH_2Br + OH^- \longrightarrow CH_3CH_2OH + Br^-$$

$$CH_3CH_2I + SH^- \longrightarrow CH_3CH_2SH + I^-$$

$$CH_3CH_2Cl + CN^- \longrightarrow CH_3CH_2CN + Cl^-$$

$$CH_3CH_2Br + I^- \longrightarrow CH_3CH_2I + Br^-$$

$$CH_3CH_2Br + NH_3 \longrightarrow CH_3CH_2\overset{+}{N}H_3 + Br^-$$

Figure 7-14. *Five examples of the displacement reaction in which a halide is replaced by a group Y. Group Y may be nearly any anion, and some neutral molecules with unbonded electrons, like NH_3, are also effective. Tertiary halides do not undergo this reaction.*

It is immediately evident that displacement reactions differ profoundly from all other reactions we have studied so far. In the first place the reagents are bases (OH^-, NH_3) or anions (CN^-, I^-) rather than the acids or radicals that have been involved in earlier reactions. Second, if we study the effect of structure on the ease of the displacement reaction, we find that a 1° halide is the most reactive, followed by a 2° halide; 3° halides do not react by displacement at all. This is exactly the opposite order of reactivity from that found among alcohols in the Lucas reaction, e.g., Fig. 6-7. Finally, we find a remarkable difference if we use a pure enantiomeric halide as the starting material for a displacement reaction. If, for instance, we start with pure S-2-bromobutane and allow it to react with hydroxide ion, we find that the alcohol formed is pure *R*-2-butanol. We say that the reaction is *stereospecific;* one pure stereo isomer has been formed from another. What is even more remarkable, a right-handed molecule has been formed from a left-handed one in the course of the transformation.

Results like these clearly rule out a carbonium-ion intermediate in the displacement reaction and indicate a different pathway, one in which the

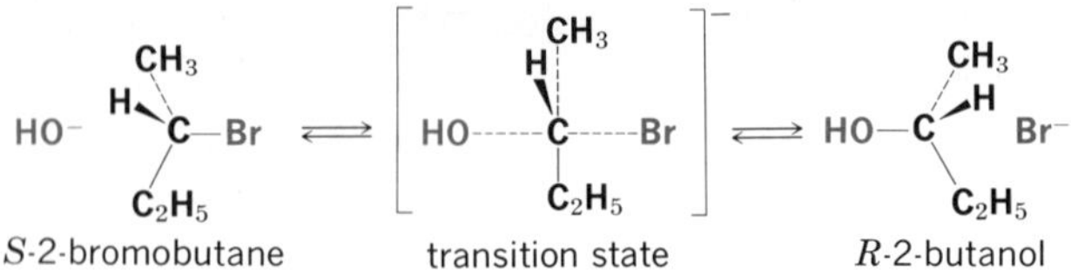

$$HO^- + \underset{C_2H_5}{\overset{CH_3}{H\!-\!C\!-\!Br}} \rightleftharpoons \left[HO \cdots \underset{C_2H_5}{\overset{CH_3}{\overset{H}{C}}} \cdots Br \right]^- \rightleftharpoons HO\!-\!\underset{C_2H_5}{\overset{CH_3}{C\!-\!H}} \quad Br^-$$

S-2-bromobutane — transition state — *R*-2-butanol

Figure 7-15. *In a displacement reaction, the entering group attacks at the rear of the C—Br bond, turning the molecule inside out in what is known as a Walden inversion. Too many large groups about the rear of the C—Br bond (as in a tertiary halide) prevent the reaction.*

molecule is literally turned inside out by a process known as Walden inversion (Fig. 7-15). The incoming group Y attacks the rear of the C—X bond and pushes X out from the rear. As the C—Y bond is formed the C—X bond is broken, and at an intermediate point in the reaction both bonds are about half-made, half-broken. The other three groups at this point lie in a plane containing the central carbon atom. This high energy state, in which the transformation of reactants to products is about half over, is called the *transition state* of the reaction. It does not represent a molecule that can be isolated, but rather the arrangement of atoms at the highest energy point along the path the reaction takes. This reaction mechanism is in good accord with the order of reactivity $1° > 2° \gg 3°$, since an increase in the number of alkyl groups around the C—Br bond will make it increasingly difficult for the chloride to approach the rear of the carbon atom. Other examples of displacement are shown in Fig. 7-16.

Prob. 7-3. Suppose you wanted to prepare ethyl *tert*-butyl ether, $C_2H_5OC(CH_3)_3$, by a reaction using the first example in Fig. 7-16 as an analogy. There are two different combinations of halide and alkoxide you might use. Explain why one combination would give the product wanted and the other would not.

Elimination Reactions of Halides

Alkyl halides sometimes react with anions to give alkenes. This type of reaction is known as an *elimination reaction.* The best yield results from treatment with a base, and hydroxide or alkoxide is frequently used for this purpose. With primary and secondary halides, the two reactions, displacement and elimination, are competitive. When isopropyl bromide reacts with ethoxide ion in ethanolic solution, for instance, 20% ethyl isopropyl ether and 80%

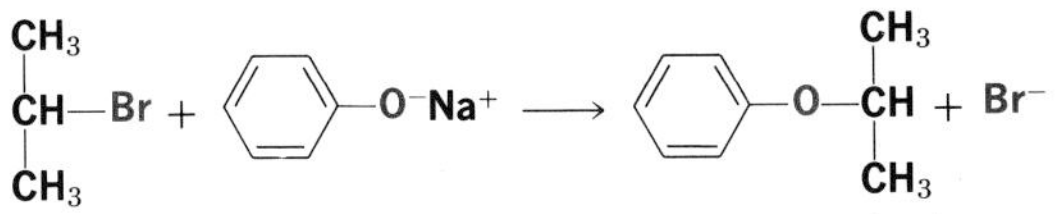

$$(CH_3)_2CH{-}Br + C_6H_5{-}O^-Na^+ \longrightarrow C_6H_5{-}O{-}CH(CH_3)_2 + Br^-$$

phenyl isopropyl ether

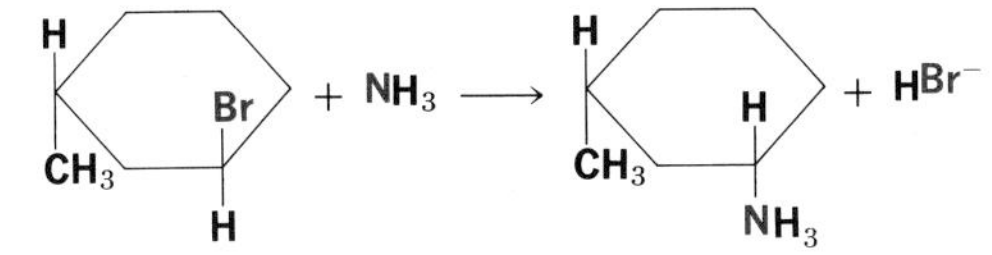

$$trans\text{-3-methyl-1-bromocyclohexane} + NH_3 \longrightarrow cis\text{-3-methylcyclohexyl ammonium bromide} + HBr^-$$

trans-3-methyl-1-bromocyclohexane

cis-3-methylcyclohexyl ammonium bromide

$$C_6H_5{-}CH_2Cl + CN^- \longrightarrow C_6H_5{-}CH_2CN + Cl^-$$

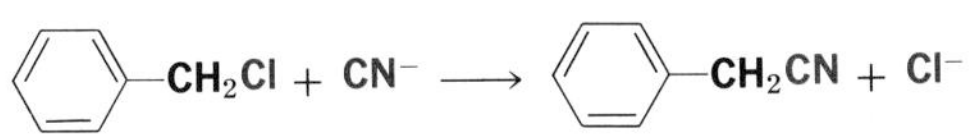

benzyl chloride

benzyl cyanide

Figure 7-16. *Three examples of the displacement reaction. All these reactions proceed with Walden inversion, although only in the second case is this obvious from the structure of the product.*

propylene are formed (Fig. 7-17). Since tertiary halides do not undergo the displacement reaction, only alkenes and no alcohols or ethers form when a tertiary halide is allowed to react with base.

The elimination and displacement reactions proceed by related pathways. In both, approach by the base precedes the reaction, but the attack in elimination is on the hydrogen atom located on the carbon atom adjacent to that bearing the halogen atom. The oxygen–hydrogen bond and the

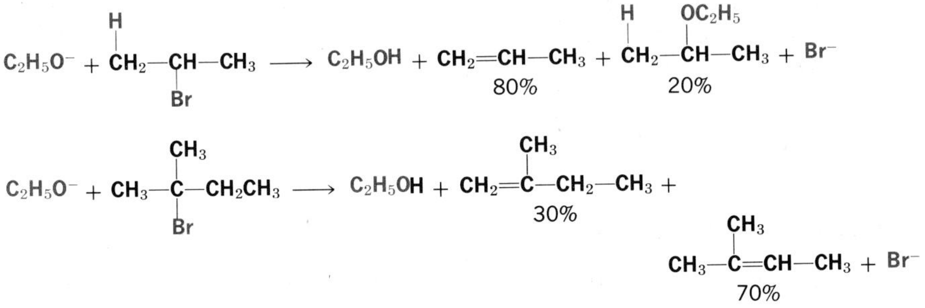

$$C_2H_5O^- + CH_2(H){-}CH(Br){-}CH_3 \longrightarrow C_2H_5OH + \underset{80\%}{CH_2{=}CH{-}CH_3} + \underset{20\%}{CH_2(H){-}CH(OC_2H_5){-}CH_3} + Br^-$$

$$C_2H_5O^- + CH_3{-}C(CH_3)(Br){-}CH_2CH_3 \longrightarrow C_2H_5OH + \underset{30\%}{CH_2{=}C(CH_3){-}CH_2{-}CH_3} + \underset{70\%}{CH_3{-}C(CH_3){=}CH{-}CH_3} + Br^-$$

Figure 7-17. *When a halide reacts with a base, elimination and displacement reactions compete. Tertiary halides form only alkenes; where a choice of alkenes exists, the olefin with the fewest hydrogens on the double bond usually predominates (Saytzeff's rule, Fig. 6-8).*

carbon–carbon double bond form, whereas the hydrogen–carbon and carbon–halogen bonds break. When elimination can take place in more than one direction, a mixture of products usually results.

Elimination reactions occur most readily if the hydrogen and halogen atoms are *trans* to one another. This stereochemical requirement has its most striking effect in cyclic systems. Suppose *trans*-2-methylcyclopentyl bromide is treated with base (Fig. 7-18). Attack from the rear of the C—Br bond can occur on carbon to give the alcohol with Walden inversion or 3-methylcyclopentene can result by attack on the *trans* hydrogen (in red) and elimination of HBr. No 1-methylcyclopentene can form because there is no *trans* hydrogen on the carbon bearing the methyl group, and elimination of HBr by removing the hydrogen *cis* to the bromine atom is very difficult. It should be clear from this example that stereochemistry can have a profound effect on chemical reactivity.

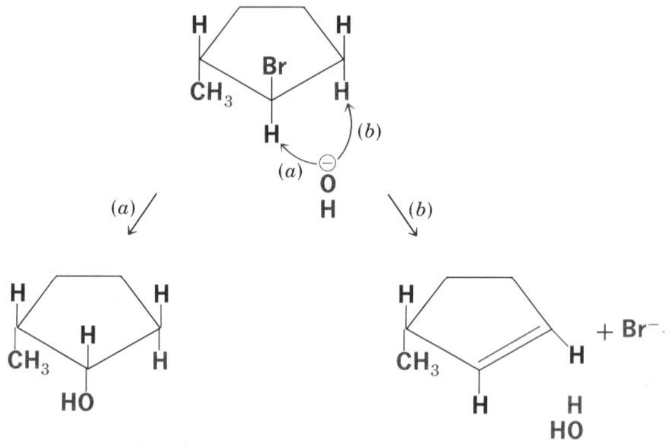

Figure 7-18. *Both displacement (path a) and elimination (path b) are initiated by attack from the side of the molecule opposite to the bromine. No elimination can occur toward the methyl group because there is no hydrogen* trans *to the bromine atom on that carbon.*

Organometallic Reagents

Reaction of organic halides with certain metals gives molecules with carbon–metal bonds. We have already seen that the reaction of ethyl bromide with lead can give the important antiknock agent tetraethyllead. For organic synthesis, more reactive organometallic compounds than tetraethyllead are needed. Organomagnesium compounds form from the reaction, in ether, of alkyl or aryl halides with metallic magnesium (Fig. 7-19). These compounds

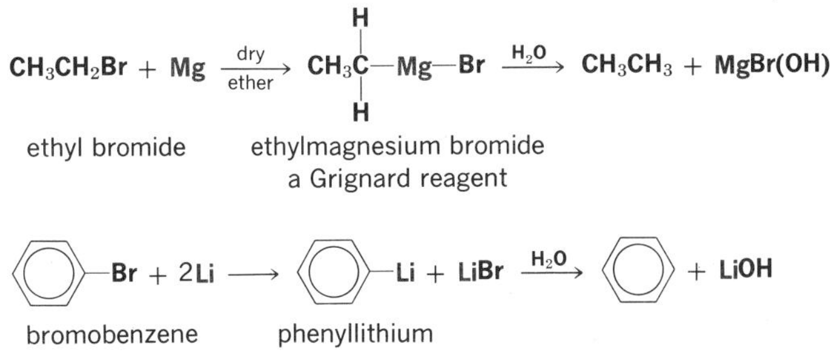

$$CH_3CH_2Br + Mg \xrightarrow[\text{ether}]{\text{dry}} CH_3\overset{H}{\underset{H}{C}}—Mg—Br \xrightarrow{H_2O} CH_3CH_3 + MgBr(OH)$$

ethyl bromide — ethylmagnesium bromide, a Grignard reagent

$$C_6H_5—Br + 2Li \longrightarrow C_6H_5—Li + LiBr \xrightarrow{H_2O} C_6H_6 + LiOH$$

bromobenzene — phenyllithium

Figure 7-19. *Most organic chlorides, bromides, and iodides react with magnesium (to give Grignard reagents) or with lithium. The resultant organometallic compounds react vigorously with water, acids, or other sources of protons to form hydrocarbons.*

are more commonly called Grignard reagents after their discoverer, the French chemist Victor Grignard (Nobel Prize, 1912). Organolithium compounds, formed by the reaction of organic halides with metallic lithium, are also useful. Most carbon–metal bonds are not particularly ionic, but to interpret their reactions it helps to think of them as reacting as if they were $\geq C^-Mg^{2+}Br^-$ or $\geq C^-Li^+$. Both of these organometallic compounds readily combine with any source of a proton (from water, acid, alcohol, etc.) to give $\geq C—H$. So, all reactions involving these intermediates must be carried out in the absence of even traces of moisture.

Grignard and organolithium compounds are among the most important reagents in organic chemistry because they react readily with carbon–oxygen double bonds (aldehydes, ketones, esters, etc.) to form large organic molecules from small ones. These reactions of Grignard reagents will be covered in later chapters.

Aryl and Vinyl Halides

A halogen atom attached directly to a carbon atom of a carbon–carbon unsaturated bond has a great deal less reactivity than an ordinary alkyl halide. Specifically, aryl and vinyl halides undergo displacement reactions (with hydroxide, alkoxide, etc.) only under such stringent conditions that these reactions are seldom attempted in the laboratory. Elimination reactions of vinyl halides are less difficult, and vinyl bromide is converted to acetylene by heating with base (Fig. 7-20). If an aromatic halide is allowed to react with a strong base like sodium amide ($NaNH_2$), elimination also takes place. The resulting intermediate is called a *benzyne* and is highly strained. Benzynes cannot be isolated, since they react readily with either excess amide or with themselves to give a dimer.

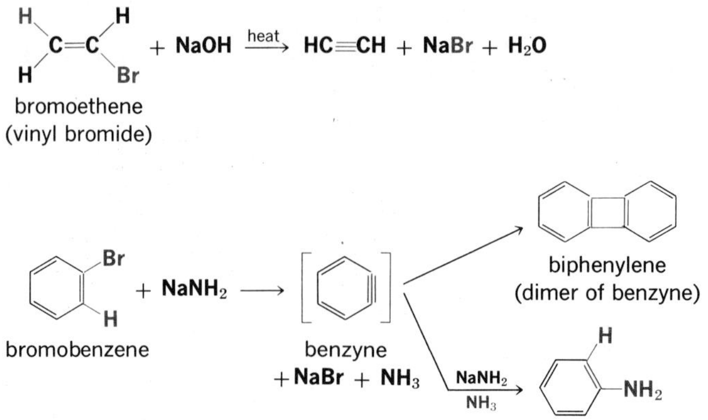

$$H_2C{=}CHBr + NaOH \xrightarrow{heat} HC{\equiv}CH + NaBr + H_2O$$

bromoethene (vinyl bromide)

$$C_6H_5Br + NaNH_2 \longrightarrow [\text{benzyne}] + NaBr + NH_3$$

bromobenzene → benzyne + $NaBr + NH_3$ → biphenylene (dimer of benzyne); benzyne $\xrightarrow[NH_3]{NaNH_2}$ $C_6H_5NH_2$

Figure 7-20. *Halogen atoms attached to a double bond or aromatic ring are quite unreactive in displacement reactions. Vinyl and aryl halides form acetylenes on treatment with strong base, but those from aryl halides, called benzynes, are quite unstable. One of the products from this elimination reaction of halobenzenes is the dimer of benzyne.*

Visual Chemical Tests for Halides

The presence of halogen in an organic molecule is usually revealed by the *sodium fusion test*. A small sample of the organic compound is heated with a piece of sodium metal about the size of a small pea until the sodium melts. Any halogen in the molecule reacts with the sodium to form sodium halide (compare with the reaction of lithium metal in Fig. 7-19). The excess sodium is carefully destroyed with alcohol, aqueous acid is added, and then a few drops of a silver nitrate solution. The appearance of a precipitate of silver halide is a positive visual test for an organic halide. The test can be made quantitative.

It is possible to distinguish among 1°, 2° and 3° halides in much the same way that the Lucas test distinguishes among 1°, 2° and 3° alcohols. A 3° halide is extremely reactive in any reaction that occurs by way of an intermediate carbonium ion. One such reaction is that with silver nitrate in ethyl alcohol solution; a 3° halide will give an immediate precipitate of silver halide. A 2° halide reacts only slowly with the same reagent, and a 1° halide does not react. Neither do aryl or vinyl halides.

A second visual chemical test used to distinguish among bromides takes advantage of the fact that 1° halides react rapidly in displacement reactions; 2° halides react only slowly; and 3°, aryl, and vinyl halides do not react at all. To carry out this test, the halide is dissolved in acetone containing sodium iodide. Primary bromides react immediately to give an alkyl iodide and a precipitate of sodium bromide; 2° bromides react only slowly; and 3°, aryl, and vinyl halides do not react.

Commercially Important Organic Halides

A number of halo compounds have important practical applications. Carbon tetrachloride (CCl_4), for example, is a dense, nonflammable liquid (b.p. 78°) that is a good solvent for oils and greases. It finds wide use in the dry cleaning industry. Carbon tetrachloride is poisonous, and so should always be used with adequate ventilation.

Two industrial methods are available for the synthesis of carbon tetrachloride, and they furnish good examples of some of the economic factors that play a role in industrial chemistry. The simplest method involves the direct chlorination of methane, while in the second method carbon and sulfur are combined to give carbon disulfide, which is then chlorinated to furnish carbon tetrachloride. Twice as much chlorine is required for chlorination of methane to carbon tetrachloride as for chlorination of carbon disulfide. Most companies have uses for the hydrogen chloride produced as a byproduct in the first reaction, or can sell it, but if they do not, the synthesis from carbon disulfide is cheaper (Fig. 7-21).

$$CH_4 + 4\,Cl_2 \longrightarrow CCl_4 + 4HCl$$

$$CS_2 + 2\,Cl_2 \longrightarrow CCl_4 + 2S$$

Figure 7-21. *Two industrial syntheses of carbon tetrachloride. If a market were not readily available for hydrogen chloride, the second reaction would be the more economical because carbon disulfide is synthesized from carbon and sulfur, two cheap raw materials.*

Carbon tetrachloride is also used in the production of chloroform and as a fire extinguisher. In the latter role it acts as a smothering agent; its dense vapors blanket the fire and keep out oxygen. There is danger, however, because carbon tetrachloride is oxidized to the poisonous gas phosgene, $COCl_2$.

Prob. 7-4. During World War II, many fire bombs were used which were primarily powdered magnesium metal. This metal burns fiercely, and the bombs caused great destruction. Why were people warned not to use carbon tetrachloride fire extinguishers on such fires?

Chloroform, a sweet colorless liquid (b.p. 61°), is prepared industrially by the reduction of carbon tetrachloride with iron. Like nearly all halo compounds, chloroform is somewhat toxic. As it is especially apt to cause damage to the liver, it is now seldom used as a general anesthetic, although it is still used as a mild sedative in cough syrups, cough drops, or liniments. Ethyl chloride (b.p. 12°) is also commonly used as a topical anesthetic. It owes its effect to its physical rather than its chemical properties: When sprayed on the skin, ethyl chloride evaporates rapidly, cooling the skin and nerve endings.

A number of halogen-containing compounds find wide use as insecticides and herbicides. The most famous of these is DDT, an abbreviation for its common name *di*chloro*di*phenyl-*tri*chloroethane. This compound is formed by the reaction of chlorobenzene with chloral (Fig. 7-22). DDT is active against a wide variety of insects and proved its usefulness in striking fashion when, after World War II, nearly all the residents of Naples were dusted with DDT to bring a typhus epidemic to a halt. It has also been used to eradicate malaria from many parts of the world and has probably saved the lives of over 100 million humans. Unfortunately, it is broken down only slowly in nature and tends to concentrate in the organs of animals exposed to it.

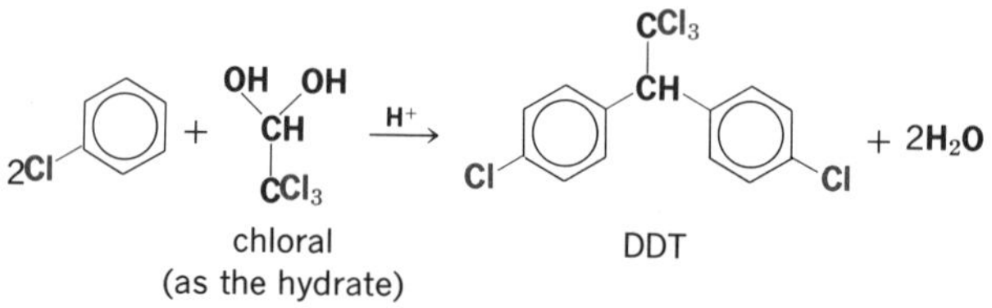

$$2\,Cl{-}C_6H_4{-}H + CH(OH)_2CCl_3 \xrightarrow{H^+} (Cl{-}C_6H_4)_2CH{-}CCl_3 + 2H_2O$$

chloral (as the hydrate) — DDT

Figure 7-22. *DDT was the first really potent synthetic organic insecticide.*

Organic Fluorine Compounds

Organic fluorides differ greatly from other organic halides, some being more reactive and others, less reactive. A detailed examination of their reactivity is beyond the scope of this book. More important to us is the knowledge that compounds containing two or more fluorine atoms on the same carbon atom are extremely resistant to chemical attack. Tetrafluoroethylene, for instance, can be polymerized to Teflon (Fig. 7-23), which is an extremely inert, high-

$$F_2C{=}CF_2 \longrightarrow {-}CF_2{-}CF_2{+}CF_2{-}CF_2{+}_nCF_2{-}CF_2{-}$$

tetrafluoroethylene — Teflon

Figure 7-23. *Polytetrafluoroethylene (Teflon) is the completely fluorinated analog of polyethylene. It is stable to 380° and completely inert to boiling acids and bases, and to many oxidizing agents.*

melting plastic. It is used to coat the insides of frying pans to prevent sticking; it also has many industrial and scientific applications, such as being made into bearings to be exposed to a corrosive atmosphere.

Completely fluorinated hydrocarbons—fluorocarbons—have unusual physical and chemical properties. In particular, their boiling points are much lower than hydrocarbons of the same molecular weight. Perfluoro-*n*-butane (C_4F_{10}, mol. wt. 138) has a boiling point even lower ($-1.7°$) than that of *n*-butane (mol. wt. 58, b.p. $-0.5°$).

Sodium monofluoroacetate is a powerful rat-killing agent, whereas Freon, CF_2Cl_2, is a completely nontoxic gas used in refrigerators and as a propellant in aerosol sprays. Satisfactory methods for preparing a variety of completely fluorinated hydrocarbons were discovered in the late 1930s, just in time to be used in the atomic energy program developed during the war years. The great unreactivity of these hydrocarbons made them valuable in contact with uranium hexafluoride, and this use gave great impetus to investigations of their syntheses and properties. Fluorinated compounds are still quite expensive, but industrial applications are increasing.

PROBLEMS

1. Identify the chiral carbon in each of the following compounds.
   (a) $CH_3CHBrCH_2CH_3$ (b) $(C_2H_5)_2CHCH(CH_3)C_2H_5$
   (c) $CH_2{=}CH{-}CH(OH)CH_3$ (d) Cl, H, CH_3, CH_3 (substituted cyclopropane)
   (e) HO H (cyclohexene ring) (f) $C_6H_5CH(CH_3)C_6H_4Cl$

2. Complete the following reactions. If more than one organic product is expected, give the structures of all. If no reaction occurs, so indicate.
   (a) $CH_3CH(CH_3)CH_2Cl + I^- \longrightarrow$
   (b) C_6H_5–$CH_2Br + OH^- \longrightarrow$
   (c) $(CH_3)_3C{-}Br + OH^- \longrightarrow$
   (d) $CH_3Br + CH_3O^- \longrightarrow$
   (e) C_6H_5–$Br + Mg \longrightarrow$
   (f) $CH_3Li + H_2O \longrightarrow$

(g) Br + OH^- $\longrightarrow$ (2 products)

(h) CH_3, H, Br, H, H, C_2H_5 + OH^- $\longrightarrow$ (2 products; include stereochemistry)

(i) $CH_3CHICH(CH_3)_2 + CH_3O^-$ $\longrightarrow$ (3 products)

(j) $(CH_3)_2CHBr + Mg$ $\longrightarrow$ $\xrightarrow{H_2O}$

(k) Br + HNO_3 $\xrightarrow{H_2SO_4}$

3. Suppose that each of the following reactions were carried out with the enantiomer indicated. Draw the three-dimensional structure(s) of the product(s) formed.
   (a) *R*-1-phenylethanol + HBr $\longrightarrow$
   (b) *S*-3-bromoheptane + CN^- $\longrightarrow$
   (c) *R*-3-methyl-3-bromohexane + OH^- $\longrightarrow$
   (d) *R*-3-bromo-1-pentene + H_2 $\xrightarrow{Pt}$
   (e) *S*-3-methylhexane + Br_2 $\xrightarrow{light}$ (3° bromide only)
   (f) *R*-4-phenyl-1-pentanol + HBr $\longrightarrow$
4. What is the smallest noncyclic saturated hydrocarbon that could exist in enantiomeric forms? The smallest noncyclic alkene? The smallest cyclic saturated hydrocarbon?
5. Draw the structure of each of the following commercially important compounds.
   (a) 1,1-di-(*p*-chlorophenyl)-ethanol (DMC, useful in killing mites)
   (b) 3-chloro-2-methylpropene (methallyl chloride, an insecticide for grain weevils)
   (c) hexachloroethane (used in smoke pots)
   (d) 1,1-dihydroxy-2,2,2-trichloroethane (chloral hydrate, a hypnotic)
   (e) 1,2,3,4,5,6 hexachlorocyclohexane (Lindane, an insecticide)
6. Production of DDT by the Friedel–Crafts reaction shown in Fig. 7-22 actually leads to three isomeric DDT molecules, only one of which is shown. Suggest structures for the other two isomers based upon your knowledge of the directive effect of chlorine in aromatic substitution reactions.
7. If pure *R*-2-bromobutane is allowed to react with sodium bromide, the product is racemic 2-bromobutane. Explain.

8. Draw the structure of each of the following:
(**a**) chloroform (**b**) *R*-3-phenyl-3-bromooctane
(**c**) S-1-hexen-3-ol (**d**) S-2-cyclohexenol
(**e**) perfluorobutane

9. How could you distinguish among the following compound by visual chemical tests?

(**a**) $CH_3(CH_2)_4CH{=}CHCH_3$ $CH_3(CH_2)_4CH(OH)CH_3$ $CH_3(CH_2)_4CH(Br)CH_3$

—Br $CH_3(CH_2)_4CH_2C{\equiv}CH$

(**b**) Br

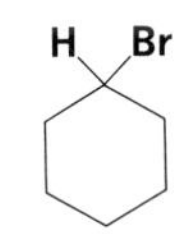

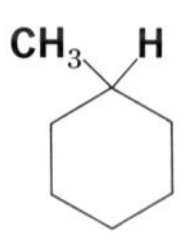

8
aldehydes and ketones

With aldehydes and ketones we meet for the first time the chemistry of the carbon–oxygen double bond, known as the *carbonyl* group. In an *aldehyde* at least one hydrogen atom is attached to the carbonyl group, whereas in a *ketone* both remaining valences of the carbonyl carbon are joined to other carbon atoms. Thus formaldehyde, $H_2C{=}O$, is the simplest aldehyde, and acetone, $(CH_3)_2C{=}O$, the simplest ketone. Among familiar compounds, glucose is an aldehyde, as are odoriferous constituents found in cinnamon, vanilla, and heliotrope, for example, while ketones contribute to the odor and flavor of peppermint, caraway, and camphor.

Acetone was known among the alchemists of the middle ages because it could be prepared by the strong heating of lead acetate, which in turn was made by dissolving lead salts in vinegar. It obtained its present common name during the 1830s, since the ending *-one* was used to mean "derived from," and acetone was derived from acetic acid. As the chemical nature of the carbon–oxygen double bond became clear in the 1850s the -one ending was used more and more to indicate a ketone; it was adopted in the IUPAC nomenclature for that purpose. Aldehydes owe their names to the fact that they can be obtained from *al*cohols by *dehy*drogenation (i.e., oxidation); in the IUPAC system the ending *-al* is used to indicate their presence. Their common names derive from the names of the acids they yield on oxidation (i.e., *acet*aldehyde gives *acet*ic acid).

Nomenclature

Both common and systematic names are widely used for aldehydes and ketones. Some aldehydes are listed in Table 8-1. The common names are derived, as we have seen, from those of the corresponding acids. Thus formaldehyde gives, on oxidation, *formic* acid (Latin: "ant," because it occurs in the stinging red

Table 8-1
Names and Physical Properties of Some Simple Aldehydes

Structure	Common Name	IUPAC Name	B.P.	M.P.
$H_2C{=}O$	formaldehyde	methanal	−19°	−92°
$CH_3\overset{H}{C}{=}O$	acetaldehyde	ethanal	21°	−123°
$CH_3CH_2\overset{H}{C}{=}O$	propionaldehyde	propanal	49°	−81°
$CH_3CH_2CH_2\overset{H}{C}{=}O$	*n*-butyraldehyde	butanal	75°	−99°
$C_6H_5\overset{H}{C}{=}O$	benzaldehyde	benzaldehyde	179°	−56°
$(H)(CH_3)C{=}C(H)(\overset{H}{C}{=}O)$	crotonaldehyde	*trans*-2-butenal	104°	−76°

ant), propionaldehyde, propionic acid (Latin: "first fat") and benzaldehyde, *benzoic* acid (since it can be obtained from the fragrant resin, gum benzoin). Because acids are so easy to isolate, they were among the earliest organic compounds known in something approaching a pure state, and so they gave their names to many of the standard series of organic chemicals. Propane was named from propionic acid, butane (and butyraldehyde) from butyric acid (Latin: "butter"), benzene from benzoic acid. The IUPAC systematic names of aldehydes are derived by using the ending *-al;* the aldehyde group takes precedence in numbering over other groups we have discussed.

Among ketones, common names are also widespread, and many have been incorporated into the IUPAC system, particularly in the aromatic series. The common and IUPAC names of some simple ketones are given in Table 8-2. In the IUPAC system the ending for a ketone is *-one.* Among groups encountered thus far, it takes precedence over all but the aldehyde group. Many common names combine the group names with the word ketone. Prac-

Table 8-2
Names and Physical Properties of Some Simple Ketones

Structure	Common Name	IUPAC Name	B.P.	M.P.
$CH_3\overset{O}{\overset{\|\|}{C}}CH_3$	acetone	propanone	56°	−94°
$CH_3COCH_2CH_3$	methyl ethyl ketone	2-butanone	79°	−86°
$CH_3COCH{=}CH_2$	methyl vinyl ketone	3-buten-2-one	81°	
$C_6H_5COC_6H_5$	diphenyl ketone	benzophenone	305°	48°
$C_6H_5COCH_3$	methyl phenyl ketone	acetophenone	202°	20°

tice in the use of IUPAC nomenclature as applied to aldehydes and ketones will be found in the problems section at the end of the chapter.

Synthesis of Aldehydes and Ketones

Aldehydes and ketones are most easily prepared by oxidation of alcohols; this is especially true for ketones of low molecular weight, which may be prepared from inexpensive secondary alcohols. For example, acetone is prepared in enormous quantities by oxidation (Fig. 8-1) of isopropyl alcohol, obtained from propylene. In the laboratory ketones may be formed from alcohols by oxidation with dichromate.

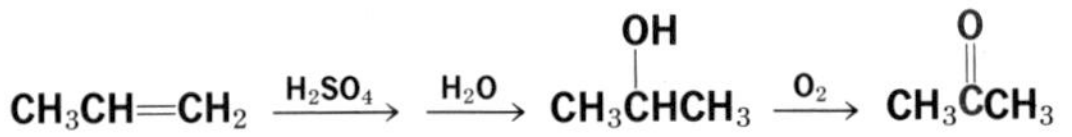

$$CH_3CH{=}CH_2 \xrightarrow{H_2SO_4} \xrightarrow{H_2O} CH_3\overset{\overset{\displaystyle OH}{|}}{C}HCH_3 \xrightarrow{O_2} CH_3\overset{\overset{\displaystyle O}{\|}}{C}CH_3$$

Figure 8-1. *The preparation of acetone from propylene via isopropyl alcohol. Industrially, air is used as the oxidant, while in the laboratory $K_2Cr_2O_7$ is used.*

Prob. 8-1. Show the preparation of methyl ethyl ketone from 1-butene by the series of reactions of Fig. 8-1. Could methyl ethyl ketone also be prepared from 2-butene?

Primary alcohols can sometimes be oxidized to aldehydes, although this is a delicate reaction, since aldehydes are themselves easily oxidized. Special conditions are usually employed, such as dehydrogenation over copper at high temperatures. Aromatic aldehydes are conveniently produced by hydrolysis of 1,1-dichloro compounds (Fig. 8-2).

Both aldehydes and ketones can be formed by the action of ozone (O_3) on alkenes. The initial product is an adduct called an *ozonide*, whose hydrolysis gives two molecules of carbonyl compounds, plus hydrogen peroxide. Since

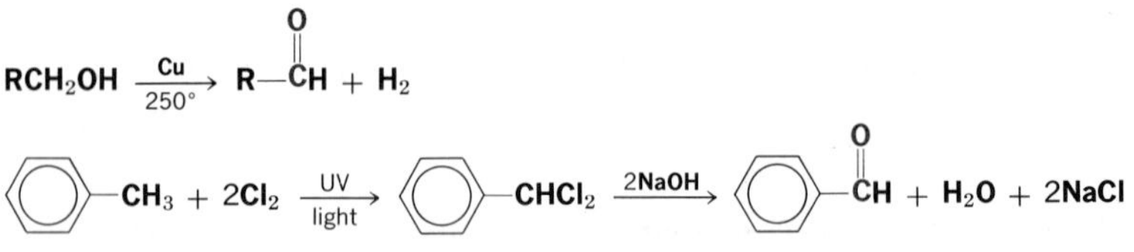

$$RCH_2OH \xrightarrow[250^\circ]{Cu} R{-}\overset{\overset{\displaystyle O}{\|}}{C}H + H_2$$

$$C_6H_5{-}CH_3 + 2Cl_2 \xrightarrow[\text{light}]{\text{UV}} C_6H_5{-}CHCl_2 \xrightarrow{2NaOH} C_6H_5{-}\overset{\overset{\displaystyle O}{\|}}{C}H + H_2O + 2NaCl$$

toluene — benzal chloride — benzaldehyde

Figure 8-2. *Two methods for the synthesis of aldehydes.*

the hydrogen peroxide may oxidize any aldehyde formed, the hydrolysis is usually carried out in the presence of a reducing agent (H_2 or Zn) to reduce H_2O_2 to water (Fig. 8-3).

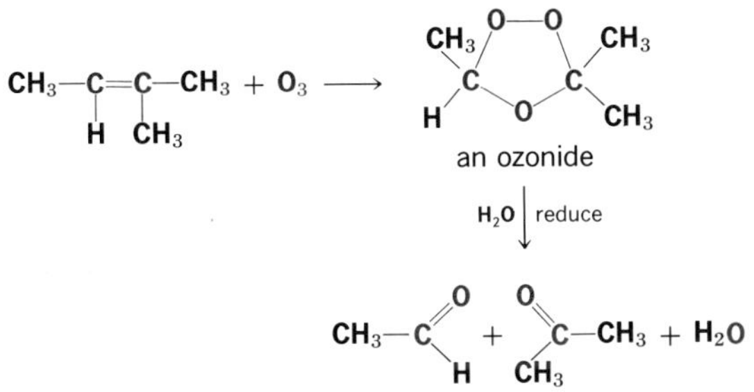

$$CH_3{-}\underset{H}{C}{=}\underset{CH_3}{C}{-}CH_3 + O_3 \longrightarrow \text{an ozonide} \xrightarrow[\text{reduce}]{H_2O} CH_3{-}\overset{O}{\overset{\|}{C}}{-}H + CH_3{-}\overset{O}{\overset{\|}{C}}{-}CH_3 + H_2O$$

Figure 8-3. Ozonolysis *cleaves a carbon–carbon double bond with the formation of two carbon–oxygen double bonds.*

Ozonolysis, as this process is called, is useful not only in synthesizing aldehydes and ketones, but also in locating carbon–carbon double bonds, since the position of the carbonyl bonds in the products defines that of the carbon–carbon double bond in the alkene.

Prob. 8-2. If an alkene gives 2-butanone and propanal when treated with ozone followed by palladium and hydrogen, what is its structure?

An important reaction for synthesizing aromatic ketones involves the use of an acid chloride like acetyl chloride in the Friedel–Crafts reaction (Fig. 8-4). Substitution nearly always occurs in the *para* position relative to an electron-donating substituent. The reaction does not occur when the ring contains *meta* directing substituents.

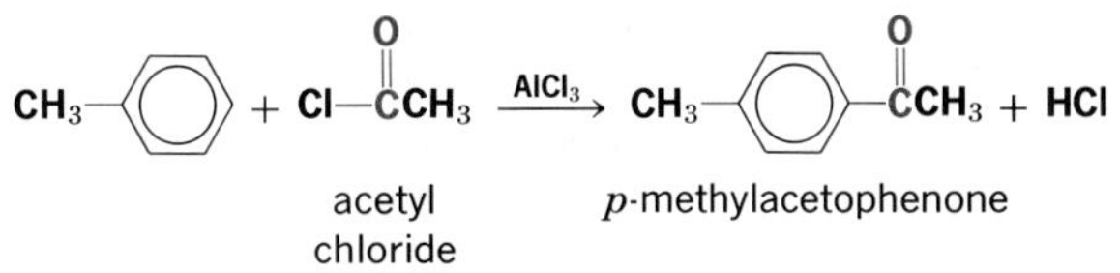

$$CH_3{-}C_6H_5 + Cl{-}\overset{O}{\overset{\|}{C}}CH_3 \xrightarrow{AlCl_3} CH_3{-}C_6H_4{-}\overset{O}{\overset{\|}{C}}CH_3 + HCl$$

acetyl chloride — *p*-methylacetophenone

Figure 8-4. *Aromatic ketones are frequently prepared by the Friedel–Crafts reaction.*

Reactions of the Carbonyl Group

To understand the reactions of aldehydes and ketones, we must understand how a carbon–oxygen double bond resembles, and how it differs from, a

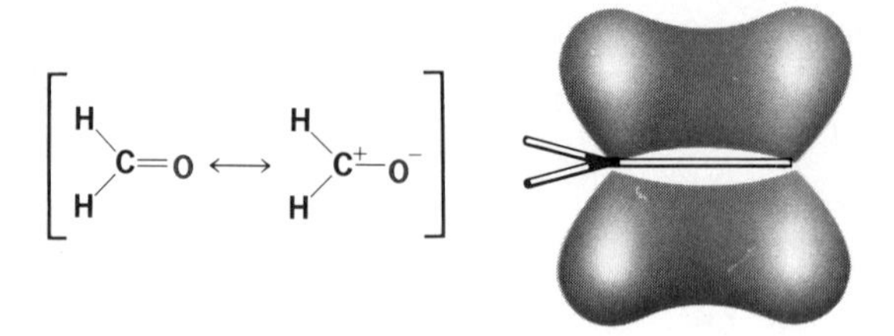

Figure 8-5. *The carbon–oxygen, like the carbon–carbon, double bond results from a combination of σ and π bonds. But the carbonyl group has a dipole moment, with the oxygen slightly negative, the carbon slightly positive, as shown by the resonance structures in the valence bond formula.*

carbon–carbon double bond. Electronically, the two are formed in the same way, by the end-on overlap of hybrid orbitals to form a σ bond and by the edgewise overlap of *p* orbitals to form a π bond (Fig. 8-5). In a carbonyl group the two atoms making up the double bonds have different *electronegativities*. The oxygen atom, with its greater nuclear charge, demands, and gets, a greater share of the four electrons making up the double bond than does the carbon atom. As a result the carbonyl group has a *dipole moment*, with the oxygen atom slightly negative and the carbon atom slightly positive. In valence bond formulas, this is shown by writing two resonance structures for the carbonyl group, in one of which the oxygen has a negative charge, and the carbon a positive charge (Fig. 8-5). Placed in an electric field, a molecule containing a carbonyl group aligns itself, with the oxygen pointing toward the positive pole, the carbon toward the negative pole.

Carbonyl groups undergo addition reactions. Acidic reagents add across the carbon–oxygen double bond, the positive part of the reagent adding to the oxygen, the negative part to the carbon. As with additions to carbon–carbon double bonds, a carbonium ion is the intermediate, and the carbonium ion that is formed by protonation of the oxygen of a carbonyl group is, relative to most other carbonium ions, a very stable one (Fig. 8-6). However, most addition products of the carbon–oxygen double bond cannot be isolated, since they quickly revert to the carbonyl group. As a general rule it is usually not possible to isolate compounds with two hydroxyl groups on the same carbon, those with one hydroxyl and a halogen on the same carbon, or those with one hydroxyl and one ether group on the same carbon.

Acetals and Hemiacetals

Addition of the elements of an alcohol across the carbon–oxygen double bond of a carbonyl group forms a *hemiacetal* from an aldehyde or a *hemiketal* from

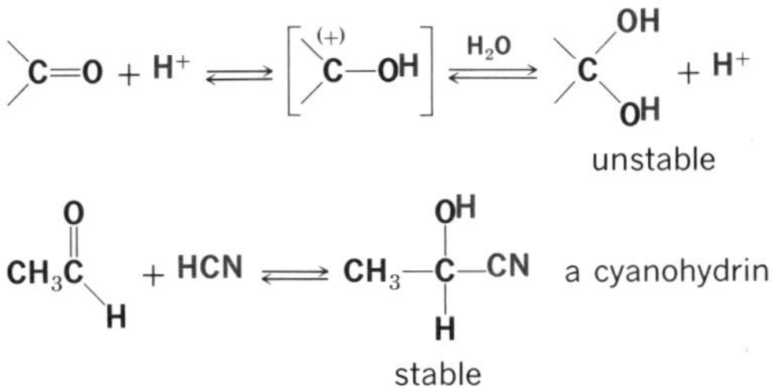

$$>C{=}O + H^+ \rightleftharpoons \left[>\overset{(+)}{C}{-}OH \right] \xrightleftharpoons{H_2O} >C(OH)_2 + H^+$$
unstable

$$CH_3C(=O)H + HCN \rightleftharpoons CH_3{-}C(OH)(H){-}CN \quad \text{a cyanohydrin}$$
stable

Figure 8-6. *Addition reactions of the carbonyl group occur readily, with the proton adding to the oxygen atom. The products are usually too unstable to be isolated, however, and quickly revert to the carbon–oxygen double bond. An exception is the HCN adduct, which can be isolated.*

a ketone. We have just stated that compounds of this type are usually too unstable to be isolated, but in alcohol solution they do exist in equilibrium with aldehydes and ketones. If the hydroxyl group of a hemiacetal or hemiketal is converted to an ether, an *acetal* or *ketal* results. Acetals and ketals, which have two ether groups on the same carbon atom, are stable and can be isolated. (Fig. 8-7).

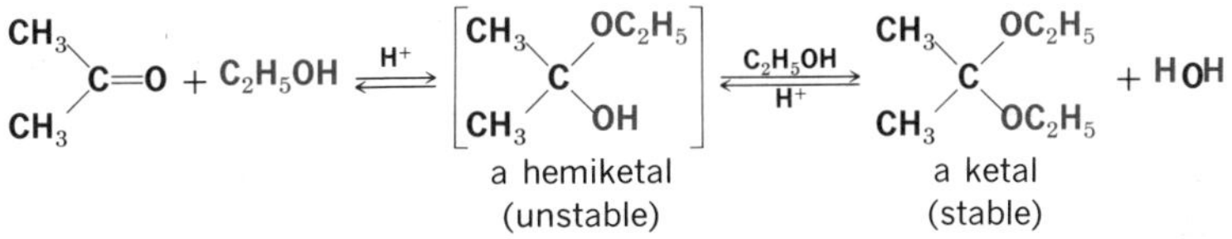

$$(CH_3)_2C{=}O + C_2H_5OH \xrightleftharpoons{H^+} \left[(CH_3)_2C(OC_2H_5)(OH) \right] \xrightleftharpoons[H^+]{C_2H_5OH} (CH_3)_2C(OC_2H_5)_2 + HOH$$
a hemiketal (unstable) — a ketal (stable)

Figure 8-7. *A ketone reacts with an alcohol to form an unstable hemiketal. Longer reaction with alcohol and acid gives the stable ketal. Aldehydes undergo similar reactions to form analogous hemiacetals and acetals.*

Prob. 8-3. Draw the structure of the hemiacetal and acetal resulting from the reaction of benzaldehyde with methanol in acid.

Hemiacetals and hemiketals play an important role in carbohydrate chemistry, to be considered in the next chapter. Among commercially important acetals is that formed by reaction between the polymer *polyvinyl alcohol* and butyraldehyde (Fig. 8-8), which is used as an adhesive between the sheets of glass in safety glass.

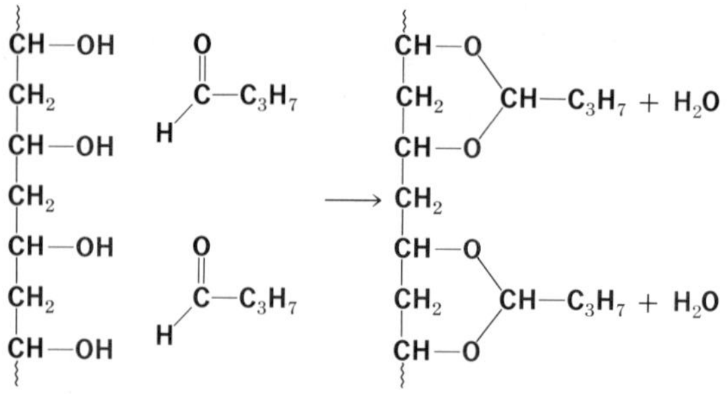

$$\begin{matrix} \text{CH—OH} \\ | \\ \text{CH}_2 \\ | \\ \text{CH—OH} \\ | \\ \text{CH}_2 \\ | \\ \text{CH—OH} \\ | \\ \text{CH}_2 \\ | \\ \text{CH—OH} \end{matrix} \quad \begin{matrix} \text{H—C(=O)—C}_3\text{H}_7 \\ \\ \text{H—C(=O)—C}_3\text{H}_7 \end{matrix} \longrightarrow \begin{matrix} \text{CH—O} \\ | \\ \text{CH}_2 \;\; \text{CH—C}_3\text{H}_7 + \text{H}_2\text{O} \\ | \\ \text{CH—O} \\ | \\ \text{CH}_2 \\ | \\ \text{CH—O} \\ | \\ \text{CH}_2 \;\; \text{CH—C}_3\text{H}_7 + \text{H}_2\text{O} \\ | \\ \text{CH—O} \end{matrix}$$

polyvinyl alcohol   butyraldehyde

Figure 8-8. *Cyclic acetals are formed between the hydroxyl groups of polyvinyl alcohol and butyraldehyde to give an adhesive used in safety glass.*

Reaction with Ammonia and Its Derivatives

Carbonyl compounds react readily with bases as well as with acids; in this they differ from alkenes, which are usually unreactive toward base. Addition of a base to the carbon–oxygen double bond is easy because the negative charge that is introduced into the molecule can be accommodated on an oxygen atom. Ketones and aldehydes react with ammonia and its derivatives to form unstable addition products analogous to hemiacetals and hemiketals (Fig. 8-9). These can generate new classes of compounds containing carbon–nitrogen double bonds by the loss of water.

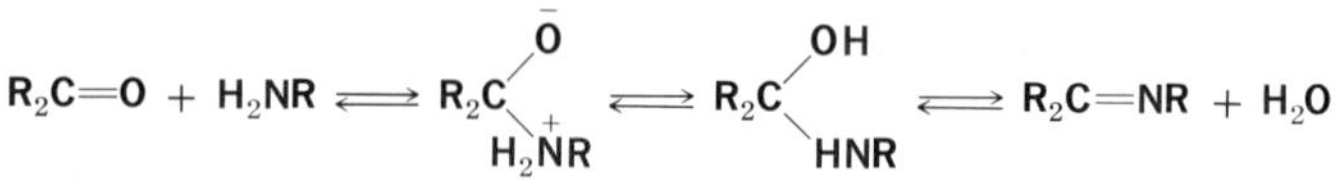

$$R_2C{=}O + H_2NR \rightleftharpoons R_2C\begin{matrix} O^- \\ H_2\overset{+}{N}R \end{matrix} \rightleftharpoons R_2C\begin{matrix} OH \\ HNR \end{matrix} \rightleftharpoons \underset{\text{imine}}{R_2C{=}NR} + H_2O$$

Figure 8-9. *Ammonia and its derivatives add to carbonyl groups to form unstable addition products. Loss of water from these gives* imines, *compounds with carbon–nitrogen double bonds.*

If the ammonia derivative is properly chosen, the product of its reaction with an aldehyde or ketone is an easily purified and identified *solid derivative,* whose ready formation and precipitation constitutes a *visual chemical test* for these functional groups. Phenylhydrazine is a typical reagent used to make solid derivatives of aldehydes and ketones. When heated together in alcoholic solution, cyclohexanone and phenylhydrazine react and precipitate the beautifully crystalline cyclohexanone *phenylhydrazone.* Semicarbazide reacts similarly to form a *semicarbazone* and hydroxylamine forms an *oxime* (Fig. 8-10).

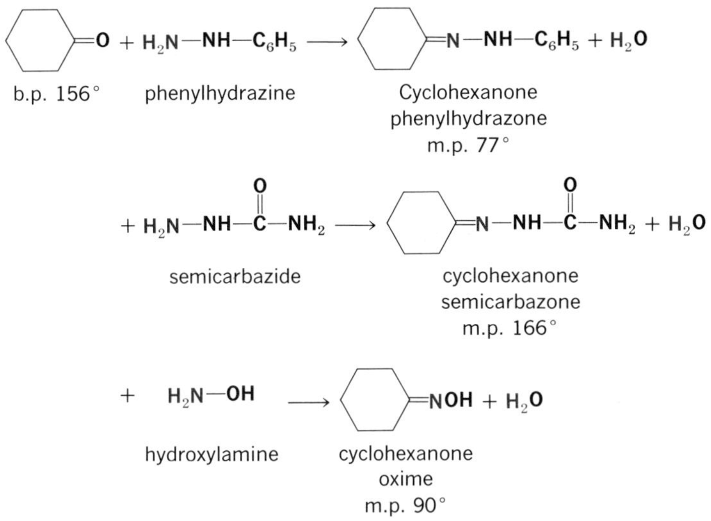

$$\text{C}_6\text{H}_{10}\text{=O} + H_2N\text{—}NH\text{—}C_6H_5 \longrightarrow \text{C}_6\text{H}_{10}\text{=N—NH—}C_6H_5 + H_2O$$

b.p. 156° phenylhydrazine Cyclohexanone phenylhydrazone m.p. 77°

$$+ H_2N\text{—}NH\text{—}\overset{O}{\overset{\|}{C}}\text{—}NH_2 \longrightarrow \text{C}_6\text{H}_{10}\text{=N—NH—}\overset{O}{\overset{\|}{C}}\text{—}NH_2 + H_2O$$

semicarbazide cyclohexanone semicarbazone m.p. 166°

$$+ H_2N\text{—}OH \longrightarrow \text{C}_6\text{H}_{10}\text{=NOH} + H_2O$$

hydroxylamine cyclohexanone oxime m.p. 90°

Figure 8-10. *These three ammonia derivatives react readily with aldehydes and ketones to form solid derivatives. The appearance of a precipitate of these derivatives constitutes a visual chemical test for aldehydes and ketones.*

Grignard Reactions of Carbonyl Groups

Carbon is much less electronegative than oxygen, and hence a carbon with a negative charge (a carbanion R_3C^-) is a much stronger base than an oxygen anion (HO^- or RO^-). We have seen that bases can add to the carbon atom of a carbonyl group; therefore Grignard reagents, which we have described as being essentially magnesium salts of carbanions, react readily with aldehydes and ketones (Fig. 8-11). The product of such a reaction is an alcohol, and

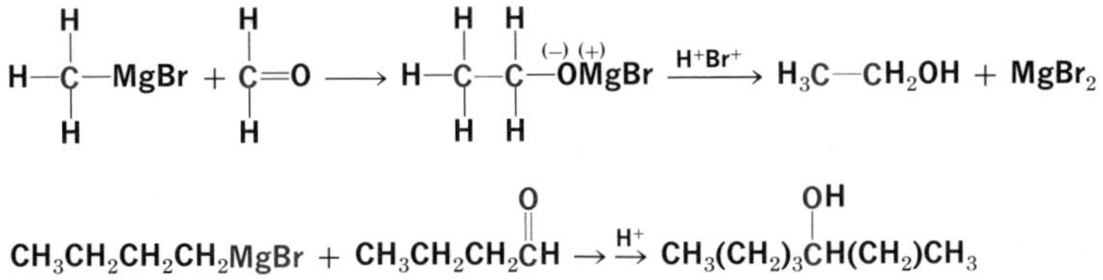

$$H_3C\text{—}MgBr + H_2C\text{=}O \longrightarrow H_3C\text{—}CH_2\text{—}\overset{(-)}{O}\overset{(+)}{Mg}Br \xrightarrow{H^+Br^+} H_3C\text{—}CH_2OH + MgBr_2$$

$$CH_3CH_2CH_2CH_2MgBr + CH_3CH_2CH_2\overset{O}{\overset{\|}{C}}H \longrightarrow \xrightarrow{H^+} CH_3(CH_2)_3\overset{OH}{\overset{|}{C}}H(CH_2)CH_3$$

Figure 8-11. *Grignard reagents add readily to carbonyl groups, with the negatively charged carbon of the organomagnesium compound adding to the carbon of the carbon–oxygen double bond. Acidification gives an alcohol. In this way a new carbon–carbon bond is formed, and large organic molecules may be prepared from smaller ones.*

a new carbon–carbon bond has been formed. In this way large organic molecules can be formed from smaller ones. For example, a four-carbon halide can be converted to a Grignard reagent and reacted with a four-carbon aldehyde to give an eight-carbon alcohol (Fig. 8-11).

Reaction of a Grignard reagent with formaldehyde gives a 1° alcohol; 2° alcohols are formed from all other aldehydes; and ketones give 3° alcohols (Fig. 8-12). This is one of the most useful reactions in all of organic chemistry and one of the few that forms new carbon–carbon single bonds. Exactly analogous reactions occur with organolithium reagents (Fig. 7-19).

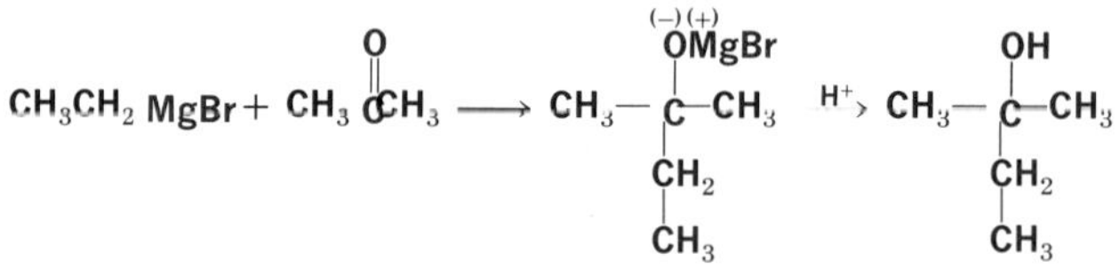

$$CH_3CH_2MgBr + CH_3\overset{\overset{O}{\|}}{C}CH_3 \longrightarrow CH_3-\underset{\underset{CH_3}{|}}{\underset{CH_2}{|}}{\overset{\overset{(-)(+)}{OMgBr}}{\overset{|}{C}}}-CH_3 \xrightarrow{H^+} CH_3-\underset{\underset{CH_3}{|}}{\underset{CH_2}{|}}{\overset{OH}{\overset{|}{C}}}-CH_3$$

Figure 8-12. Tertiary alcohols result from the reaction of a Grignard reagent with a ketone.

Prob. 8-4. What aldehyde and Grignard reagent would react to form $CH_3CH_2CH_2CH_2OH$? $CH_3CH_2CH(OH)CH_2CH_3$? Show another combination of Grignard reagent and ketone that could be used to synthesize the 3° alcohol in Fig. 8-12.

Carbanions from other sources also add to the carbonyl group. Acetylenes easily form salts in which the carbon atom bears a negative charge and hence is a carbanion. Sodium acetylide is used commercially to prepare unsaturated alcohols (Fig. 8-13).

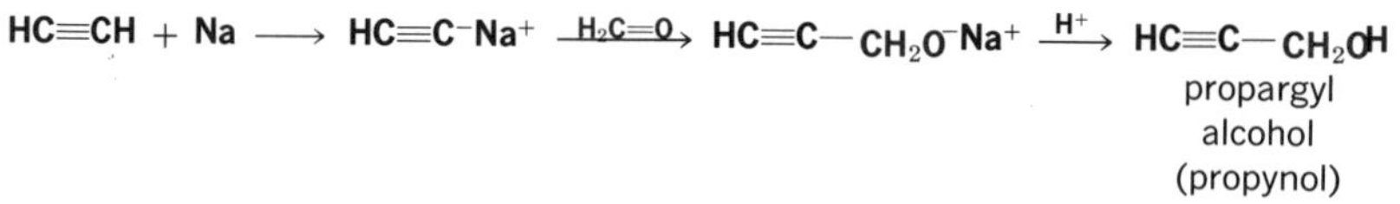

$$HC{\equiv}CH + Na \longrightarrow HC{\equiv}C^-Na^+ \xrightarrow{H_2C=O} HC{\equiv}C-CH_2O^-Na^+ \xrightarrow{H^+} HC{\equiv}C-CH_2OH$$

propargyl alcohol (propynol)

Figure 8-13. Sodium acetylide forms readily from acetylene and sodium. The carbanion adds to a carbonyl group analogously to a Grignard reagent.

Reductions of the Carbon-Oxygen Double Bond

Aldehydes and ketones may be reduced to alcohols by hydrogen in the presence of a catalyst (Fig. 8-14). The reaction is more difficult than reduction of a carbon–carbon double bond and usually requires heating with hydrogen gas under high pressure. For laboratory reductions certain *metal hydrides* are more convenient. In compounds like lithium aluminum hydride ($LiAlH_4$) or sodium borohydride ($NaBH_4$), the hydrogen is negatively charged (H^-) and, like other bases, is capable of adding to the carbon of a carbonyl group. When the reduction is complete, acidification gives the alcohol (Fig. 8-14).

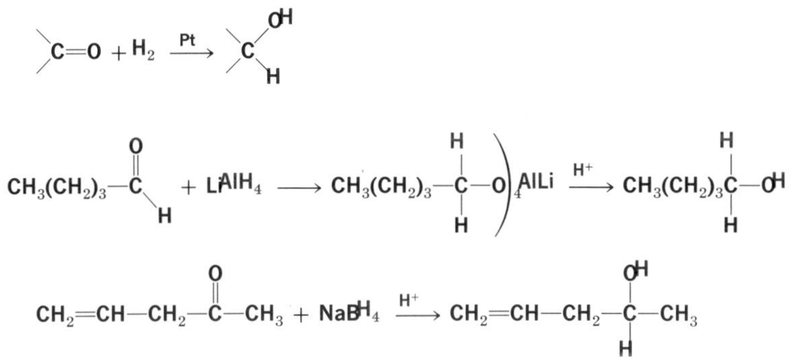

$$\text{>C=O} + H_2 \xrightarrow{\text{Pt}} \text{>C(OH)H}$$

$$CH_3(CH_2)_3\text{—}\overset{O}{\overset{\|}{C}}\text{—}H + LiAlH_4 \longrightarrow \left(CH_3(CH_2)_3\text{—}CH_2\text{—}O\right)_4AlLi \xrightarrow{H^+} CH_3(CH_2)_3CH_2\text{—}OH$$

$$CH_2\text{=}CH\text{—}CH_2\text{—}\overset{O}{\overset{\|}{C}}\text{—}CH_3 + NaBH_4 \xrightarrow{H^+} CH_2\text{=}CH\text{—}CH_2\text{—}CH(OH)\text{—}CH_3$$

Figure 8-14. *Carbonyl groups may be reduced by hydrogen or, more conveniently, by* hydrides. *Since metal hydrides reduce by the addition of H^-, they ordinarily do not react with alkenes.*

Oxidation of Aldehydes

The carbonyl–hydrogen bond of an aldehyde is subject to relatively easy oxidation; a ketone, lacking a carbonyl–hydrogen bond, is not oxidized as readily. The difference is similar to that between oxidizable primary and secondary alcohols and oxidation-stable tertiary alcohols. Even the oxygen of air can oxidize an aldehyde to the corresponding acid. Powerful oxidizing agents like potassium permanganate or potassium dichromate perform the job very well, but milder reagents are often used to avoid oxidation of other parts of the molecule. Testing for oxidation by one of these mild reagents can show whether a compound that reacts with phenylhydrazine is an aldehyde or a ketone. If a reaction with ammoniacal silver nitrate (Tollens reagent) is carried out in a clean test tube, the silver metal precipitates in a thin film on the walls and produces a mirror (Fig. 8-15). For this reason the reaction is often called the silver mirror test for an aldehyde.

Aldehydes and Ketones in Nature

Simple and complex aldehydes and ketones are found widely in nature and many have odors that we associate with their sources (Fig. 8-16). Among

$$R{-}\overset{O}{\overset{\|}{C}}{-}H + 2\overset{+}{Ag}(NH_3)_2 \xrightarrow[H_2O]{OH^-} R{-}\overset{O}{\overset{\|}{C}}{-}O^- + 2Ag\downarrow \quad \text{Tollens test}$$

Figure 8-15. *Aldehydes are easily oxidized to salts of acids by silver ions. Because the reaction only proceeds in basic solution, the silver ion must be complexed with ammonia to prevent its precipitation. This reaction is a visual chemical test for an aldehyde and distinguishes it from a ketone.*

aldehydes *citronellal*, for example, occurs in lemon oil, *cinnamaldehyde* in cinnamon and *vanillin* in vanilla. Ketones, too, occur in plants and flowers. *Camphor* (from the camphor tree) and *carvone* (found in caraway and dill oils) will be familiar by odor if not by name. Note that citronellal, camphor, and carvone (and also pinene and limonene, Fig. 3-29) all have the same C_{10} carbon skeleton, sometimes noncyclic and sometimes joined into one or more rings. These compounds all belong to the general class of plant products known as terpenes (Chapter 16). *Cortisone* is an example of a *steroid hormone;* it is used in the treatment of arthritis. The same biosynthetic pathways that lead to terpenes in plants produce steroids in animals (Chapter 16).

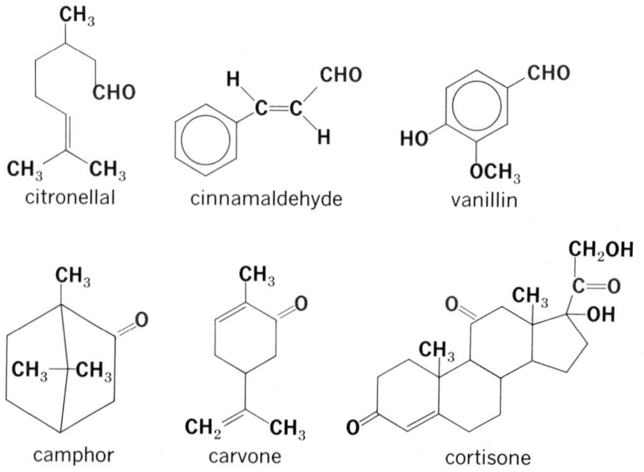

citronellal cinnamaldehyde vanillin

camphor carvone cortisone

Figure 8-16. *Examples of naturally occurring aldehydes and ketones.*

Industrially Important Aldehydes and Ketones

Acetone (b.p. 56°), the simplest ketone, is an inexpensive and widely employed industrial solvent. One industrial synthesis uses the oxidation of isopropyl alcohol shown earlier in Fig. 8-1. It can also be formed by fermentation of sugars or starch, and is found on the breath and in the urine of many sufferers of diabetes.

Methyl ethyl ketone (b.p. 80°) shows many of acetone's desirable solvent properties and is used when a solvent with a higher boiling point is needed. It is approximately twice as expensive as acetone; however, in many processes the solvent is recovered and reused, so that the greater initial expense is not especially important.

Acetaldehyde (b.p. 20°) is formed readily by the hydration of acetylene and is used mainly as a raw material for the synthesis of other organic compounds (Fig. 8-17), among them acetic acid by reaction with oxygen. Chlorination of acetaldehyde gives chloral, which is the active ingredient of "knockout" drops and is also used in the synthesis of DDT.

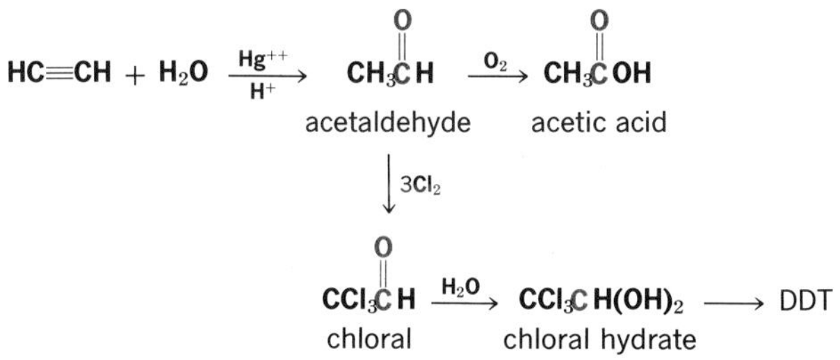

$$HC{\equiv}CH + H_2O \xrightarrow[H^+]{Hg^{++}} \underset{\text{acetaldehyde}}{CH_3\overset{O}{\overset{\|}{C}}H} \xrightarrow{O_2} \underset{\text{acetic acid}}{CH_3\overset{O}{\overset{\|}{C}}OH}$$

$$\downarrow 3Cl_2$$

$$\underset{\text{chloral}}{CCl_3\overset{O}{\overset{\|}{C}}H} \xrightarrow{H_2O} \underset{\text{chloral hydrate}}{CCl_3CH(OH)_2} \longrightarrow DDT$$

Figure 8-17. *Acetaldehyde is an intermediate in the synthesis of a variety of useful products.*

n-Butyraldehyde (b.p. 75°), formed from oxidation of *n*-butyl alcohol, is widely used in the production of safety glass (Fig. 8-8).

From a commercial standpoint, the simplest aldehyde, *formaldehyde,* is also the most important. Pure formaldehyde is a gas (b.p. −21°) that can be neither readily isolated nor handled in this state. It is prepared from controlled oxidation of methanol over a metal catalyst, and is sold as a 37% solution in water (formalin) and also as a solid polymer (paraformaldehyde), from which it can be recovered by heating (Fig. 8-18). Candles of paraformaldehyde are used as fumigating agents.

Since formaldehyde can be produced inexpensively and polymerizes readily (to paraformaldehyde), a great deal of effort has gone into investigations of its use in the preparation of commercial polymers. The problem in using pure polyformaldehyde has been to prevent its decomposition to formaldehyde, which is poisonous. It can be shown that the decomposition begins at the ends of the chains, but when these free hydroxyl groups are converted into ester groups, a stable polymer results. Polyformaldehyde is now being produced and sold under the trade name *Delrin,* used as a wood substitute for many applications.

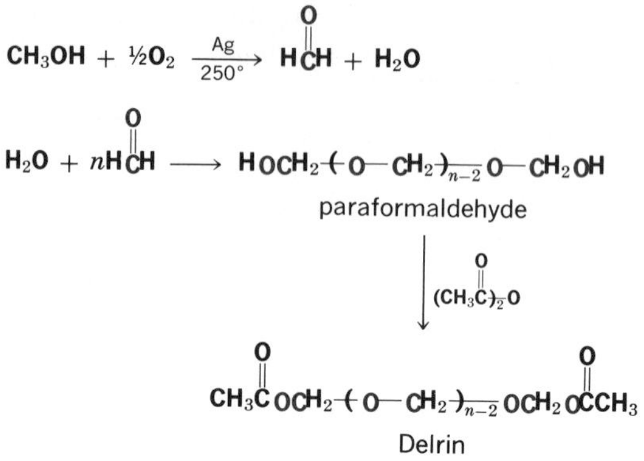

$$CH_3OH + \tfrac{1}{2}O_2 \xrightarrow[250^\circ]{Ag} HCH(=O) + H_2O$$

$$H_2O + nHCH(=O) \longrightarrow HOCH_2\text{-}(O\text{-}CH_2)_{n-2}\text{-}O\text{-}CH_2OH$$

paraformaldehyde

$$\downarrow (CH_3C(=O))_2O$$

$$CH_3C(=O)OCH_2\text{-}(O\text{-}CH_2)_{n-2}\text{-}OCH_2OC(=O)CH_3$$

Delrin

Figure 8-18. *Formaldehyde is prepared by oxidation of methanol. A formaldehyde polymer (Delrin) is stabilized by capping the ends of the chain with acetate groups to prevent decomposition to formaldehyde.*

Formaldehyde and ammonia react to form the cagelike compound hexamethylenetetramine, which liberates formaldehyde upon treatment with acid (Fig. 8-19). Hexamethylenetetramine is medicinally useful as a urinary antiseptic (urotropine) and is also oxidized by nitric acid to the important military explosive cyclonite (RDX).

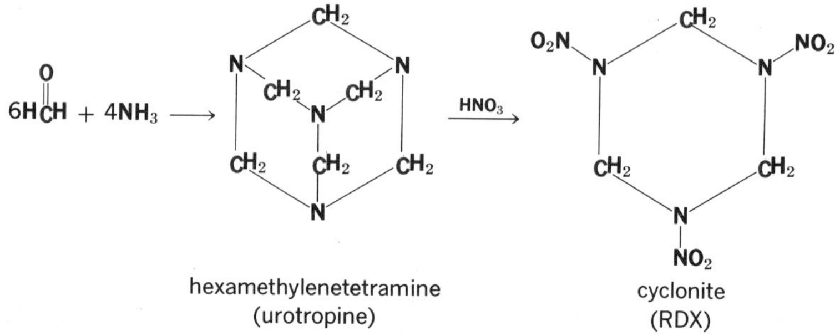

$$6HCH(=O) + 4NH_3 \longrightarrow \text{hexamethylenetetramine} \xrightarrow{HNO_3} \text{cyclonite}$$

hexamethylenetetramine (urotropine): four N atoms linked by six CH_2 groups in a cage.

cyclonite (RDX): six-membered ring of alternating CH_2 and $N\text{-}NO_2$ groups (O_2N, NO_2, NO_2 on N).

hexamethylenetetramine
(urotropine)

cyclonite
(RDX)

Figure 8-19. *Formaldehyde and ammonia form hexamethylenetetramine, a source of the explosive cyclonite.*

The largest commercial use of formaldehyde is in the preparation of phenol-formaldehyde resins for molded articles (Bakelite) and for use in plywood (Fig. 8-20). These resins are *thermosetting;* this means that in the course of the manufacture of an article the phenol-formaldehyde resin melts, and then, under the influence of heat and pressure, resolidifies to become permanently infusible. The properties of thermosetting polymers contrast with those of thermoplastic polymers. The latter, which include most vinyl polymers,

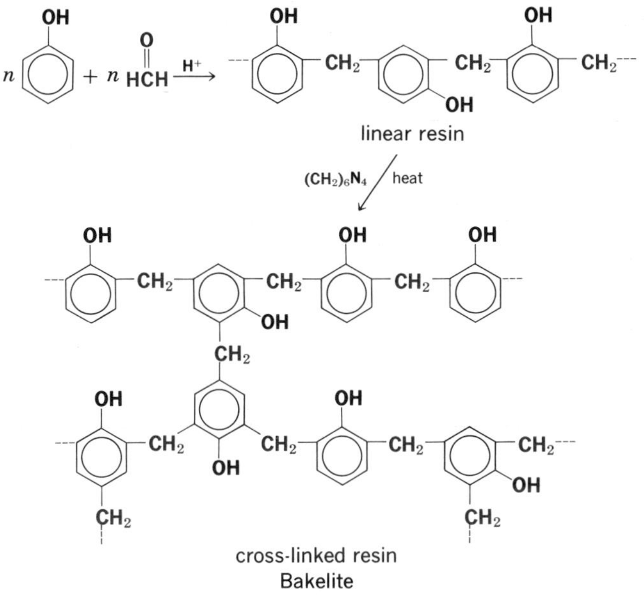

linear resin

$(CH_2)_6N_4$ / heat

cross-linked resin
Bakelite

Figure 8-20. *Bakelite can be made by heating a linear phenol-formaldehyde resin with hexamethylenetetramine.*

are solids at room temperature but can be molded at higher temperatures. This fusion-solidification process can be repeated numerous times.

In one method for the preparation of Bakelite, formaldehyde is heated with an excess of phenol to form a linear polymer whose structure approximates that shown in Fig. 8-20. This thermoplastic polymer is available in powdered form for the molding of articles. In the molding process this powder is mixed with hexamethylenetetramine, which serves as a source both of more formaldehyde and of ammonia, a basic catalyst. The mixture is melted in a mold where further polymerization takes place, this time *between* the chains. This converts the individual linear polymer molecules into a single gigantic molecule, by a process known as *cross-linking*. The plastic resolidifies in the mold, since the molecular weight has increased enormously.

Spectroscopy of the Carbon–Oxygen Double Bond

Simple aldehydes and ketones show only very weak absorption of ultraviolet light near 260 nm. This absorption does not involve the electrons of the double bond but rather the nonbonding electrons on the carbonyl oxygen; because these electrons are not present in a carbon–carbon double bond, alkenes do not show a comparable absorption. Carbonyl groups that are conjugated to a double bond or an aromatic ring absorb much more strongly in the uv (Fig. 8-21).

Carbonyl groups show extremely characteristic infrared absorption bands, generally absorbing in the region between 1600 and 2000 cm^{-1}. The ir spectrum of benzaldehyde is shown in Fig. 8-22; the C=O stretching vibration can be seen very clearly (in red). The exact position of the carbonyl absorption can give a great deal of structural information. Among cyclic ketones, for example, it is possible to tell whether the carbonyl group is in a four-, five- or six-membered ring by the wavelength of the infrared radiation it absorbs, and conjugated unsaturated carbonyl compounds absorb at a different point within the carbonyl region than unconjugated carbonyl groups.

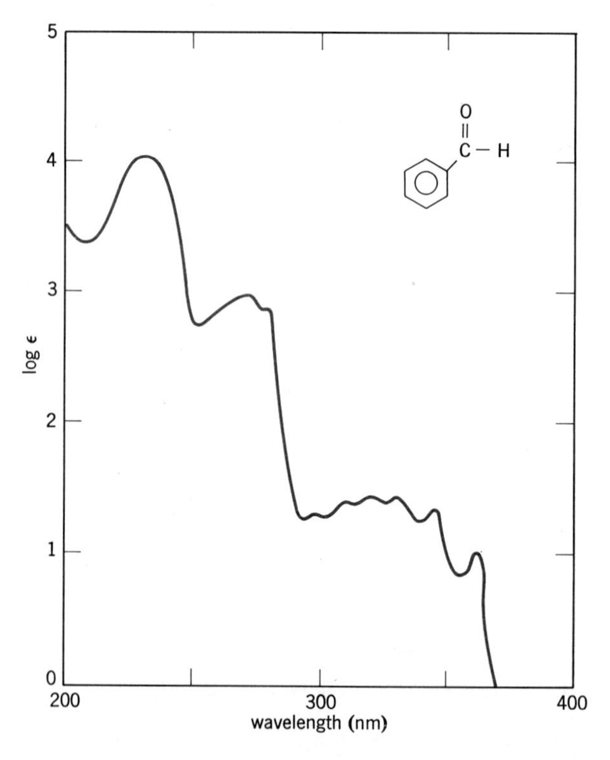

Figure 8-21. *Unconjugated aldehydes and ketones show only weak ultraviolet absorption, but when conjugated to a benzene ring they absorb strongly.*

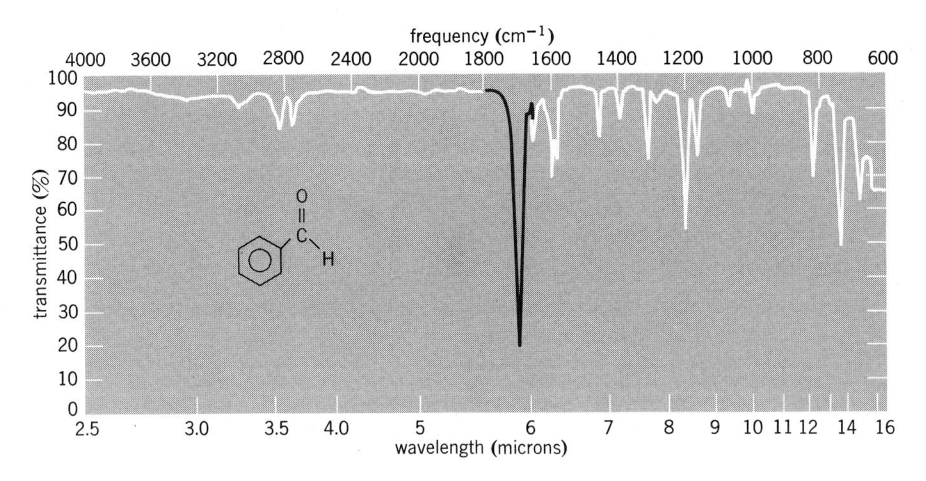

Figure 8-22. *The infrared absorption spectrum of benzaldehyde. The C=O stretching absorption is usually the most intense in the spectrum and always occurs in a relatively small region of the spectrum. Its exact absorption position within this region can give a great deal of information about the structure of the molecule.*

PROBLEMS

1. The IUPAC ending for the aldehyde group is *al* and for the ketone group is *one*. Draw the structures of:

(a) methanal
(b) cyclohexanone
(c) 2-pentanone
(d) 3-methylhexanal
(e) 2-methyl-2-phenyl-3-octanone
(f) 2-cyclopentenone
(g) 3-cyclopentenone
(h) *cis*-2-butenal
(i) 2,4-pentanedione
(j) *p*-hydroxybenzaldehyde

2. Give the IUPAC name for:

(a) $C_2H_5COC_2H_5$
(b) $(CH_3)_2CHCHO$
(c) $(CH_3)_3CCOCH_3$
(d) $C_6H_5CH_2COCH_3$

(e) 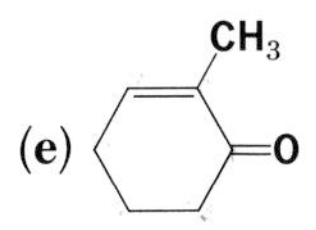

(f) CH$_3$ / CH$_3$ / O (cyclobutanone ring)

(g) $CH_3CH{=}CH{-}COCH_3$

(h) Cl–C$_6$H$_4$–COCH$_3$

(i) 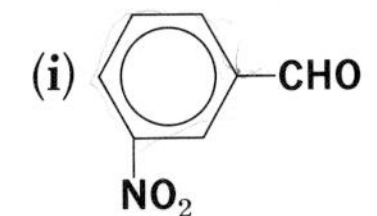

(j) $CH_3CH_2CH_2CHO$

3. Give common names for 1(a,c), 2(a,b,c,d,j).
4. Draw structures for:
   (a) benzaldehyde oxime
   (b) acetone phenylhydrazone
   (c) the diethylacetal of propionaldehyde
   (d) the hemiacetal of formaldehyde and methanol
   (e) the semicarbazone of 2-pentanone
   (f) the ozonide of cyclopentene
   (g) the dialdehyde formed by hydrolysis of (f) in the presence of zinc
   (h) the cyanohydrin of acetophenone
5. Complete the following reactions:
   (a) $(CH_3)_2C{=}O + NH_2OH \longrightarrow$

   (b) $C_6H_5CHO + LiAlH_4 \longrightarrow \xrightarrow{H^+}$

   (c) $CH_3CH_2CHO + Ag^+(NH_3)_2 \xrightarrow{H_2O}$

   (d) $C_6H_5COCH_3 + CH_3MgBr \longrightarrow \xrightarrow{H^+}$

   (e) cyclopentanone ($C_5H_8{=}O$ ring) $+ CH_3OH \underset{heat}{\overset{H^+}{\rightleftharpoons}}$

   (f) $C_6H_5CH{=}CH{-}CHO + H_2 \xrightarrow{Pd}$

   (g) $C_6H_5CH{=}CH{-}CHO + NaBH_4 \longrightarrow \xrightarrow{H^+}$

   (h) $Cl{-}C_6H_4{-}CHO + H_2NNHCONH_2 \longrightarrow$

   (i) $(CH_3)_2C{=}O + C_6H_5Li \longrightarrow \xrightarrow{H^+}$

   (j) $C_6H_6 + C_6H_5COCl \xrightarrow{AlCl_3}$
6. What would be the structures of the alkenes or alkadienes that yield the following products on ozonolysis?
   (a) formaldehyde and acetone
   (b) 2,7-octadione
   (c) equal amounts of acetone, propionaldehyde, and 1,4-butandial
   (d) 2,3-butanedione and 2,5-hexanedione
7. How could each of the following compounds be prepared by making use of a Grignard reagent? More than one way may be possible. Use as starting material any halide and any aldehyde or ketone.
   (a) $C_6H_5CH_2OH$ (b) 2-propanol

   (c) $(C_6H_5)_3COH$ (d) *n*-hexane

   (e) $CH_3{-}C(OH)(C_2H_5){-}C_6H_5$ (3 ways)

8. How could you distinguish among the following compounds by simple visual chemical tests? Tell what you would do and see.

(a) C_6H_5CHO, $C_6H_5COCH_3$, $C_6H_5CH_2OH$, $C_6H_5C{\equiv}CH$,$C_6H_5CH{=}CH_2$

(b) $C_2H_5COCH_3$, $C_2H_5\underset{\substack{|\\ OH}}{C}HCH_3$, $(CH_3)_3COH$, C_3H_7CHO, $CH_3CH{=}CH{-}COCH_3$

9. How could you carry out the following transformations? More than one step will be required in each case. You may use any compounds or reagents you wish as long as you start with the indicated compound.

(a) $CH_3COCH_3 \longrightarrow CH_3CH{=}CH_2$

(b) $C_6H_5CHO \longrightarrow C_6H_5CH_2Br$

(c) $CH_3CH_2CHO \longrightarrow CH_3\underset{\substack{|\\ OH}}{C}HCH_3$

(d) ⬡=O $\longrightarrow H\overset{O}{\overset{\|}{C}}(CH_2)_4\overset{O}{\overset{\|}{C}}H$

(e) $CH_3CH_2CHO \longrightarrow CH_3CH_2\underset{\substack{|\\ OH}}{C}(CH_3)_2$

(f) $CH_3CH{=}CH_2 \longrightarrow (CH_3)_2C{=}NOH$

9 carbohydrates, complex stereochemistry

Among naturally occurring substances the most important aldehydes and ketones are the *carbohydrates,* which include sugars and starch, cellulose, and numerous antibiotics and other related compounds. The name "carbohydrate" is derived from the molecular formula of these compounds, which in many, but not all, cases may be written $C_m(H_2O)_n$ (e.g., glucose, $C_6H_{12}O_6 = C_6(H_2O)_6$). Carbohydrates are *polyfunctional molecules,* that is to say, each molecule contains several functional groups. For example, glucose contains an aldehyde group as well as five hydroxyl groups. Our study of organic chemistry has been based on the idea that a functional group will react in nearly the same way no matter in which molecule it finds itself; therefore, we expect to find glucose exhibiting the reactions of both an aldehyde and an alcohol. So it does. At the same time, when two or more functional groups are held in close proximity to one another, as of course they must be when they are nearby within the same molecule, there is inevitably some interaction between them, interaction which leads to a modification of functional group reactivities. In this chapter we shall explore the chemistry of the carbohydrates and see how functional groups influence each other in polyfunctional molecules.

Among the carbohydrates we shall also encounter for the first time molecules that contain more than one chiral center. Table sugar (sucrose), for example, contains 9 chiral carbons, and 512 stereoisomers are possible. Remarkably, only one of these possible isomers occurs naturally. We shall also discover compounds that contain several chiral carbons and yet are identical with their mirror images, and so they do not rotate the plane of polarized light. Compounds of this type play an especially important role when we set about to determine the stereochemical configuration of a molecule that contains several chiral centers.

Structure of Glucose

Glucose, also known as dextrose because it rotates the plane of polarized light in the (+) direction, i.e., is *dextrorotatory*, is the most widely distributed carbohydrate. Chemically, it is a pentahydroxyaldehyde (Fig. 9-1) and has the chemical properties of the carbonyl, as well as the hydroxyl, groups. It differs importantly from most simple aldehydes, however, because it forms stable hemiacetals, making use of one of its own hydroxyl groups as the alcohol. In fact, the cyclic hemiacetal form of glucose (Fig. 9-1) is more stable than the open-chain carbonyl form, and a sample of crystalline glucose will not, for example, show any infrared absorption band in the carbonyl region. In this way it differs from a typical aldehyde, whose hemiacetal form is usually too unstable to be isolated. Nevertheless, when placed in solution an equilibrium is established between the hemiacetal and free aldehyde forms of glucose.

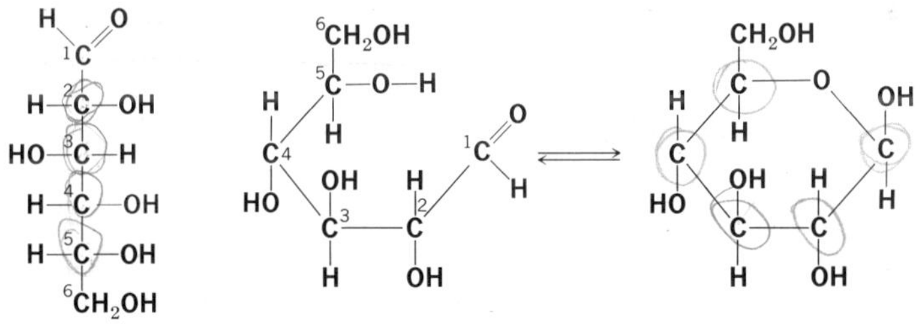

glucose

glucopyranose

Figure 9-1. *Glucose is a pentahydroxy aldehyde. Its hemiacetal with one of its own alcohol groups is stable, in contrast to most hemiacetals. In fact, glucose exists mainly* ($>$ *99.9%) in the* glucopyranose *form, a cyclic hemiacetal.*

Prob. 9-1. Give the IUPAC name for glucose. How many chiral carbons does it contain? How many in the cyclic form?

Even though the amount of aldehyde present at equilibrium is small, many reactions of glucose and similar sugars are those of the free aldehyde group. As the small amount of aldehyde reacts, the equilibrium operates so as to restore the original ratio of hemiacetal to acetal (Le Chatelier's principle), thus constantly providing a fresh (though small) supply of the reactive aldehyde.

Glucose gives all of the typical reactions of the aldehyde group, including the formation of an oxime with hydroxylamine (Fig. 8-10), the reduction of silver ions to metallic silver (silver mirror test, Fig. 8-15) or the precipitation

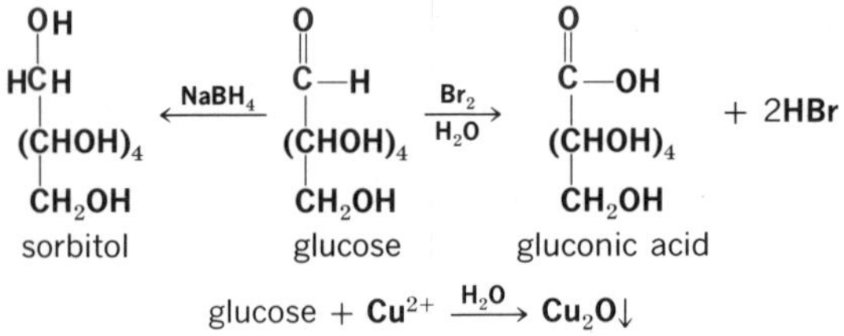

$$\underset{\text{sorbitol}}{\begin{array}{c}\text{OH}\\|\\\text{HCH}\\|\\(\text{CHOH})_4\\|\\\text{CH}_2\text{OH}\end{array}} \xleftarrow{\text{NaBH}_4} \underset{\text{glucose}}{\begin{array}{c}\text{O}\\\|\\\text{C—H}\\|\\(\text{CHOH})_4\\|\\\text{CH}_2\text{OH}\end{array}} \xrightarrow[\text{H}_2\text{O}]{\text{Br}_2} \underset{\text{gluconic acid}}{\begin{array}{c}\text{O}\\\|\\\text{C—OH}\\|\\(\text{CHOH})_4\\|\\\text{CH}_2\text{OH}\end{array}} + 2\text{HBr}$$

$$\text{glucose} + \text{Cu}^{2+} \xrightarrow{\text{H}_2\text{O}} \text{Cu}_2\text{O}\downarrow$$

Figure 9-2. *Glucose gives typical aldehyde reactions of oxidation and reduction because the aldehyde form is in equilibrium with the more abundant hemiacetal form. Glucose is a* reducing sugar *because it reduces cupric ion.*

of bright red cuprous oxide from a solution of cupric ion (Fig. 9-2). This last reaction is used as a visual chemical test for the presence of glucose in urine, a common symptom of diabetes. As we shall see, not all sugars will reduce Cu^{2+} ion; if one does, we call it a *reducing* sugar. The aldehyde group of glucose may itself be reduced to an alcohol, using one of the metal hydrides (Fig. 9-2).

Among reagents used to form crystalline adducts with sugars, phenylhydrazine is the most important, at least historically. The great German chemist Emil Fischer, who won the Nobel Prize in 1902 for his brilliant work in elucidating the structures of many of the sugars, called phenylhydrazine his “first and most lasting chemical love,” even though he became violently allergic to it. The reaction of phenylhydrazine with a sugar is more complex than its reaction with a simple carbonyl compound, because sugars are easily oxidized and phenylhydrazine is a mild oxidizing agent. The initially formed phenylhydrazone consumes a second mole of phenylhydrazine and is oxidized at carbon-2. A third mole of phenylhydrazine adds to form a di-phenylhydrazone. Fischer coined the name *osazone* for these compounds, which are nearly always nicely crystalline (Fig. 9-3).

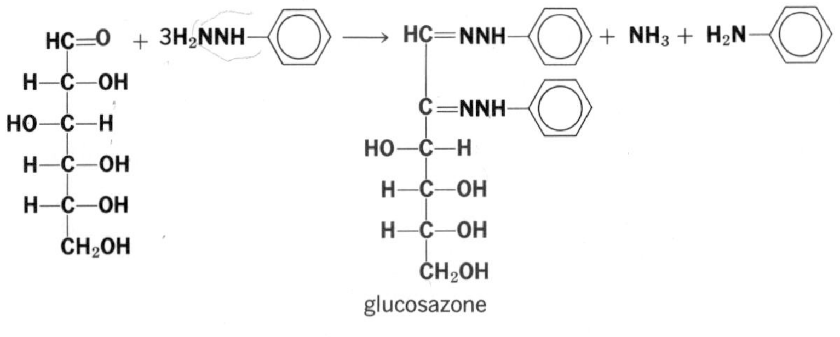

$$\begin{array}{c}\text{HC=O}\\|\\\text{H—C—OH}\\|\\\text{HO—C—H}\\|\\\text{H—C—OH}\\|\\\text{H—C—OH}\\|\\\text{CH}_2\text{OH}\end{array} + 3\text{H}_2\text{NNH—C}_6\text{H}_5 \longrightarrow \underset{\text{glucosazone}}{\begin{array}{c}\text{HC=NNH—C}_6\text{H}_5\\|\\\text{C=NNH—C}_6\text{H}_5\\|\\\text{HO—C—H}\\|\\\text{H—C—OH}\\|\\\text{H—C—OH}\\|\\\text{CH}_2\text{OH}\end{array}} + \text{NH}_3 + \text{H}_2\text{N—C}_6\text{H}_5$$

Figure 9-3. *When treated with an excess of phenylhydrazine, glucose forms an* osazone *(a di-phenylhydrazone). Carbon-2 of the sugar is oxidized by a molecule of phenylhydrazine, which is reduced to aniline and ammonia.*

Nomenclature of Sugars

The names of carbohydrates all have the ending *-ose,* e.g., gluc*ose,* fruct*ose,* rib*ose.* Glucose contains six carbons and is an aldehyde; it therefore belongs to the class of sugars known as *aldohexoses,* being both an *aldose* and a *hexose;* since fructose also has six carbons but is a ketone rather than an aldehyde, it is a *ketohexose* (Fig. 9-13).

As we have already shown for glucose (Fig. 9-1), sugars exist mainly in the form of cyclic hemiacetals. If the ring formed in this cyclization is six-membered, as it is in glucose, the sugar is in the *pyranose* form (glucopyranose). If the ring is five-membered (as, for instance, is fructose when combined in table sugar), it is said to be in the *furanose* form (fructofuranose, Fig. 9-13). Sugars with more than six carbons, as well as those with fewer than six, are well known. Sedoheptulose, a *heptose* (seven-carbon sugar), plays a key role in photosynthesis, and the *pentose* ribose is involved in the biological synthesis of proteins (RNA, Chapt. 14).

Prob. 9-2. Draw the structure of an aldopentose; a ketotetrose. Draw the former in both a pyranose and a furanose form.

Stereochemistry of Sugars: Diastereomers

If a molecule contains a single chiral carbon, two enantiomers exist. Carbohydrates, and many other synthetic and naturally occuring organic molecules, contain more than a single chiral carbon; sucrose we have said contains nine. Before considering further the reactions of carbohydrates, we must examine the complications introduced into the structure of sugars by this multitude of chiral centers.

First consider sugars with two chiral centers, the *aldotetroses.* Each chiral carbon (C-2 and C-3) can exist in either the *R* or *S* configuration. Four stereoisomers are therefore possible, *R,R;* S,S; S,*R;* and *R,*S. These four isomers are shown in Fig. 9-4. Note that the isomers fall into pairs of enantiomers, 2*R*,3*R* being the mirror image of 2S,3S and 2*R*,3S the mirror image of 2S,3*R*. Because they are mirror images, these pairs obviously have identical physical properties and rotate the plane of polarized light by equal amounts in opposite directions. But the 2*R*,3*R*-isomer is not a mirror image of either the 2*R*,3S- or 2S,3*R*-isomer, and will differ from them in physical and chemical properties. Stereoisomers that are not mirror images are called *diastereomers.* So each erythrose is a diastereomer of both threose isomers. The aldotetroses then fall into two pairs of sugars, erythrose and threose, with different physical properties. Each of these sugars exists as enantiomers, (+)-erythrose and (−)-erythrose, (+)-threose and (−)-threose.

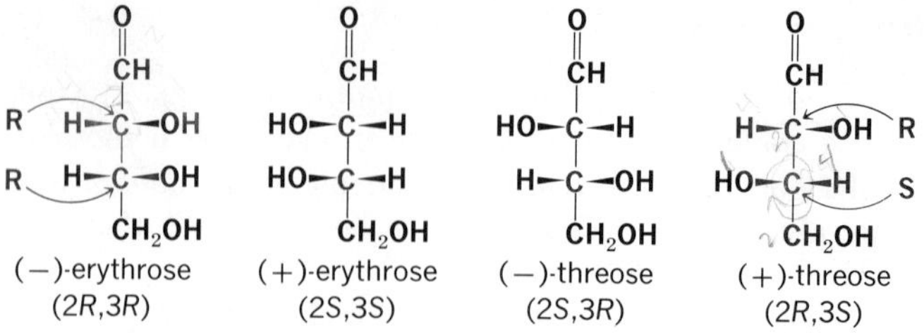

(−)-erythrose (2*R*,3*R*) (+)-erythrose (2*S*,3*S*) (−)-threose (2*S*,3*R*) (+)-threose (2*R*,3*S*)

Figure 9-4. *The aldotetroses contain two chiral carbon atoms, each of which may exist in an* R- *or* S- *form, giving rise to four stereoisomers. The two erythrose isomers are mirror images (enantiomers), as are the two threose isomers. Each erythrose is a diastereomer of the threoses, and vice versa.*

To draw these stereoisomers quickly and conveniently, we use *Fischer projection formulas* (Fig. 9-5). In this convention the carbon backbone is drawn vertically and the hydroxyl and hydrogen substituents are placed horizontally to the right or left. These horizontal bonds by definition project outward, toward the viewer, while the vertical bonds are behind the plane of the paper, away from the viewer. The carbon atoms making up the backbone lie at the intersections of the vertical and horizontal lines. The Fischer convention is illustrated in Fig. 9-5 for 2*R*,3*R*-erythrose.

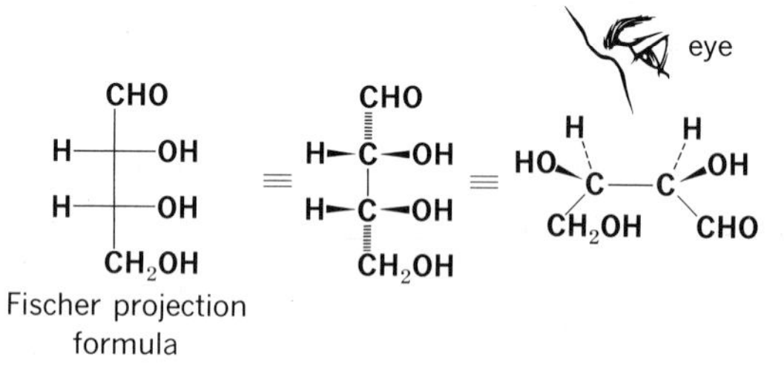

Fischer projection formula

Figure 9-5. *In the Fischer convention the molecule is viewed from above with horizontal groups projecting out of the page, vertical groups behind the page.*

Each time we add a chiral center to a molecule, we double the possible number of stereoisomers. With 1 chiral center, there are 2 isomers, 2 chiral centers, 4 possible isomers, 3 centers, 8 isomers and 4 centers, 16 possible stereoisomers. For an arbitrary number (n) of chiral centers in a molecule there are 2^n possible stereoisomers. Sucrose, with nine chiral carbons, has $2^9 = 512$ stereoisomers. Glucose has four chiral carbons in its aldehyde form, and so there are $2^4 = 16$ possible stereoisomers of this formula, only one of which is dextrose [(+)-glucose]. These 16 isomers are shown in Fig. 9-6. We could, of course, designate the stereochemistry at each chiral carbon as *R* or S; in this way carbons 2 through 5 in dextrose could be labeled 2*R*,3S,4*R*,5*R*.

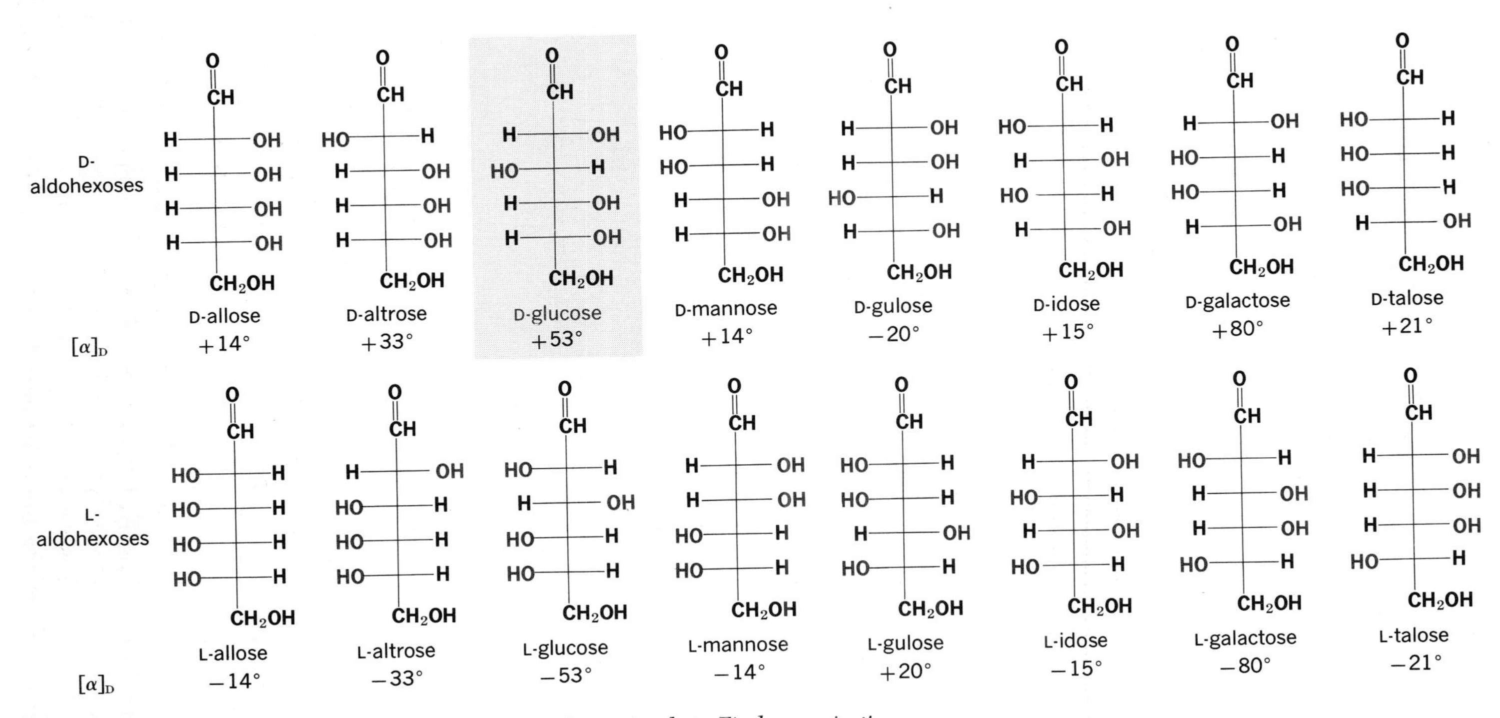

Figure 9-6. *The 16 stereoisomers of glucose are shown in their Fischer projection formulas. Those isomers with their bottom hydroxyl group on the right are called* D-, *those with it on the left are* L- *isomers. Each aldohexose is an enantiomer of one other sugar and a diastereomer of the other 14.*

More commonly, the sugars are divided into two broad classes, D- or L-, depending upon whether their bottom hydroxyl group is on the right (Dextro) or left (Levo) in the Fischer convention. This has been done in Fig. 9-6.

Prob. 9-3. How many stereoisomers are possible for an aldoheptose? A ketoheptose?

Meso Isomers: The Assignment of Configuration to Sugars

A molecule may contain several chiral carbons and still be identical with its mirror image; we call such isomers *meso,* and of course they do not rotate the plane of polarized light. Since only symmetrical objects are identical with their mirror images, meso isomers must contain some internal symmetry despite the presence of chiral centers. Usually, this is a *plane of symmetry,* some imaginary plane which divides the molecule into two halves such that one half is the mirror image of the other. For simplicity consider the isomers of 2,3-butanediol (Fig. 9-7). Because there are two chiral carbons, we might expect four stereoisomers, two pairs of enantiomers. Instead, only three isomers exist; one stereoisomer is identical with its mirror image. This meso isomer can be written in such a way that it is divided by an internal plane of symmetry; the two chiral carbons that it contains are mirror images of one another.

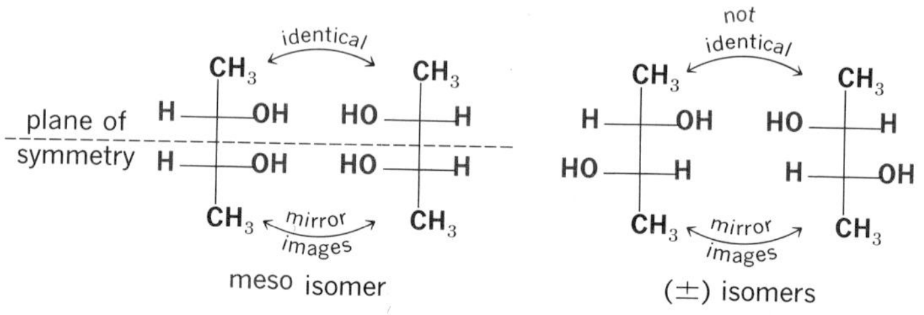

Figure 9-7. *Only three stereoisomers of 2,3-butanediol exist, one pair of enantiomers and one meso isomer. The meso isomer is identical with its mirror image. It can be recognized by the presence of a plane of symmetry that divides the molecule into two halves, each the mirror image of the other.*

The existence of meso isomers can be extremely useful in assigning stereochemistry to sugars. Let us return to the four aldotetroses (Fig. 9-4). Suppose we have on our desk four bottles, each containing one of these four sugars. Two of them we have called erythrose, two threose. Our problem is to assign a structure to the compound in each bottle. First, we reduce the aldehyde group in all four compounds to the alcohol (with $NaBH_4$, Fig. 9-2). From two of the bottles, those corresponding to the two threose isomers, enantiomeric tetraols will be formed which rotate the plane of polarized light in equal and opposite directions. But reduction of either of the two erythrose isomers will give the same, optically inactive meso tetraol (Fig. 9-8). From these results we can assign two of the four aldotetroses as (+)- and (−)-erythrose, as shown in Fig. 9-8, leaving the other two structures for the two threose enantiomers.

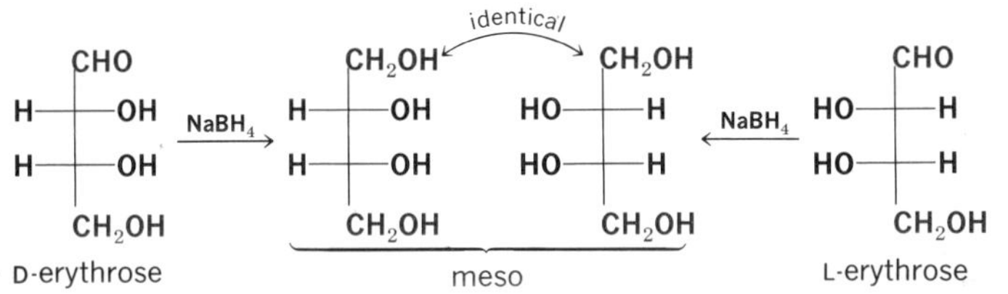

Figure 9-8. *If we take a sample of either stereoisomer of* erythrose *and reduce it, we get the same* meso *tetraol. In order for this to be true, the erythrose isomers must have the structures shown. Reduction of the two* threose *isomers gives enantiomeric tetraols that still rotate the plane of polarized light.*

Prob. 9-4. Draw the structures of the alcohols that would result from the reduction of D- and L-threose and show that they are enantiomers.

Prob. 9-5. Show that reduction of the aldehyde group of D- or L-allose gives the same *meso* alcohol. What other pair of aldohexose isomers also gives a meso alcohol on reduction?

The Absolute Configuration of the Aldotetroses

One problem still remains. We have two bottles of erythrose, (+)-erythrose and (−)-erythrose, and we have two structural formulas, D-erythrose and L-erythrose.[1] We need now to determine which structural formula corresponds to which direction of rotation of polarized light. To do this, we must start with a molecule whose absolute configuration at a chiral carbon is already known and synthesize threose and erythrose from this by reactions that do

[1] We could also call them 2*R*, 3*R*-erythrose and 2*S*, 3*S*-erythrose.

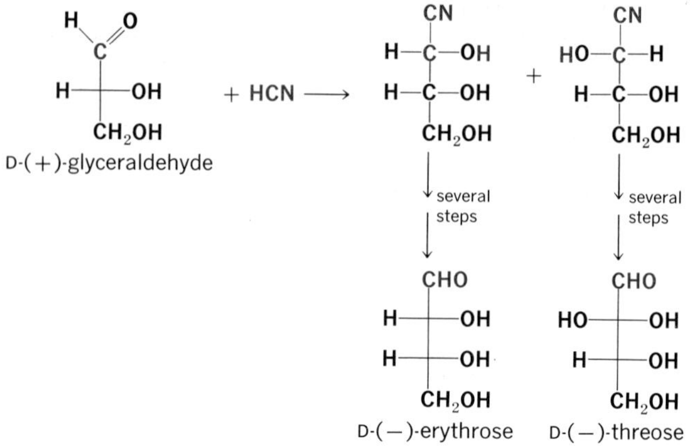

D-(+)-glyceraldehyde + HCN ⟶ (H—C—OH isomer) + (HO—C—H isomer)

several steps

D-(−)-erythrose D-(−)-threose

Figure 9-9. *Beginning with* D-(+)-*glyceraldehyde, those isomers of erythrose and threose which rotate in the* (−) *direction are formed. These must, like* D-*glyceraldehyde, have the bottom hydroxyl on the right, and so be the* D-*isomers. Note especially that, in the first step of HCN addition, a new chiral center is formed (in red), and so two isomers* (R- *and* S- *at this center) are formed.*

not affect the chiral carbon. In the sugar series the standard is D-(+)-glyceraldehyde (Fig. 9-9). This aldehyde might be considered to be the simplest sugar, a triose. It has a single chiral carbon, and the (+) isomer is known from X-ray data to have the absolute configuration with the hydroxyl on the right as shown. It then corresponds to the D-definition we have given for sugars. If we add HCN to the aldehyde group of D(+)glyceraldehyde, we form a *cyanohydrin* (Figs. 9-9 and 8-6). Note that a new chiral center is formed in this reaction, so that *two* new compounds are formed. The cyanide (CN) group in these adducts can be converted to an aldehyde group by a series of steps that we need not detail here, and we find experimentally that the (−) isomers of both threose and erythrose result. Since we started with a compound with the bottom hydroxyl on the right, and since our synthesis did nothing to change this, then (−) threose and (−) erythrose must also have the bottom hydroxyl groups on the right and so be D-isomers. Therefore, we know the absolute configuration of all four isomers: D-(−)-erythrose, L-(+)-erythrose, D-(−)-threose, and L-(+)-threose. We emphasize here again that the direction of rotation of the plane of polarized light, (+) or (−), is purely an experimental quantity and has nothing to do with the symbols D- or L-, or *R*- or S-, which we use to represent the stereochemical relationships of groups to one another in three-dimensional formulas or models. Only when we connect (+) and (−) with D- and L- or *R*- and S- experimentally do we know the absolute configuration.

Stereochemistry of Glucose

By extension of the methods outlined in the last section, the absolute configuration of glucose may be determined and that single structural formula among the 16 possible for an aldohexose (Fig. 9-6) which shows the three-dimensional arrangement of D-(+)-glucose can be identified. All the details for this structure proof cannot be given here, but some are included among the problems at the end of the chapter. One can see, however, that if the synthetic steps shown in Fig. 9-9 are repeated using D-(−)-erythrose as a starting material rather than D-(+)-glyceraldehyde, then two aldopentoses will be formed; if the same steps are repeated on these aldopentoses, four aldohexoses will result. Experimentally, one of these four is found to be (+)-glucose, so that it, too, must belong to the D-series. The three-dimensional structure of glucose is shown in several views in Fig. 9-10.

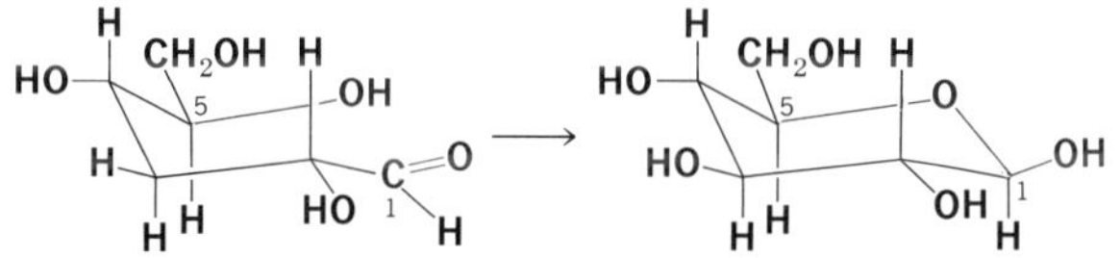

Figure 9-10. *In one conformation of glucose the hydroxyl group of carbon-5 is extremely close to the carbonyl group and can easily form a hemiacetal.*

Mutarotation of Glucose

Note from Fig. 9-10 that glucose may adopt a conformation in which substituents attached to carbon 5 are close in space to the aldehyde group of carbon 1. Because of this proximity, addition of the C-5 hydroxyl group across the carbon–oxygen double bond of glucose is especially favorable, and a stable six-membered ring, the hemiacetal form of glucose, results. In this addition a new chiral carbon atom is formed (carbon 1), so that two isomers of the pyranose form of glucose exist, α-D-glucopyranose and β-D-glucopyranose (usually called α- and β-glucose) (Fig. 9-11). Since these two isomers are identical at four of their chiral carbons and mirror images at the fifth, they are diastereomers of one another, have different physical and chemical properties, and can therefore be separated by crystallization. The pure α form has $[\alpha]_D + 113°$ and the pure β form $[\alpha]_D + 19°$. If either form is dissolved in water, it is found that its initial rotation gradually changes to that of an equilibrium rotation of $+52.5°$. This gradual change of rotation until equilib-

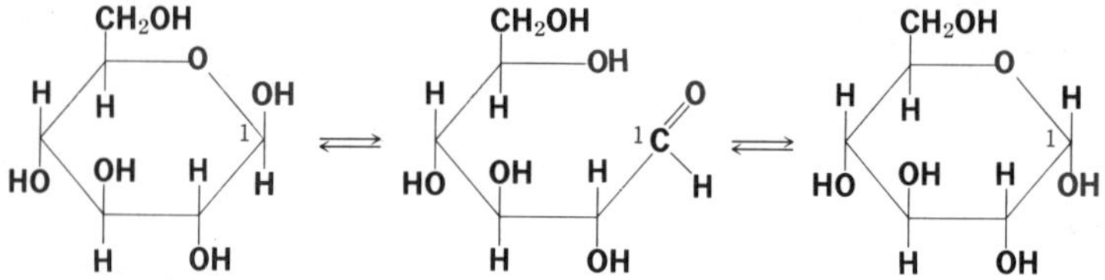

β-D-glucopyranose $[\alpha]_D^{20^\circ} + 19^\circ$ ⇌ D-glucose (aldehyde form) ⇌ α-D-glucopyranose $[\alpha]_D^{20^\circ} + 113^\circ$

Figure 9-11. *At equilibrium,* D*-glucose consists of 37%* α-D*-glucopyranose and 63%* β-D*-glucopyranose, with less than 0.05% of the free aldehyde present.*

rium is reached is known as *mutarotation* (Fig. 9-11). Both α- and β-glucoses are stable as solids, but in solution an equilibrium is gradually established between them. Simple arithmetic shows that in water the equilibrium mixture, with $[\alpha]_D + 52.5^\circ$, is 37% α-D-glucose (rotation $+113^\circ$) and 63% β-D-glucose (rotation $+19^\circ$).

Glucosides

We have seen previously that a hemiacetal may be converted to an acetal, a molecule in which two ether groups are attached to the same carbon. A similar reaction may be carried out with either α- or β-glucose (Fig. 9-12)

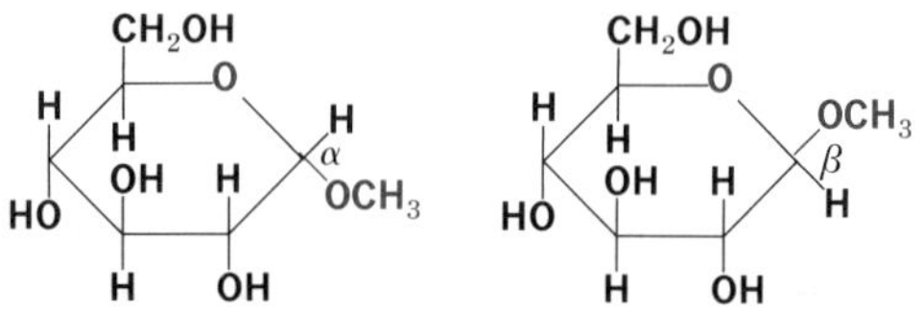

methyl α-D-glucoside   methyl β-D-glucoside

Figure 9-12. *Upon treatment with methyl alcohol and an acid catalyst, glucose is converted to a mixture of its acetals, known as glucosides. In contrast to the hemiacetal form of glucose, these acetals are not in ready equilibrium with the free aldehyde, and glucosides do not give typical aldehyde reactions.*

by heating with an alcohol and a trace of acid catalyst. Glucose acetals are called *glucosides*, and they do not convert back to the free aldehyde form unless heated with aqueous acid. Consequently glucosides do not mutarotate, nor do they give typical aldehyde reactions. Most important, glucosides (and the acetal forms of other sugars) are *nonreducing sugars*. They do not reduce copper solutions, because these reductions are characteristic of the aldehyde group, which is in equilibrium with the hemiacetal but not the acetal form of an aldehyde. The structure of a glucoside (α or β) can be determined by whether its hydrolysis back to glucose is catalyzed by the enzyme maltase (an α-glucosidase) or emulsin (a β-glucosidase). These enzymes are so specific that they easily distinguish between these seemingly minor structural features.

Fructose, a Keto Sugar

Not all sugars are aldehydes. Fructose, an important sugar isomeric with glucose, is a ketone, a ketohexose. Because α-hydroxy ketones are easily oxidized, fructose gives a positive test as a reducing sugar. Fructose exists primarily in the form of cyclic hemiketals and readily forms ketals with alcohols in acid solution (Fig. 9-13). For fructose the five-membered *furanose*

fructose

β-D-fructofuranose

α-D-fructofuranose

methyl β-D-fructofuranoside

methyl α-D-fructofuranoside

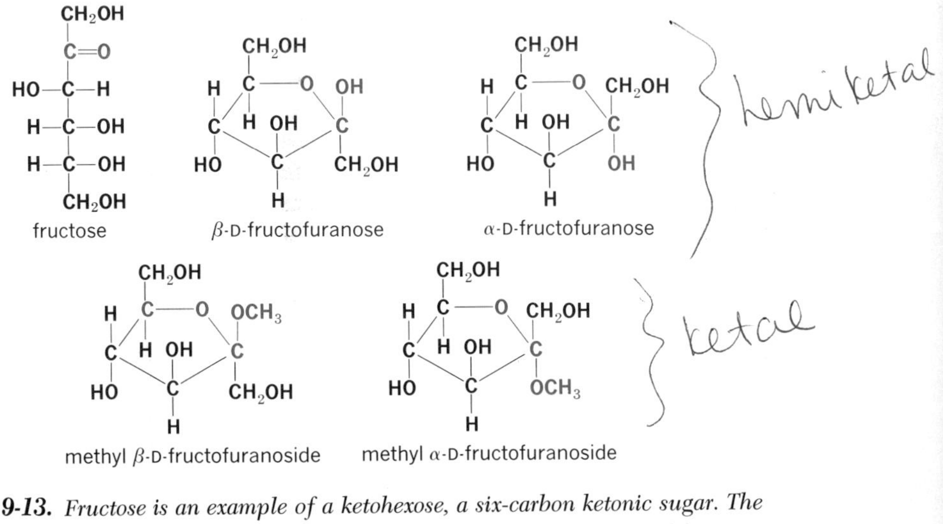

Figure 9-13. *Fructose is an example of a ketohexose, a six-carbon ketonic sugar. The hemiketal and ketal forms of fructose are illustrated here in the five-membered furanose ring form. Pyranose (six-membered ring) forms of fructose are also known.*

form is shown, since this form is important in table sugar, sucrose. Because they are hemiketals, the fructofuranoses are in equilibrium with fructose itself, show typical carbonyl reactions, and undergo mutarotation. The ketal forms of fructose, however, are diethers, do not show carbonyl reactions and do not mutarotate. In particular they are *nonreducing* derivatives of fructose.

Prob. 9-6. Draw the structures of α- and β-D-glucofuranose and α- and β-D-fructopyranose.

Disaccharides

Glucose and fructose are *monosaccharides* because each is composed of a single sugar molecule. Carbohydrates containing two monosaccharides joined together with the loss of a molecule of water are called *disaccharides.* A simple example is *maltose,* formed on partial hydrolysis of starch. Maltose has the formula $C_{12}H_{22}O_{11}$ and can be further hydrolyzed to two molecules of glucose. *Cellobiose* is an isomer of maltose obtained by partial hydrolysis of cellulose, and also gives two molecules of glucose on further hydrolysis. In both maltose and cellobiose the hydroxyl group on carbon 4 of one glucose molecule serves as the alcohol group to convert a second molecule of glucose into its acetal. The two disaccharides differ only in the stereochemistry of the acetal linkage, maltose being an α-glucoside and cellobiose a β-glucoside. Their structures are given in Fig. 9-14. In both maltose and cellobiose one of the two aldehyde groups is still present as a hemiacetal, so these carbohydrates are reducing sugars.

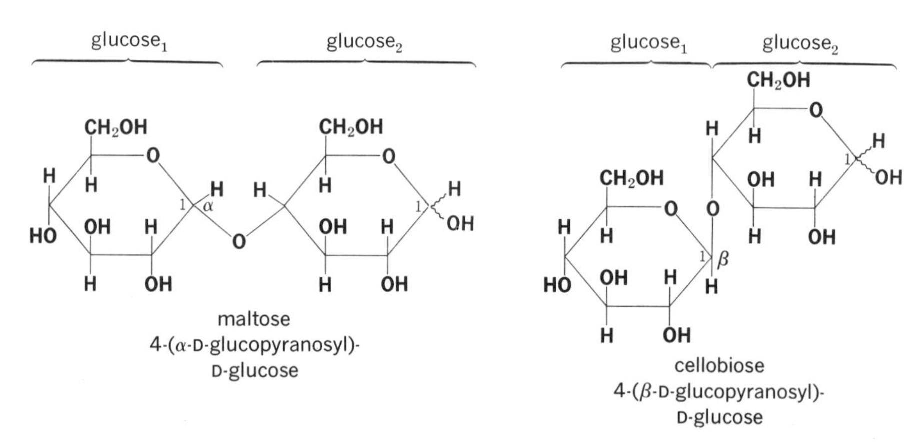

***Figure 9-14.** Two reducing disaccharides. Note that in both sugars glucose$_1$ is present in the acetal form, while glucose$_2$ is still a hemiacetal and so in equilibrium with the free aldehyde. Maltose is hydrolyzed to glucose by maltase, showing that it is an α-glucoside, while cellobiose is hydrolyzed by emulsin and therefore is a β-glucoside. Both maltose and cellobiose exist as mixtures of α- and β- forms at the hemiacetal group of glucose$_2$.*

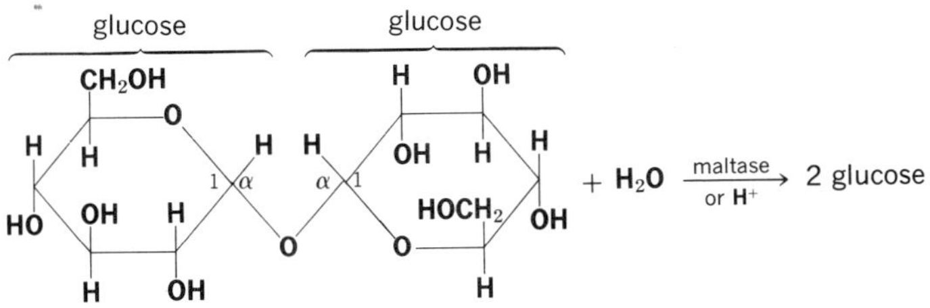

glucose glucose + H_2O $\xrightarrow[\text{or } H^+]{\text{maltase}}$ 2 glucose

α,α-trehalose
(α-D-glucopyranosyl-α-D-glucopyranoside)

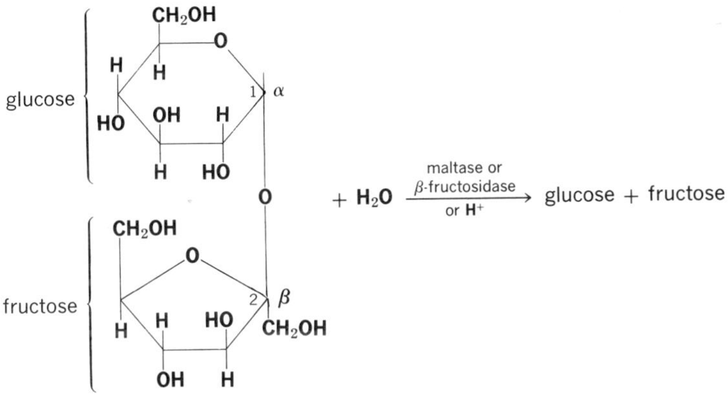

glucose / fructose + H_2O $\xrightarrow[\text{or } H^+]{\text{maltase or } \beta\text{-fructosidase}}$ glucose + fructose

sucrose
(α-D-glucopyranosyl-
β-D-fructofuranoside)

Figure 9-15. *Two nonreducing disaccharides. In each sugar both carbonyl groups are present as a ketal or acetal and so do not show carbonyl reactions.*

α,α-Trehalose is a disaccharide of glucose found in yeast, young mushrooms, and seaweed (Fig. 9-15). It does not react with cupric ion (Fehling's solution). Neither does it form derivatives with phenylhydrazine or hydroxylamine, as glucose, fructose, maltose, and cellobiose do. Evidently, trehalose contains neither a free carbonyl group nor its equivalent, a hemiacetal or hemiketal grouping. Since it is nonreducing, both aldehyde groups must be present as acetals, and this can occur only if the two glucose molecules are joined by an ether (acetal) linkage at their respective C-1 positions.

Sucrose, ordinary table sugar (cane or beet), is the pure organic chemical sold in greatest quantity. It is a disaccharide, which on hydrolysis gives one mole of glucose and one of fructose. Like trehalose, sucrose is a nonreducing sugar. Since the hydrolysis of sucrose is catalyzed by maltase, it is evidently an α-glucoside. The hydroxyl group of the hemiketal (at C-2) of fructose must also be involved in the glucoside linkage, because if any other hydroxyl group were used, fructose would be a reducing sugar. Fructose can also be hydrolyzed by a β-fructosidase found in yeast. The complete structure of sucrose is given in Fig. 9-15.

Carbohydrate Antibiotics

Several widely used antibiotics either contain unusual sugars or are themselves examples of complex tri- and tetrasaccharides. The latter category includes streptomycin, used in the treatment of tuberculosis and other bacterial infections, and neomycin, used topically as a component of burn ointments. An interesting feature of each of these antibiotics is the presence of one or more aminosugars in which NH_2 groups replace hydroxyl groups, and of highly substituted cyclohexane derivatives (Fig. 9-16).

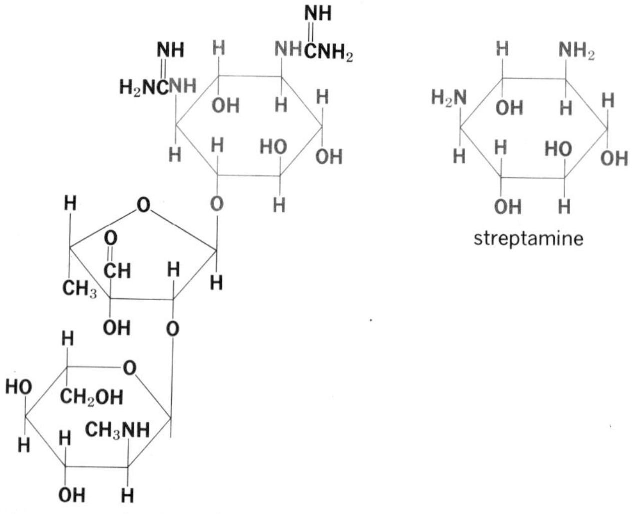

streptomycin

streptamine

Figure 9-16. *Several important antibiotics are carbohydrate derivatives. Streptomycin contains the cyclohexane derivative streptamine (in red) which is related to compounds present in several other antibiotics.*

Polysaccharides

Starch and *cellulose* are polysaccharides, high molecular weight polymers of glucose. Partial hydrolysis of cellulose produces cellobiose as the only disac-

cellulose

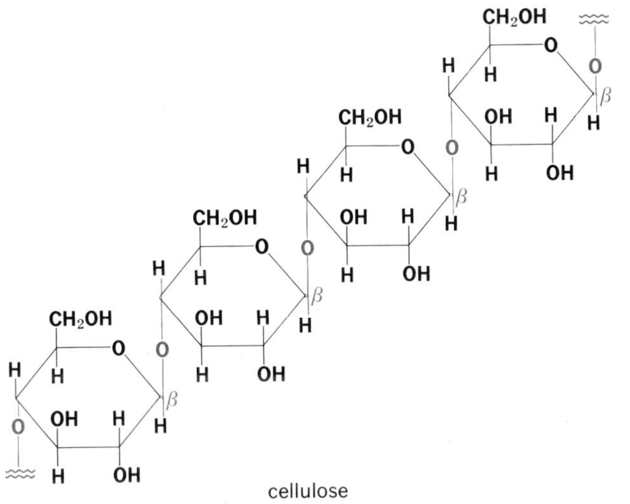

Figure 9-17. *Cellulose is a polymer of 2000 to 3000 β-glucose units.*

charide; thus cellulose must be a polymer of β-glucose. Physical studies have shown that cellulose has a molecular weight of 300,000 to 500,000, so that each molecule has 2000 to 3000 glucose units in it (Fig. 9-17). Cotton and linen are good examples of relatively pure cellulose. Wood also contains large amounts of cellulose. Humans and animals do not produce the necessary enzymes for its hydrolysis, so that cellulose cannot be used as food. However, many microorganisms can hydrolyze cellulose. The termite, which does make a meal of wood, cannot digest cellulose either; however, he keeps a friendly group of bacteria in his alimentary canal that performs the hydrolysis for him. The cow keeps similar bacteria in its rumen, a "predigester"; cellulose is hydrolyzed there by the bacteria and then transferred to the stomach.

Starch, a polymer of α-glucosides, is distributed widely throughout the plant kingdom—in potatoes, corn, etc. Most starch is actually a mixture of two polymers, amylose and amylopectin. Both are poly-α-glucosides; since the human body contains enzymes that can hydrolyze α-linkages, starch, unlike cellulose, is useful as food.

Amylose, the starch polymer usually present in a smaller amount, is a linear poly-α-glucoside exactly analogous to cellulose, though its chains are not as long, usually containing only 50 to 300 glucose molecules. *Amylopectin,* the more abundant form of starch, usually contains 10 to 20 times more glucose molecules than does amylose. It is also a polymer of α-glucosides, but, in addition to the usual acetals involving the C-4 hydroxyl group, contains numerous branches formed from acetals involving the primary C-6 hydroxyl. As a consequence, the amylopectin molecule looks more like a bush than a chain (Fig. 9-18).

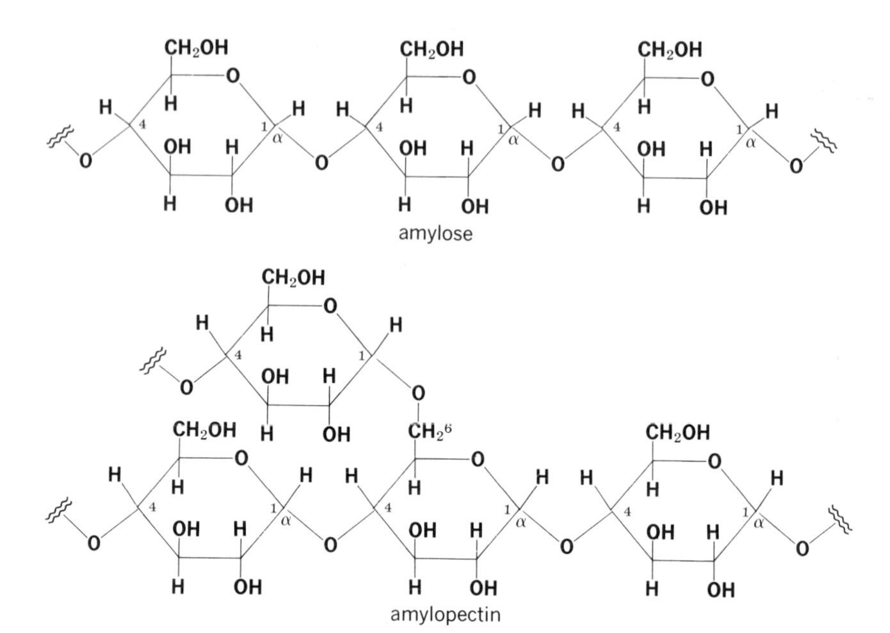

Figure 9-18. *Amylose, a component of starch, is, like cellulose, a polymer of glucose except that each acetal link is α instead of β. Amylopectin is a polymer of α-glucose that contains 1,6-acetal linkages in addition to the more common 1,4-linkages.*

Dextrins are polymers of α-glucose intermediate in molecular weight between glucose and starch. They are formed by the heating of dry starch or by the partial hydrolysis of starch. The shiny finish imparted to a starched shirt by ironing is caused by the starch breaking down into dextrins as a result of the heat of the iron.

Glycogen is another polysaccharide whose structure is quite similar to that of amylopectin shown in Fig. 9-18, but it is still more branched and of higher molecular weight. Glycogen is the principal storage form for carbohydrates in animals. Although its structure is like that of amylopectin, its formation from starch requires a complete breakdown of the latter and a synthesis of the former from glucose (Fig. 9-19).

$$\text{starch} \xrightleftharpoons{\text{amylase, } H_2O} \text{maltose} \xrightleftharpoons{\alpha\text{-glucosidase, } H_2O} \text{D-glucose}$$

$$\text{D-glucose} \rightleftharpoons \text{activated glucose} \rightleftharpoons \text{glycogen}$$

Figure 9-19. *Starch is hydrolyzed enzymatically in the body to glucose, which is converted to glycogen, a polymer whose structure resembles that of amylopectin (Fig. 9-18). Glycogen is stored and can be reconverted to glucose as needed for energy.*

First, the enzyme *amylase* converts the polysaccharide starch to the disaccharide maltose; this process is begun in the mouth by the amylase present in saliva and is responsible for the sweet taste of bread chewed for a few minutes. Additional amylase in the pancreatic juices of the intestine carries the polysaccharide hydrolysis to completion, with the assistance of another enzyme that hydrolyzes the occasional 1,6-linkages of starch. Maltose is hydrolyzed, in turn, to glucose, with the aid of an α-glucosidase of the intestines. Glucose is then absorbed from the intestine into the blood stream, in which it is transported to the liver, muscles, and other sites. At these sites it is stored as glycogen until it is needed again as glucose for conversion to energy. Formation and destruction of glycogen do not involve glucose directly, but rather activated forms of glucose. Glycogen provides a great reservoir of readily available energy for animals.

Rayon and Other Cellulose Derivatives

In cotton and linen, the cellulose molecules occur as long fibers that are suitable for conversion into thread and fabric. In wood the cellulose does not

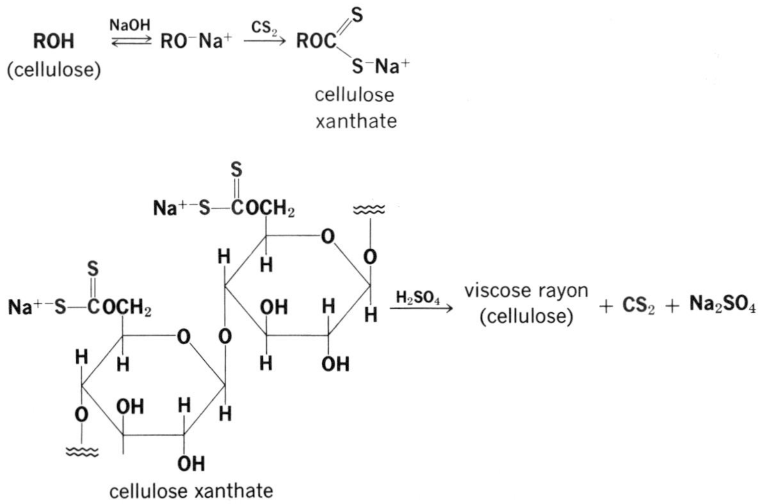

$$\underset{\text{(cellulose)}}{ROH} \overset{NaOH}{\rightleftharpoons} RO^-Na^+ \xrightarrow{CS_2} \underset{\text{cellulose xanthate}}{ROC(=S)S^-Na^+}$$

$$\underset{\text{cellulose xanthate}}{[Na^+\,{}^-S{-}C(=S)OCH_2\text{-glucose units}]} \xrightarrow{H_2SO_4} \underset{\text{(cellulose)}}{\text{viscose rayon}} + CS_2 + Na_2SO_4$$

Figure 9-20. *In the preparation of viscose rayon, cellulose (Fig. 9-17) is brought into solution by means of its xanthate salt. Cellulose is regenerated in long threads as viscose rayon by forcing the xanthate through small holes into an acid solution.*

differ chemically from that in cotton and linen, but the cellulose molecules are tangled and bunched. Wood cellulose, therefore, cannot be directly used for making cloth, but it is so cheap that much effort has been expended to convert it into useful forms.

The main difficulty with the modification of cellulose is its insolubility. Cellulose contains too many hydroxyl groups to be soluble in organic solvents, but it is too nonpolar to dissolve in water. In the synthesis of *viscose rayon*, this insolubility is overcome. First cellulose is mixed with strong alkali solution and then allowed to react with carbon disulfide (Fig. 9-20). A xanthate derivative of cellulose is formed by a reaction exactly analogous to the reaction of base with carbon dioxide. This xanthate salt is ionic; therefore, cellulose xanthate dissolves in water. The thick viscous xanthate solution of cellulose is forced through tiny holes into a solution of sulfuric acid. Just as sodium bicarbonate, when treated with acid, regenerates carbon dioxide, so cellulose

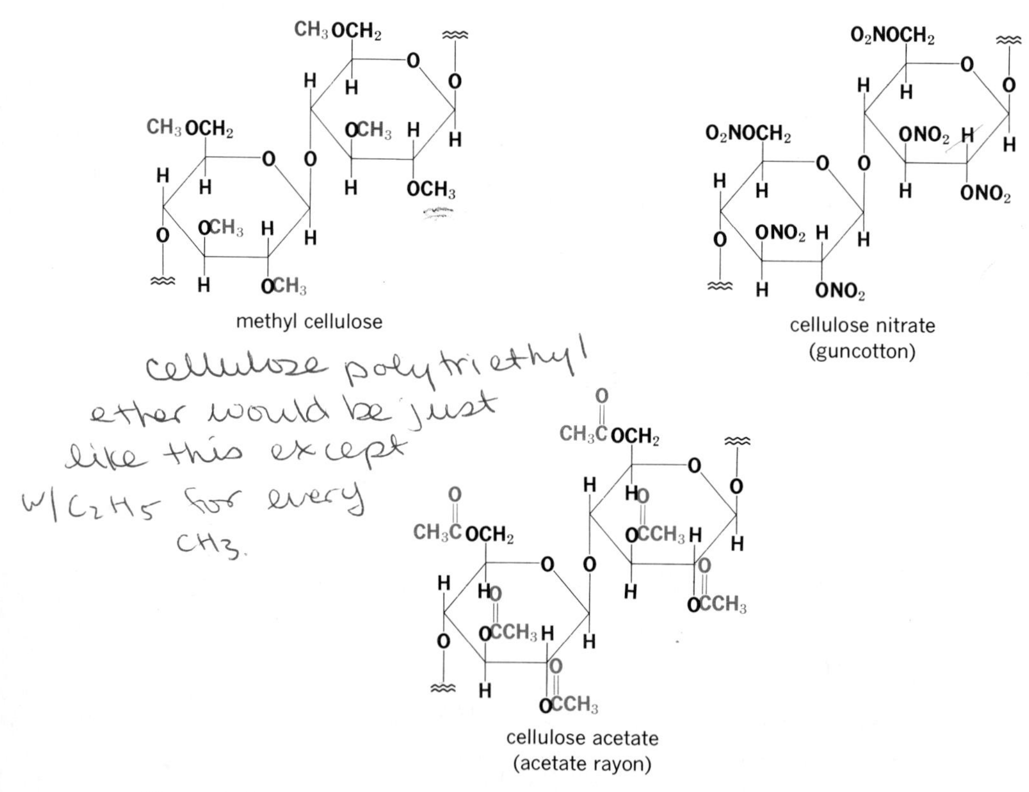

Figure 9-21. *In methyl cellulose, the hydroxyl groups are converted to ethers by reaction with methyl chloride and sodium hydroxide. Guncotton is the nitrate ester of cellulose and acetate rayon is the acetate ester.*

xanthate regenerates carbon disulfide and precipitates cellulose in the form of a continuous thread that is drawn off and dried. Viscose rayon is then chemically identical to ordinary cellulose, although the treatment with acid and base usually results in shortening the polymer chain.

Cellulose can be made soluble in organic solvents by converting its hydroxyl groups to less polar derivatives. A cellulose ether is the simplest derivative that meets this condition. The sodium salt of cellulose can be treated, for instance, with methyl chloride to form methyl cellulose (Fig. 9-21). The resultant product does not resemble cotton or linen but is used to impart oil and grease resistance to paper. Acetate rayon (cellulose acetate) is formed by the reaction of cellulose with acetic anhydride to yield esters of the hydroxyl groups. Guncotton is an ester of cellulose with nitric acid.

PROBLEMS

1. Give an example of each of the following:
   (**a**) a ketopentose (**b**) a methyl glucoside
   (**c**) an osazone (**d**) the xanthate salt of an alcohol
   (**e**) a meso compound (**f**) a nonreducing disaccharide
   (**g**) the hemiacetal form of an aldotetrose
   (**h**) a glucoside that is hydrolyzed by maltase
   (**i**) cellulose polytriethyl ether

2. What would be the product of the reaction of glucose with each of the following reagents?
   (**a**) $NaBH_4$ (**b**) NH_2OH (**c**) CH_3OH, H^+
   (**d**) C_6H_5—$NHNH_2$ (**e**) HCN (**f**) Br_2/H_2O

3. Draw a structure for any aldopentose. Draw its furanose hemiacetal in both the α- and β-forms. Draw a disaccharide of either form that would be a reducing disaccharide and another that would be a nonreducing disaccharide.

4. Draw all the stereoisomers (configurational and geometrical) of each of the following compounds. Indicate which are enantiomers and which meso compounds.
   (**a**) $CH_3CH(OH)CH(OH)C_6H_5$ (**b**) $C_2H_5CH(CH_3)CH_2CH(CH_3)C_6H_5$
   (**c**) $HOCH_2CH(OH)CH(OH)CH_2OH$ (**d**) $CH_3CH{=}CHCH(OH)CH_3$
   (**e**) $C_2H_5CH(CH_3)CH(CH_3)C_2H_5$ (**f**) 1,2-cyclopentanediol
   (**g**) 1,3-cyclobutanediol (**h**) 2,4-hexadiene

5. Draw the structures of the four isomeric D-aldopentoses. Which two would be reduced to meso alcohols by $NaBH_4$? Which two would be reduced to optically active alcohols?

6. Draw the structures of the osazones that would be produced by reaction of each of the four D-aldopentoses with excess phenylhydrazine. How many *different* osazones are formed from the four sugars?
7. Which pairs of D-aldohexoses (Fig. 9-6) would give the same osazones on reaction with phenylhydrazine?
8. The structures of four disaccharides are given below.
   (a) Which contain glucose?
   (b) Which contain fructose?
   (c) Which are reducing sugars?
   (d) Which contain a 1,6-linkage?
   (e) Which would be hydrolyzed by emulsin?
   (f) Which contain a furanose ring? 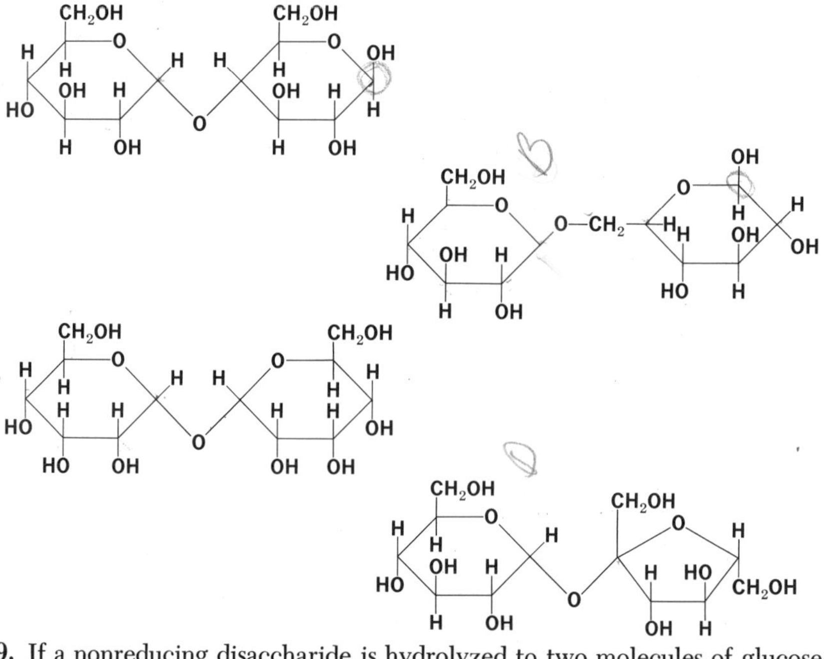
9. If a nonreducing disaccharide is hydrolyzed to two molecules of glucose by emulsin but not by maltase, what is its structure? What is its structure if it is hydrolyzed by maltase but not by emulsin? If it is hydrolyzed by both maltase and emulsin?
10. Reduction of D-glucose with $NaBH_4$ gives an optically active hexahydroxy compound. Which six structures among the D-sugars in Fig. 9-6 are possible for D-glucose on the basis of this information alone? An aldohexose can be converted to an aldopentose by a series of reactions known as the Ruff degradation:

$$\begin{array}{l}\text{HC=O} \\ | \\ \text{CHOH} \\ | \end{array} \xrightarrow[H_2O]{Br_2} \begin{array}{l}\text{COOH} \\ | \\ \text{CHOH} \\ | \end{array} \xrightarrow[H_2O_2]{Ca(OH)_2} \text{H—}\underset{|}{\text{C}}\text{=O} + CO_2$$

Application of the Ruff degradation to glucose gives an aldopentose that, on reduction with $NaBH_4$ gives an optically active penta alcohol. Show that this is expected for the structure we have given for D-glucose in Fig. 9-6. Which two other sugars among the six would give the same results?

10
amines

Amines are organic derivatives of ammonia; like ammonia, they are weak bases and form soluble salts in acid solution. Amines are notable for their physiological effects, and many of the most potent drugs and medicines, among them morphine, strychnine, and LSD, are ammonia derivatives. Amino groups are also a part of amino acids that, as the building blocks of proteins, are obviously of extreme chemical and biological importance. Amino groups, by their ability to form strong hydrogen bonds, play a key role in transmitting hereditary information in the DNA molecule. We shall discuss the latter two topics in detail in later chapters. Here we summarize the general properties of amines, discuss how they are synthesized, and give some of their more important reactions.

We classify amines as aliphatic or aromatic, depending upon whether the nitrogen is attached directly to an aromatic ring or not. This is an important distinction because the difference in basicity between aliphatic and aromatic amines matches the difference in acidity between alcohols and phenols, and for the same reason: The benzene ring interacts strongly enough with a nitrogen atom attached to it to make an appreciable difference in properties. Aliphatic amines, e.g., methylamine, have nearly the same strength as bases as ammonia, whereas a typical aromatic amine such as aniline is a million times weaker as a base. The explanation for this difference in basicity parallels exactly that which we have given for the greater basicity of an alkoxide ion than a phenoxide ion (Fig. 6-17). In an aliphatic amine, as in ammonia, the electron pair responsible for the basic properties is localized on the nitrogen atom. In an aromatic amine the electron pair is delocalized, shared by the *ortho* and *para* carbons of the benzene ring. This delocalization lowers the energy of the nonbonded electrons and makes them less available for combining with a proton. This is shown in the resonance structures of Fig. 10-1.

Nomenclature of Amines

Simple aliphatic amines are generally named by joining the name of the alkyl groups they contain to the word *amine*. Hence we can have methylamine, ethylamine, and cyclohexylamine. In each of these compounds, only one

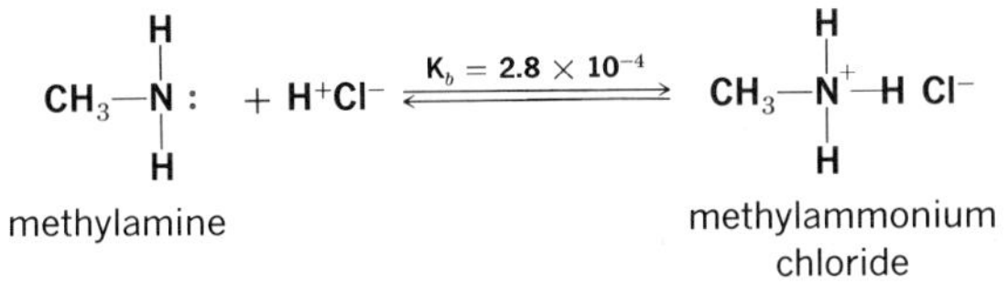

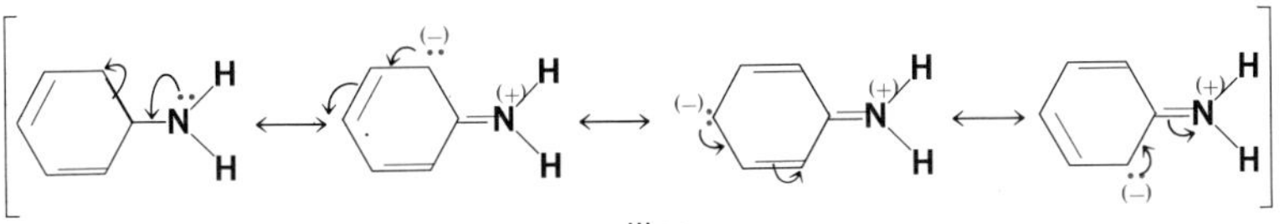

$$CH_3{-}\ddot{N}H_2 + H^+Cl^- \underset{}{\overset{K_b = 2.8 \times 10^{-4}}{\rightleftharpoons}} CH_3{-}\overset{+}{N}H_3\ Cl^-$$

methylamine — methylammonium chloride

aniline
$K_b = 2.6 \times 10^{-10}$

Figure 10-1. *The nonbonding electron pair of an amine accounts for its basicity. In aliphatic amines these electrons are localized on the nitrogen, while in aromatic amines they are delocalized and hence are less available for combination with a proton.*

organic group is attached to the nitrogen, and we say that they are *primary amines* (1° amines). If two organic groups are attached, as in dimethylamine or ethylisopropylamine, the amine is said to be *secondary* (2°) and trimethylamine and methylethylcyclohexylamine are *tertiary* (3°) amines. Tetramethylammonium bromide is a *quaternary ammonium* salt. Note carefully the difference between the use of the words primary, secondary, and tertiary in speaking of amines and of alcohols. In amines these words designate the number of carbon atoms attached to the nitrogen atom, in alcohols the number joined to the carbon to which the hydroxyl group is attached. So *tert*-butyl alcohol is a 3° alcohol, while *tert*-butylamine is a 1° amine. In the IUPAC system the amino group (NH_2) is treated as a substituent, and therefore the IUPAC name for isopropylamine is 2-aminopropane.

Among simple aromatic amines, aniline is the parent; many aromatic amines are named as its derivative. Since confusion might arise over whether a substituent is attached to the nitrogen atom or to the aromatic ring, the prefix *N*- precedes the name for nitrogen substituents; for example, $C_6H_5NHCH_3$ is called *N*-methylaniline. The structures and names for some representative amines and ammonium salts are given in Fig. 10-2.

Synthesis and Reactions of Amines

Primary amines are usually made by reaction of ammonia with alkyl halides (Fig. 7-14, 7-16) or by reduction of nitro or cyano compounds. A primary amine will react further with alkyl halides to give secondary or tertiary amines, or quaternary ammonium compounds. The synthetic utility of this reaction

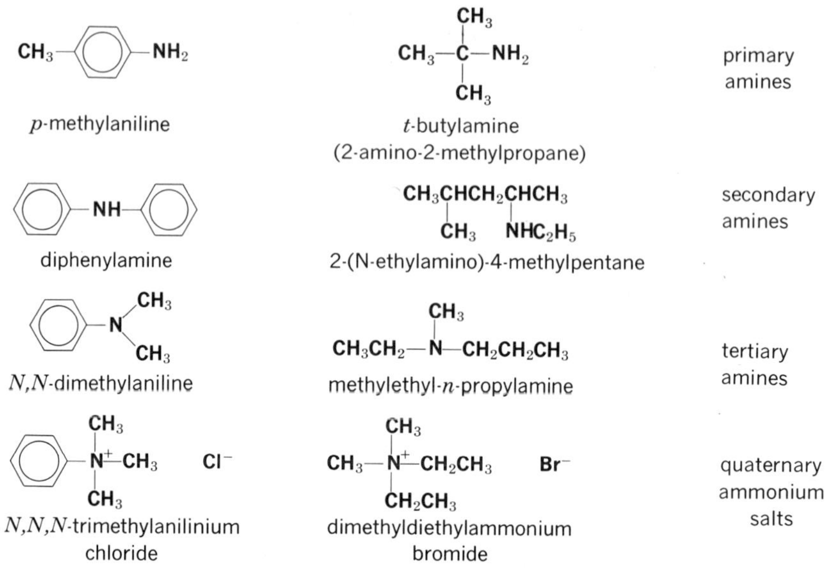

$CH_3-C_6H_4-NH_2$ *p*-methylaniline	$(CH_3)_3C-NH_2$ *t*-butylamine (2-amino-2-methylpropane)	primary amines
$C_6H_5-NH-C_6H_5$ diphenylamine	$CH_3CH(CH_3)CH_2CH(NHC_2H_5)CH_3$ 2-(N-ethylamino)-4-methylpentane	secondary amines
$C_6H_5-N(CH_3)_2$ *N,N*-dimethylaniline	$CH_3CH_2-N(CH_3)-CH_2CH_2CH_3$ methylethyl-*n*-propylamine	tertiary amines
$C_6H_5-N^+(CH_3)_3\ Cl^-$ *N,N,N*-trimethylanilinium chloride	$CH_3-N^+(CH_3)(CH_2CH_3)-CH_2CH_3\ Br^-$ dimethyldiethylammonium bromide	quaternary ammonium salts

Figure 10-2. *Names of some typical amines and quaternary ammonium salts.*

is severely limited because it is difficult to stop the reaction at the intermediate stages and because, since the reaction is a *displacement,* tertiary and aromatic halides do not react (Fig. 10-3).

$$H_3N\colon \xrightarrow{C_2H_5Br} C_2H_5NH_2 + HBr \xrightarrow{C_2H_5Br} (C_2H_5)_2NH + HBr \xrightarrow{C_2H_5Br} (C_2H_5)_3N + HBr \xrightarrow{C_2H_5Br} (C_2H_5)_4\overset{+}{N}\ Br^-$$

Figure 10-3. *Primary, secondary, and tertiary amines and quaternary ammonium salts can all be prepared by the displacement reaction of an alkyl halide and an amine. Mixtures of products are usually formed in this reaction.*

Aromatic amines are nearly always prepared by reduction of a nitro group, a reaction that occurs readily and in high yields (Fig. 10-4). Metals like zinc or tin serve as the source of electrons for the reduction, which is ordinarily carried out in acidic solution. Aliphatic amines may also be prepared by reduction of cyano or other nitrogen-containing functional groups.

Substitution Reactions of Aromatic Amines

Aniline and its derivatives undergo aromatic substitution reactions with great ease; ordinarily, it is difficult to stop such a reaction until all available *ortho*

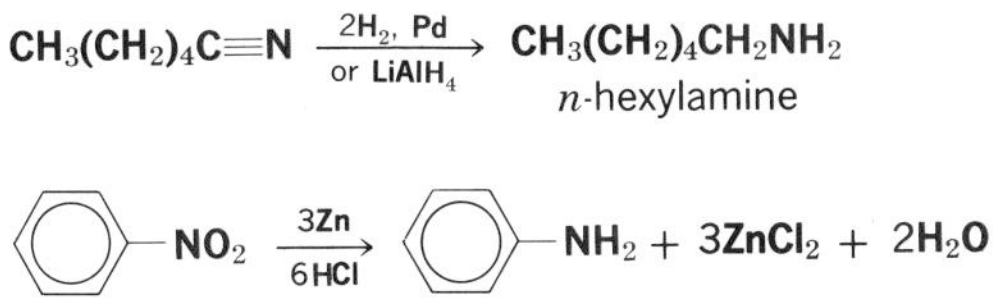

$$CH_3(CH_2)_4C{\equiv}N \xrightarrow[\text{or } LiAlH_4]{2H_2,\ Pd} CH_3(CH_2)_4CH_2NH_2$$

n-hexylamine

$$C_6H_5{-}NO_2 \xrightarrow[6HCl]{3Zn} C_6H_5{-}NH_2 + 3ZnCl_2 + 2H_2O$$

Figure 10-4. *Reductive methods are important for the preparation of pure amines. Aromatic amines are nearly always made by the reduction of aromatic nitro compounds. Other reducing agents, such as iron or tin in acid, or hydrogen gas, can be employed for the latter reduction.*

and *para* positions have been substituted. This reactivity results from the great stability that the electron pair on nitrogen imparts to an adjacent positive charge (Fig. 5-19). The reactivity of aniline found early industrial use in the recovery of bromine from the ocean, when bromine was still a rare element. Sea water contains about 0.015% bromine in the form of bromide ion. The sea water was oxidized to bromine by the addition of chlorine and then aniline was added. Even at this low concentration, bromine forms the insoluble solid 2,4,6-tribromoaniline; this compound can be easily isolated, so that the bromine can be recovered from it. Although this process has now been supplanted by less expensive methods, it illustrates the great reactivity of the aromatic ring in aniline toward electrophilic reagents like bromine (Fig. 10-5).

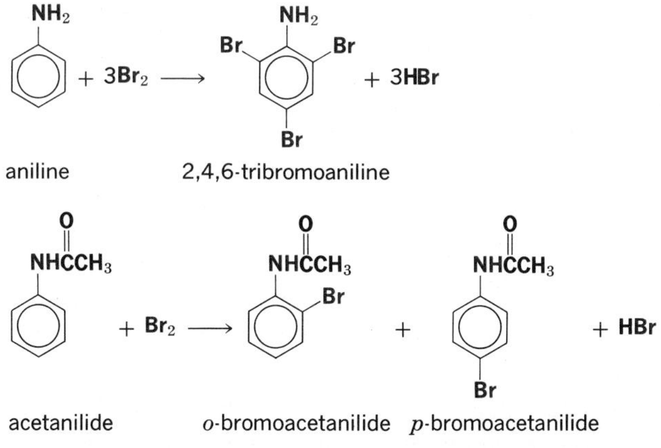

$$C_6H_5NH_2 + 3Br_2 \longrightarrow 2,4,6\text{-}Br_3C_6H_2NH_2 + 3HBr$$

aniline 2,4,6-tribromoaniline

$$C_6H_5NHCOCH_3 + Br_2 \longrightarrow o\text{-}BrC_6H_4NHCOCH_3 + p\text{-}BrC_6H_4NHCOCH_3 + HBr$$

acetanilide *o*-bromoacetanilide *p*-bromoacetanilide

Figure 10-5. *Aniline is so reactive toward aromatic substitution that the reaction is difficult to stop before trisubstitution has occurred. Acetanilide, obtained by treating aniline with acetic anhydride, is less reactive, and monosubstitution products are easily obtained.*

Prob. 10-1. Would you expect aniline to react more or less readily with bromine in a solution of hydrogen bromide? Why?

Diazonium Salts

One of the most useful reactions of primary amines is that with nitrous acid, HNO_2. With a primary aliphatic amine, nitrogen gas is eliminated rapidly and quantitatively, and if the gas is collected and its volume measured under standard conditions of temperature and pressure, the reaction serves as a valuable method for the analysis of amines (the Van Slyke method). If the primary amine is aromatic, and the reaction is conducted at ice-bath temperatures, a *diazonium salt* is formed (Fig. 10-6). These salts are highly reactive

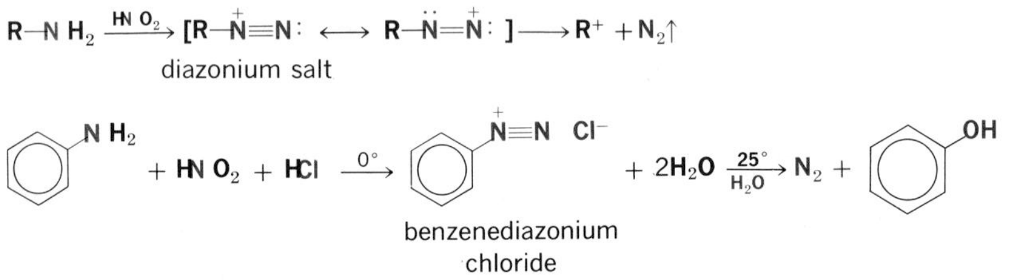

$$R{-}NH_2 \xrightarrow{HNO_2} [R{-}\overset{+}{N}{\equiv}N: \longleftrightarrow R{-}\ddot{N}{=}\overset{+}{N}:] \longrightarrow R^+ + N_2\uparrow$$

diazonium salt

$$C_6H_5NH_2 + HNO_2 + HCl \xrightarrow{0^\circ} C_6H_5\overset{+}{N}{\equiv}N\ Cl^- + 2H_2O \xrightarrow[H_2O]{25^\circ} N_2 + C_6H_5OH$$

benzenediazonium chloride

Figure 10-6. *Treatment of a primary amine with nitrous acid gives a diazonium salt. Such salts of alkyl amines are unstable and immediately decompose with the evolution of nitrogen. Aromatic diazonium salts are stable at 0° but eliminate N_2 at room temperature.*

and explode when dry, but their great reactivity makes them ideal for the synthesis of other compounds. We have already seen that aryl halides are totally unreactive toward displacement reactions. Diazonium salts play the same role in aromatic chemistry that halides do in aliphatic chemistry, for the N_2^+ group can be replaced by a great many other groups in the *Sandmeyer reaction* as shown in Fig. 10-7.

The importance of the Sandmeyer reaction is even greater than may appear at first glance because diazonium salts are obtained from amines, which in turn are best formed by reduction of a nitro group. By way of a diazonium salt, then, a *meta* directing nitro group may be transformed into an *ortho-para* directing hydroxyl or halo group. Suppose, for example, that one wishes to prepare *m*-chlorobromobenzene. Since both chlorine and bromine are *o-p* directors, neither chlorination of bromobenzene nor bromination of chlorobenzene would give the desired isomer. But bromination of nitrobenzene to give *m*-bromonitrobenzene, followed by conversion of the nitro group to

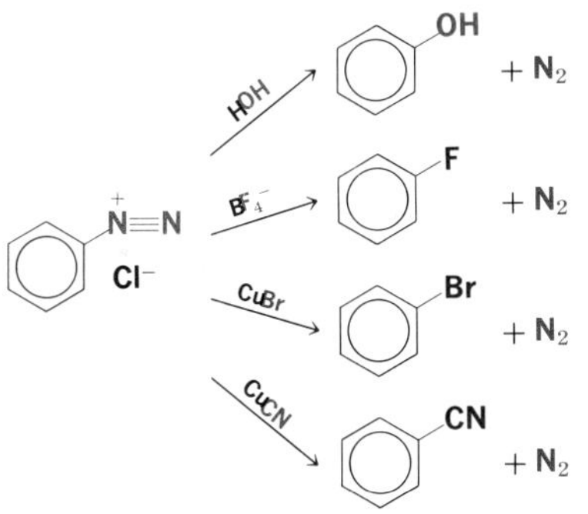

Figure 10-7. Aromatic diazonium salts are readily transformed into a variety of substituted aromatic compounds.

chlorine by way of the Sandmeyer reaction, would give the desired compound (Fig. 10-8). Although the path seems a long one, yields are high and some of the intermediates do not have to be isolated.

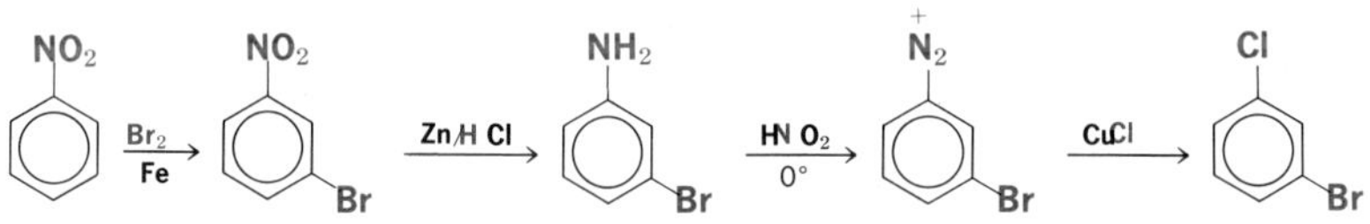

Figure 10-8. The Sandmeyer reaction is especially useful in preparing benzene derivatives with ortho-para *directing groups meta to one another* (or meta-*directing groups* ortho *or* para *to one another*).

Prob. 10-2. Show how the Sandmeyer reaction can be used to prepare *p*-cyanonitrobenzene.

Azo Dyes

Diazonium salts undergo another important reaction, *azocoupling*, a reaction that leads to a large class of dyestuffs, the *azo dyes*. The unsubstituted nitrogen of the diazonium group is a weak Lewis Acid and will carry out an aromatic substitution on a highly reactive aromatic compound (aniline or phenol derivative). The resulting azo compounds are highly colored and find wide use as dyes. Aniline yellow, the first azo dye synthesized (in 1863) is prepared by treating benzenediazonium ion with aniline. Orange II, prepared similarly

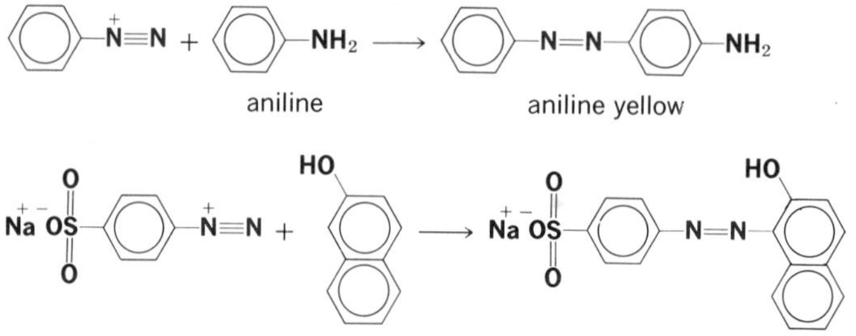

aniline

aniline yellow

β-naphthol

orange II

Figure 10-9. *Azo dyes result from the reaction of diazonium salts with aromatic amines or phenols. The azo group is shown in red in the products.*

from the diazonium salt of sulfanilic acid and β-naphthol, is rendered soluble in water by the presence of the sulfonic acid salt (Fig. 10-9).

Quaternary Ammonium Salts

These salts may be considered to be derivatives of inorganic ammonium salts, in which the nitrogen atom is covalently bonded to four groups and no longer has a free electron pair. Thus a quaternary ammonium ion is not itself a base, and any alkaline properties of its solutions are determined by the nature of the associated anions. In this respect, a quaternary ammonium ion resembles an alkali metal ion. Quaternary halides are neutral, but quaternary hydroxides are strong bases like sodium hydroxide. Quaternary ammonium salts are used in some soaps and antiseptics and they also play special roles in human physiology. One of the most important of these is that played by choline and acetylcholine. Acetylcholine is present in a combined form in nerve cells and is released when a nerve cell is stimulated; the acetylcholine in turn stimulates an adjacent nerve cell, which releases acetylcholine and so transmits the nerve impulse (Fig. 10-10).

After it has done its job of stimulating an adjacent nerve cell, acetylcholine must be quickly deactivated. This deactivation is accompanied by hydrolysis of the acetylcholine by the enzyme acetylcholinesterase into choline and acetic acid. Anything that interferes with this deactivation short-circuits nerve impulse transmission and leads quickly to paralysis or death. Some drugs contain quaternary ammonium groups that simply compete with acetylcholine and block the sites of nerve-impulse transmission. One of the most effective of these is hexamethyldecamethylenediammonium ion (decamethonium), which is used as a muscle relaxant. Another is curare, a complex base (an alkaloid) used in medicine to promote muscular relaxation and for arrow poisons by some South American Indians.

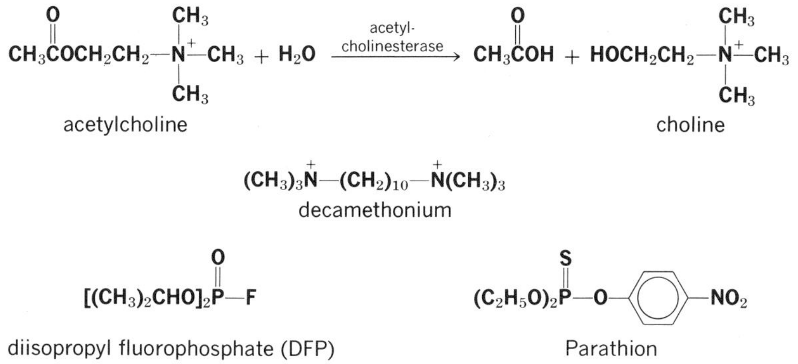

$$CH_3\overset{O}{\overset{\|}{C}}OCH_2CH_2-\overset{CH_3}{\underset{CH_3}{\overset{|}{\underset{|}{N^+}}}}-CH_3 + H_2O \xrightarrow[\text{cholinesterase}]{\text{acetyl-}} CH_3\overset{O}{\overset{\|}{C}}OH + HOCH_2CH_2-\overset{CH_3}{\underset{CH_3}{\overset{|}{\underset{|}{N^+}}}}-CH_3$$

acetylcholine | choline

$$(CH_3)_3\overset{+}{N}-(CH_2)_{10}-\overset{+}{N}(CH_3)_3$$

decamethonium

$$[(CH_3)_2CHO]_2\overset{O}{\overset{\|}{P}}-F$$

diisopropyl fluorophosphate (DFP)

$$(C_2H_5O)_2\overset{S}{\overset{\|}{P}}-O-C_6H_4-NO_2$$

Parathion

Figure 10-10. *Acetylcholine is active in the transmission of nerve impulses. Decamethonium, DFP, and Parathion interfere with this transmission.*

Other compounds function differently, by reacting with acetylcholinesterase to prevent it from hydrolyzing acetylcholine. These compounds have structures completely unrelated to that of choline, and include the venom of some snakes and also nerve gases like diisopropyl fluorophosphate. When humans take these compounds, death follows from respiratory failure. Insecticides like Parathion have a similar effect on insects (and can also kill farmers who apply them carelessly).

Quaternary ammonium salts undergo an important synthetic reaction: When heated with alkali, they form alkenes. The reaction, known as the *Hofmann elimination,* is analogous to the formation of alkenes from alkyl halides by reaction with base (see Fig. 7-17). In the case of the Hofmann elimination the alkene that is formed in greatest amount is the one which is *least substituted* with alkyl groups (Fig. 10-11).

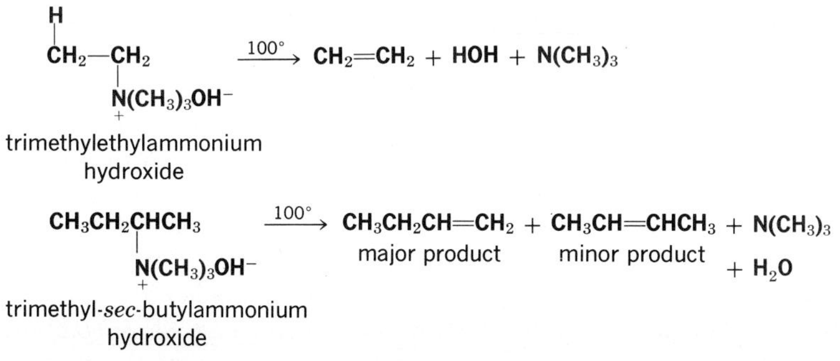

$$\overset{H}{\overset{|}{C}}H_2-\underset{\overset{+}{N}(CH_3)_3OH^-}{\underset{|}{C}H_2} \xrightarrow{100^\circ} CH_2{=}CH_2 + HOH + N(CH_3)_3$$

trimethylethylammonium hydroxide

$$CH_3CH_2\underset{\overset{+}{N}(CH_3)_3OH^-}{\underset{|}{C}H}CH_3 \xrightarrow{100^\circ} \underset{\text{major product}}{CH_3CH_2CH{=}CH_2} + \underset{\text{minor product}}{CH_3CH{=}CHCH_3} + N(CH_3)_3 + H_2O$$

trimethyl-*sec*-butylammonium hydroxide

Figure 10-11. *A Hofmann elimination results when a quaternary ammonium compound is heated with base.*

Heterocyclic Amines

If the nitrogen atom of an amine is incorporated as part of a ring, the resulting amine is said to be a *heterocyclic compound,* one of a group of compounds whose rings contain at least one element other than carbon. Heterocyclic compounds containing oxygen, sulfur, or other elements are also known; however, heterocyclic nitrogen compounds are the most common.

Saturated heterocyclic amines behave much the same as their acyclic analogs, just as n-hexane and cyclohexane are similar; *pyrrolidine* and *piperidine* (Fig. 10-12) both undergo all the reactions of typical aliphatic secondary amines. If a heterocyclic ring is highly unsaturated, however, it may acquire

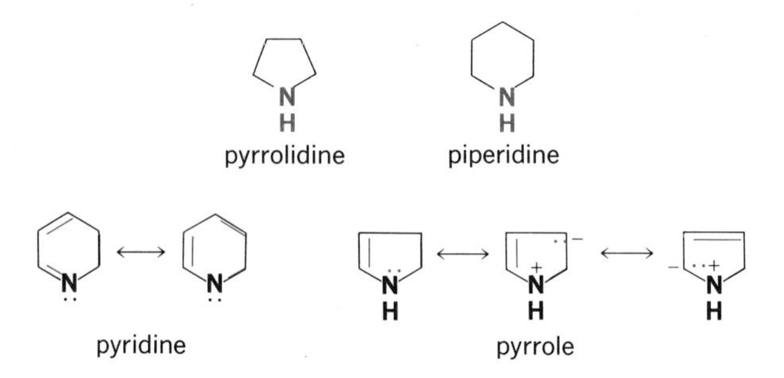

Figure 10-12. *Four heterocyclic compounds. The electrons making up the aromatic sextets in pyridine and pyrrole are shown in red.*

unusual properties all its own. For this reason, heterocyclic chemistry is often treated as a special subject. Often heterocyclic compounds have aromatic properties reminiscent of those of benzene. *Pyridine* and *pyrrole* are the most common examples (Fig. 10-12). Pyridine is the direct heterocyclic analog of benzene, in which one carbon atom and its attached hydrogen atom have been replaced by a nitrogen atom. The aromatic sextet of electrons, the closed π-electron system, is complete in pyridine without using the unshared electrons on the nitrogen atom. Pyridine is still basic, since it has this free electron pair on the nitrogen atom. Similarly, alkylation takes place on the nitrogen atom to give a quaternary salt.

Pyrrole also displays some of the properties of an aromatic compound because the two electrons that it possesses on the nitrogen atom can be used to complete an aromatic sextet, when combined with the four electrons of the double bonds (Fig. 10-12). In pyrrole, the unshared pair of electrons on the nitrogen atom is very much a part of the aromatic system, and so the compound is the heterocyclic analog of the cyclopentadiene anion (see Fig. 5-28). As a consequence, pyrrole does not behave as a base, since combination of this electron pair with a proton would disrupt the aromatic system.

A vast number of heterocyclic compounds are known, because nearly any carbon atom in any ring system can be replaced by a nitrogen atom, and oxygen, sulfur and other atoms can be introduced similarly; in addition, oxygen, sulfur, and other elements can similarly be introduced. Four heterocyclic compounds whose ring systems are of special importance in life proc-

imidazole   indole   pyrimidine   purine

Figure 10-13. *Four heterocyclic systems of particular biological significance.*

esses are shown in Fig. 10-13. The first two, imidazole and indole, are found widely in proteins; the last two, pyrimidine and purine, are the basic ring systems found in genes.

The *alkaloids,* nitrogen-containing plant products, are the largest class of naturally occurring heterocyclic amines. Many alkaloids have powerful physiological actions and are important drugs. Morphine and other opium alkaloids fall into this class of compounds. Biosynthesis of a number of alkaloids is discussed in Chapter 16.

PROBLEMS

1. Draw the structure of the following compounds:

(a) methylethyl-*t*-butylamine
(b) *N,N*-dimethylaniline
(c) *p*-nitroaniline
(d) diethyldimethylammonium hydroxide
(e) anilinium chloride
(f) 2-octylamine
(g) 2-amino-4-chlorooctane
(h) *p*-chlorobenzenediazonium chloride
(i) *N*-methylpyrrolidine
(j) 4-methylpyridine (the nitrogen atom is numbered 1)

2. Name each of the following compounds:

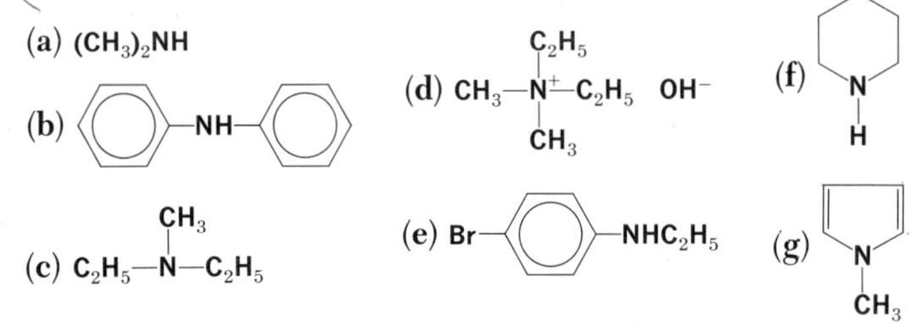

3. Complete the following reactions:

(a) $(CH_3)_3N + CH_3CH_2Br \longrightarrow$

(b) NC—⟨benzene ring⟩—CN + $LiAlH_4$ (excess) $\longrightarrow$

(c) $C_6H_5{-}NH_2 + HNO_2 \xrightarrow{0^\circ} \xrightarrow{CuBr}$

(d) $C_6H_5{-}NO_2 + Zn \xrightarrow{HCl}$

(e) $C_6H_5{-}\overset{+}{N}(CH_3)_2{-}CH_2{-}CH_3 \xrightarrow[\text{heat}]{OH^-}$

(f) $CH_3CH_2CH(CH_3){-}CH(\overset{+}{N}(CH_3)_3){-}CH_3 \xrightarrow[\text{heat}]{OH^-}$

(g) $C_6H_5{-}N(CH_3)_2 + Br_2$ (excess) $\longrightarrow$

(h) $C_6H_5{-}\overset{+}{N}{\equiv}N + C_6H_5{-}OH \longrightarrow$

4. Most amines are insoluble in water but do dissolve in aqueous acid. Most phenols are insoluble in water but dissolve in aqueous base. Both amines and phenols dissolve in ether. Suppose you had a mixture of aniline and phenol dissolved in ether. How could you separate them using only acid, base, and ether?

5. How could you convert nitrobenzene into
(a) aniline **(b)** phenol **(c)** *m*-dichlorobenzene
(d) aniline yellow

6. Starting with benzene, how could you prepare
(a) *m*-bromophenol **(b)** *p*-diaminobenzene
(c) *m*-diaminobenzene **(d)** *N,N,N*-trimethylanilinium bromide

7. How could you prepare choline (Fig. 10-10), starting with ethylene and any other reagents you choose?

8. Muscarine is a constituent of a number of poisonous mushrooms. Suggest a reason why this compound might be so deadly.

$CH_3CH_2CH(OH){-}CH(CH{=}O){-}N^+(CH_3)_3\ \ OH^-$

muscarine

9. Indicate whether each of the compounds in Problem 2 is a 1, 2, or 3° amine or a 4° ammonium salt.

10. Benzedrine (1-phenyl-2-aminopropane) is the active ingredient of many stimulants. Suggest a method of synthesis starting from benzene and any desired aliphatic compound. Several steps will be required.

---

short

11 acids and their derivatives

Organic acids contain the *carboxyl group*, —COOH, which ionizes in water to give a proton and which is neutralized by bases to give a salt. Because their salts are usually soluble in water, organic acids were among the first organic compounds to be isolated and purified; we have seen how they have given their names to many other organic molecules (p. 154). Of course, as the principal constituents of vinegar (Latin: *acetum*) acetic acid has been known since antiquity.

Acetic acid and its chloro derivatives played a key role in the attempts during the years 1830 to 1860 to work out a theory of structure and bonding in organic molecules. The idea of the ionic bond, in which many inorganic compounds are held together by the attraction of positive and negative charges, had just been worked out with the help of the newly discovered battery. Naturally, it was supposed that organic molecules were similarly constituted. Yet when acetic acid is chlorinated, chloroacetic, dichloroacetic, and trichloroacetic acids are formed, and except in acidity these three chlorinated acids hardly differ from acetic acid in appearance, odor, or chemical properties. Since chlorine was "known" to be negative and hydrogen "known" to be positive, how was it possible to replace a positive atom by a negative one without drastically changing the properties of the compound? The problem was further complicated by the fact that many chemists thought carbon had an atomic weight of 6 rather than 12. With the advent of the structural theory around 1860, it was recognized that chlorine could indeed "play the part of" hydrogen in an organic molecule, although a rational explanation of "why" had to await the development of the idea of the covalent bond in this century.

Compared to mineral acids, most organic acids are weak. In a 1 *N* solution of acetic acid in water only three molecules out of a thousand are ionized on the average. Yet as compared to the hydroxyl group in an alcohol, a carboxyl group has a much greater tendency to give up a proton, because the carboxylate anion is stabilized by resonance (Fig. 11-1). Since an anion

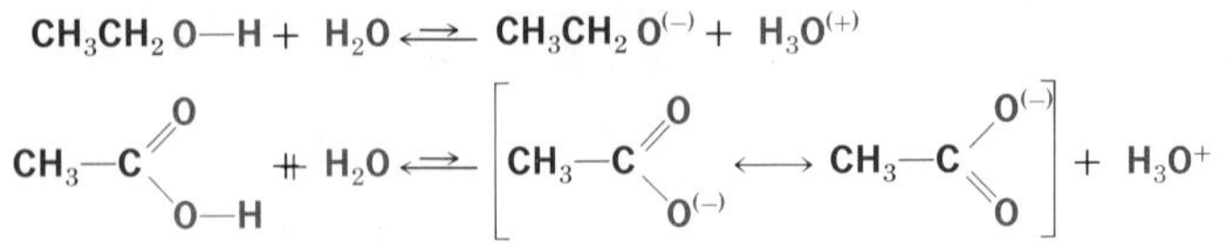

$$CH_3CH_2O{-}H + H_2O \rightleftharpoons CH_3CH_2O^{(-)} + H_3O^{(+)}$$

$$CH_3{-}C\begin{matrix}{\diagup}O\\{\diagdown}O{-}H\end{matrix} + H_2O \rightleftharpoons \left[CH_3{-}C\begin{matrix}{\diagup}O\\{\diagdown}O^{(-)}\end{matrix} \longleftrightarrow CH_3{-}C\begin{matrix}{\diagup}O^{(-)}\\{\diagdown}O\end{matrix}\right] + H_3O^+$$

Figure 11-1. *Ionization of an alcohol gives an anion in which the negative charge is localized on an oxygen atom. In the carboxylate ion the negative charge is delocalized over two oxygen atoms and so is more stable. Hence, ionization proceeds to a greater extent from an acid than from an alcohol.*

in which the negative charge is shared by two oxygens is more stable than one in which the charge is localized on a single oxygen, a carboxylate ion has less tendency to recombine with a proton than does an alkoxide ion. Hence, acetate ion is a weaker base than ethoxide ion, and acetic acid a stronger acid than ethanol. It is important to recognize this inverse relationship between an acid HX and the base derived from it by the loss of a proton (its conjugate base X^-). If the acid is strong (say HBr), then the conjugate base (Br^-) is extremely weak. If HX is very weak (say HOH), then X^- (OH^-) is a strong base. Acetic acid is about in the middle; it is a moderately weak acid and the acetate ion is a moderately weak base.

Nomenclature of Acids

In IUPAC systematic nomenclature of acids the ending *-oic acid* is added to the name of the parent hydrocarbon, and the carboxyl group is always given the number 1. Common names are also widely used for acids, and these can be used for substituted acids by locating the substituents with Greek letters. In this system the carbon adjacent to the carboxyl group is known as the α-carbon, the β-carbon is next, etc. Examples of common and systematic names for some acids are given in Fig. 11-2.

Synthesis of Acids

The most general laboratory method for the preparation of an acid is the reaction of a Grignard reagent with carbon dioxide (Fig. 11-3). Since nearly every halide can be converted to a Grignard reagent, this reaction is of great generality and the yields are usually high. The initially formed magnesium salt of the acid must be acidified to liberate the free acid.

Acids also arise from the hydrolysis of the cyanide group (—C≡N, a

	Common Names	*Systematic Names*
$HCOOH$ (HC(=O)OH)	formic acid	methanoic acid
CH_3COOH	acetic acid	ethanoic acid
CH_3CH_2COOH	propionic acid	propanoic acid
$CH_3CH_2CH_2COOH$	*n*-butyric acid	butanoic acid
$CH_3CH_2CH_2CH_2COOH$	*n*-valeric acid	pentanoic acid
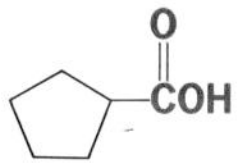 C_5H_9COOH (cyclopentyl–COOH)	cyclopentanecarboxylic acid	cyclopentylmethanoic acid
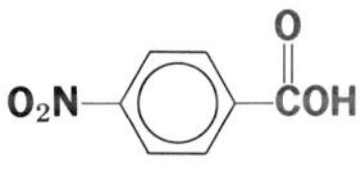 $O_2N{-}C_6H_4{-}COOH$	*p*-nitrobenzoic acid	4-nitrophenylmethanoic acid
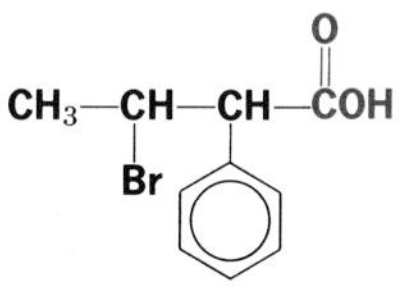 $CH_3{-}CH(Br){-}CH(C_6H_5){-}COOH$	α-phenyl-β-bromobutyric acid	2-phenyl-3-bromobutanoic acid

Figure 11-2. *Some common and systematic names for acids.*

nitrile). Acrylic acid, a useful monomer in vinyl polymerization, is made by the hydrolysis of acrylonitrile (see Table 4-1), which in turn is produced from acetylene and HCN (Fig. 11-4). In the laboratory, nitriles are easily made from 1° and 2° halides by reaction with NaCN, and their hydrolysis provides a convenient synthesis of acids.

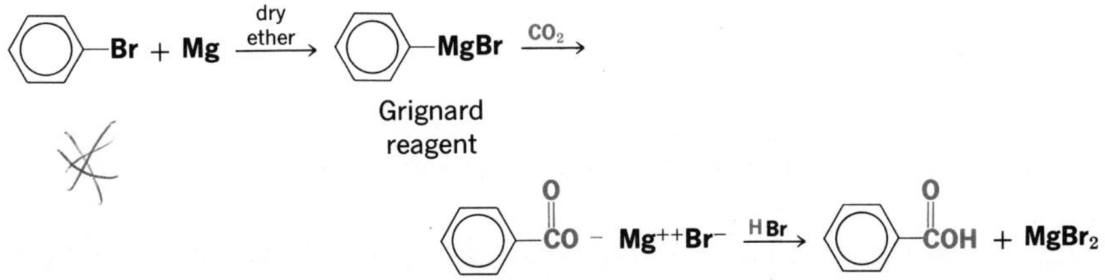

$$C_6H_5{-}Br + Mg \xrightarrow{\text{dry ether}} \underset{\text{Grignard reagent}}{C_6H_5{-}MgBr} \xrightarrow{CO_2} C_6H_5{-}\overset{O}{\overset{\|}{C}}O^-\ Mg^{++}Br^- \xrightarrow{HBr} C_6H_5{-}\overset{O}{\overset{\|}{C}}OH + MgBr_2$$

Figure 11-3. *The reaction of a Grignard reagent with carbon dioxide is a general method for the synthesis of carboxylic acids.*

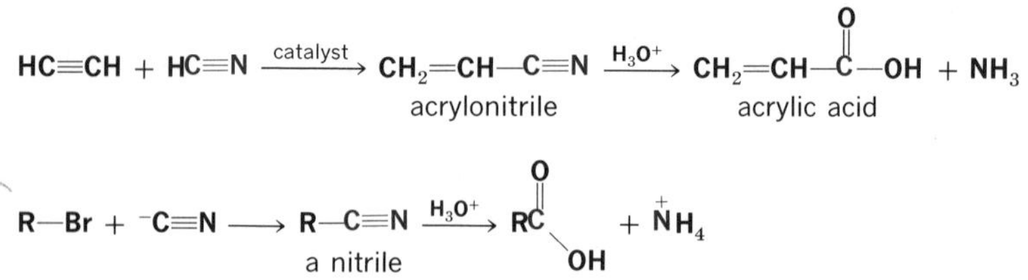

$$HC{\equiv}CH + HC{\equiv}N \xrightarrow{\text{catalyst}} \underset{\text{acrylonitrile}}{CH_2{=}CH{-}C{\equiv}N} \xrightarrow{H_3O^+} \underset{\text{acrylic acid}}{CH_2{=}CH{-}\overset{O}{\overset{\|}{C}}{-}OH} + NH_3$$

$$R{-}Br + {}^{-}C{\equiv}N \longrightarrow \underset{\text{a nitrile}}{R{-}C{\equiv}N} \xrightarrow{H_3O^+} R\overset{O}{\overset{\|}{C}}{-}OH + \overset{+}{N}H_4$$

Figure 11-4. *Hydrolysis of a nitrile is a useful method of synthesis for a carboxylic acid.*

Prob. 11-1. Why cannot acrylonitrile be obtained from the reaction of vinyl bromide with NaCN. (Ans.: See Fig. 7-20.)

Acetic acid (m.p. 17°, b.p. 118°) is the main constituent of vinegar, obtained by the fermentation of sugar in the presence of air. Industrially, acetic acid is made by the air oxidation of hydrocarbons or from acetylene by hydration to acetaldehyde and oxidation. A great deal of the acetic acid produced is converted to acetic anhydride (see below), which is used to make cellulose acetate (Fig. 9-21) and other esters. Chlorination of acetic acid gives chloroacetic acid which, upon treatment with the sodium salt of 2,4-dichlorophenol gives the weed killer 2,4-D (Fig. 11-5).

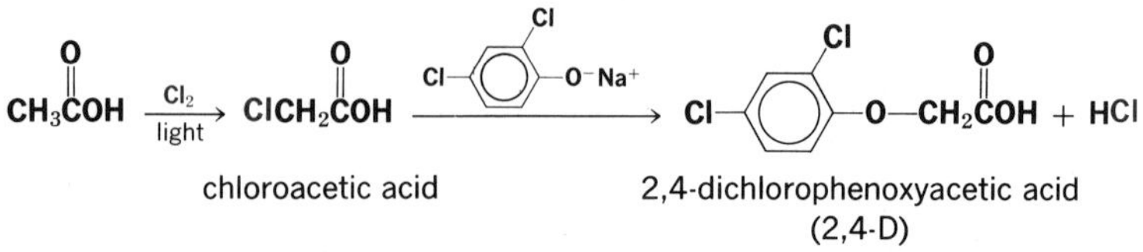

$$CH_3\overset{O}{\overset{\|}{C}}OH \xrightarrow[\text{light}]{Cl_2} \underset{\text{chloroacetic acid}}{ClCH_2\overset{O}{\overset{\|}{C}}OH} \xrightarrow{\text{2,4-}Cl_2C_6H_3O^-Na^+} \underset{\text{2,4-dichlorophenoxyacetic acid (2,4-D)}}{\text{2,4-}Cl_2C_6H_3{-}O{-}CH_2\overset{O}{\overset{\|}{C}}OH} + HCl$$

Figure 11-5. *Chloroacetic acid is used in the industrial synthesis of the weed killer 2,4-D.*

Prob. 11-2. Into what class of reactions does the formation of 2,4-D from chloroacetic acid fall? (Ans.: See Fig. 7-16.)

Benzoic acid (m.p. 121°), the simplest aromatic acid, is produced by the oxidation of toluene (Fig. 11-6). Industrially, air serves as the oxidizing agent, while in the laboratory potassium permanganate is usually used. Even

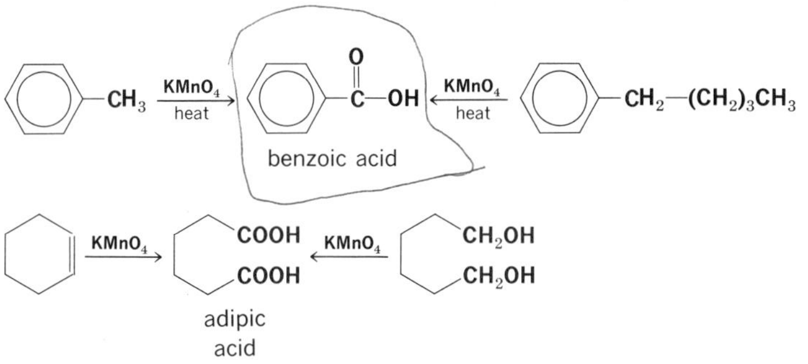

C_6H_5–CH_3 $\xrightarrow[\text{heat}]{KMnO_4}$ C_6H_5–C(=O)–OH $\xleftarrow[\text{heat}]{KMnO_4}$ C_6H_5–CH_2–$(CH_2)_3CH_3$

benzoic acid

cyclohexene $\xrightarrow{KMnO_4}$ COOH / COOH $\xleftarrow{KMnO_4}$ CH_2OH / CH_2OH

adipic acid

Figure 11-6. *Oxidation of a benzene ring with an alkyl side chain attached, no matter how long, leads to benzoic acid. Alkenes and alcohols may also be oxidized to acids.*

if the alkyl group attached to the benzene ring is longer than methyl, benzoic acid is still the product of oxidation with $KMnO_4$, and the side chain is lost. Acids may also be synthesized by the oxidation of alkenes or primary alcohols.

Acid Salts: Resolution of Enantiomers

The most characteristic reaction of a carboxylic acid is salt formation; a visual chemical test for an organic acid is to treat it with sodium bicarbonate and observe the evolution of bubbles of carbon dioxide. This serves to distinguish carboxylic acids from the only other common acids, phenols, which will form salts with sodium hydroxide but not with sodium bicarbonate (see Fig. 11-7).

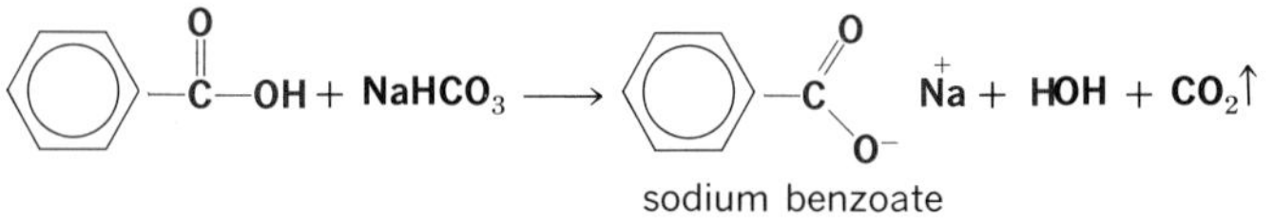

C_6H_5–C(=O)–OH + $NaHCO_3$ $\longrightarrow$ C_6H_5–C(=O)–O^- $\overset{+}{Na}$ + HOH + $CO_2\uparrow$

sodium benzoate

Figure 11-7. *The liberation of CO_2 from sodium bicarbonate serves as a visual chemical test for the carboxyl group. Sodium benzoate is widely used as a preservative, especially in fruit beverages.*

Salts formed from carboxylic acids and organic amines provide the most common means of separating a racemic mixture into its optical enantiomers, a process known as *resolution* (see p. 141). To use this method to resolve a racemic acid, one starts with a pure amine that is already optically active; such molecules are widely distributed in nature as alkaloids (morphine, strychnine, etc.) To keep our example simple, suppose that we have the pure S isomer of α-phenylethylamine and we wish to use it to resolve *R,S*-α-

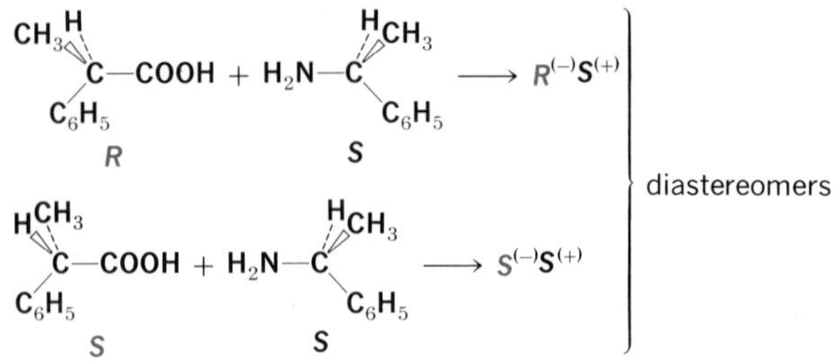

$$\underset{R}{CH_3(H)(C_6H_5)C{-}COOH} + \underset{S}{H_2N{-}C(H)(CH_3)(C_6H_5)} \longrightarrow R^{(-)}S^{(+)}$$

$$\underset{S}{H(CH_3)(C_6H_5)C{-}COOH} + \underset{S}{H_2N{-}C(H)(CH_3)(C_6H_5)} \longrightarrow S^{(-)}S^{(+)}$$

diastereomers

Figure 11-8. *Resolution of a racemic mixture of an acid can be carried out by treating the acid with an optically active amine to form diastereomeric salts that can be separated by crystallization.*

phenylpropionic acid (Fig. 11-8). Since a carboxylic acid and an amine react together to form a salt, two salts will form if we mix equimolar amounts of the S amine and the *R*,S acid, (R^-S^+ and S^-S^+). Notice that the salts are diastereomers since they both contain two chiral centers, one of which is the same in each salt (that from the amine) and one of which is different (that from the acid). Since diastereomers have different solubilities, they can usually be separated from one another by crystallization. Once the two salts have been separated, both the optically active acid and the optically active amine may be recovered in pure form. For example, suppose the pure R^-S^+ salt is added to hydrochloric acid solution. The amine will form a new, water-soluble salt with the stronger acid HCl, while the α-phenylpropionic acid will precipitate and can be removed by extraction (Fig. 11-9). In this way the amine

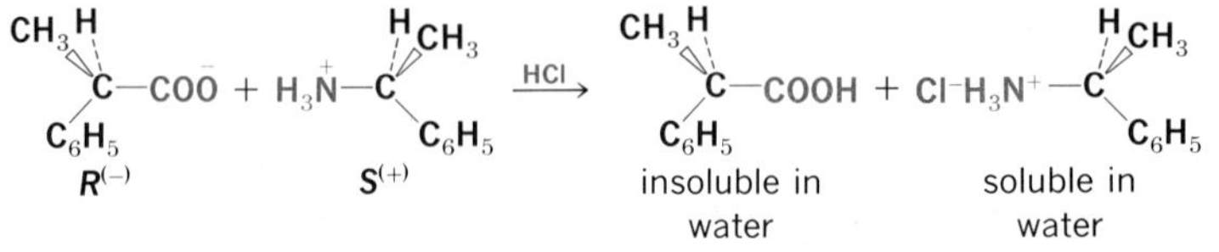

$$\underset{R^{(-)}}{CH_3(H)(C_6H_5)C{-}CO\bar{O}} + \underset{S^{(+)}}{H_3\overset{+}{N}{-}C(H)(CH_3)(C_6H_5)} \xrightarrow{HCl} \underset{\text{insoluble in water}}{CH_3(H)(C_6H_5)C{-}COOH} + \underset{\text{soluble in water}}{Cl^-H_3N^+{-}C(H)(CH_3)(C_6H_5)}$$

Figure 11-9. *Once the diastereomeric salts have been separated by crystallization, the optically active acid can be separated from the amine by acidification with mineral acid.*

can be recovered and used to resolve more acid; of course, the acid could be used to resolve more amine. If it is desired to resolve a compound that is not an acid or amine, it is often converted into a derivative which has one of these functional groups.

Prob. 11-3. Write an equation like that of Fig. 11-9 to show what would happen if the salt were added to NaOH solution. Which component would be insoluble in this solution, and which soluble?

Esters

Esters are *derivatives* of acids in which the —OH group of the acid is replaced by an —OR group derived from an alcohol. We have already discussed esters formed from alcohols and inorganic acids (Fig. 6-5). The most common way to form an ester is to heat an acid in an excess of the alcohol with a mineral acid as the catalyst (Fig. 11-10). Note carefully that *esterification* occurs by

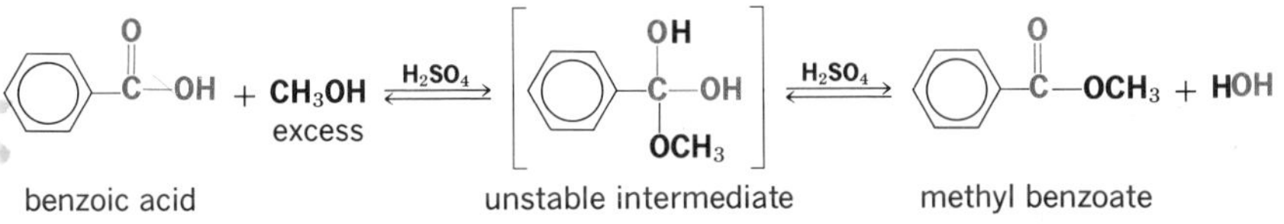

$$C_6H_5\text{—}\overset{O}{\overset{\|}{C}}\text{—OH} + \underset{\text{excess}}{CH_3OH} \xrightleftharpoons{H_2SO_4} \left[C_6H_5\text{—}\underset{OCH_3}{\overset{OH}{C}}\text{—OH} \right] \xrightleftharpoons{H_2SO_4} C_6H_5\text{—}\overset{O}{\overset{\|}{C}}\text{—OCH_3} + HOH$$

benzoic acid — unstable intermediate — methyl benzoate

Figure 11-10. *An ester is formed by heating an acid with an excess of alcohol and a mineral acid catalyst. The reaction is reversible. Note that oxygen of the water comes from the acid.*

addition of a molecule of the alcohol across the carbon–oxygen double bond of the acid, followed by the loss of water. For each molecule of water formed in the synthesis of an ester, the acid furnishes the hydroxyl group and the alcohol furnishes the proton. Esterification reactions are reversible, and the equilibrium may be shifted in either direction under appropriate conditions. By carrying out the reaction in the alcohol as a solvent, the reaction is driven to the right by the law of mass action. An ester can be hydrolyzed, i.e., split into its acid and alcohol components, by heating it in water containing an acid catalyst. In this case the law of mass action shifts the equilibrium to the left.

A better way to hydrolyze an ester is to heat it in aqueous base. The same equilibrium is established in base as in acid, since we have seen that both acids and bases catalyze addition to a carbon–oxygen double bond. But now one of the components of the equilibrium, the carboxylic acid, is completely removed by formation of a salt with the NaOH so that hydrolysis proceeds to completion (Fig. 11-11).

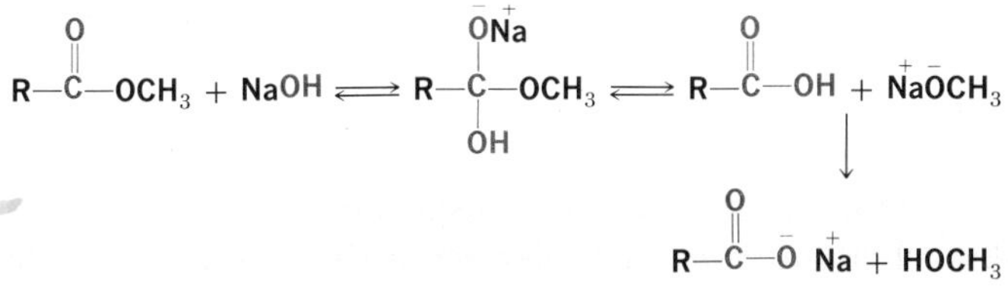

$$R\text{—}\overset{O}{\overset{\|}{C}}\text{—OCH_3} + NaOH \rightleftharpoons R\text{—}\underset{OH}{\overset{\overset{-}{O}\overset{+}{Na}}{C}}\text{—OCH_3} \rightleftharpoons R\text{—}\overset{O}{\overset{\|}{C}}\text{—OH} + \overset{+}{Na}\overset{-}{O}CH_3$$

$$\downarrow$$

$$R\text{—}\overset{O}{\overset{\|}{C}}\text{—}\overset{-}{O}\ \overset{+}{Na} + HOCH_3$$

Figure 11-11. *Esters are usually hydrolyzed in basic solution because salt formation between the base and the acid forces the reaction to completion.*

Prob. 11-4. Suppose an ester were prepared with an isotope of oxygen (^{18}O) in the position shown. Where would the isotope appear among the products of hydrolysis?

$$R-\overset{\overset{\displaystyle O}{\|}}{C}-{}^{18}OCH_3$$

Nomenclature of Esters

Esters are named after the alcohols and acids from which they are derived. For example, the ester formed from methyl alcohol and acetic acid is *methyl acetate*. In naming an ester, then, the first step is to determine its components and name them. Some examples of this procedure are given in Fig. 11-12.

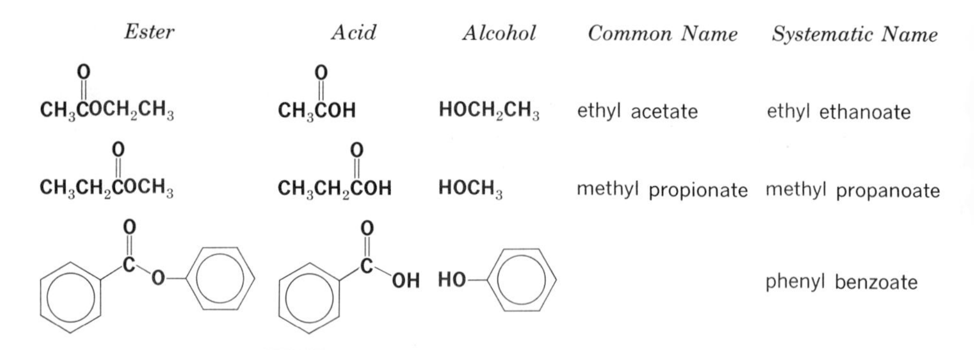

Ester	*Acid*	*Alcohol*	*Common Name*	*Systematic Name*
$CH_3\overset{O}{\overset{\|}{C}}OCH_2CH_3$	$CH_3\overset{O}{\overset{\|}{C}}OH$	$HOCH_2CH_3$	ethyl acetate	ethyl ethanoate
$CH_3CH_2\overset{O}{\overset{\|}{C}}OCH_3$	$CH_3CH_2\overset{O}{\overset{\|}{C}}OH$	$HOCH_3$	methyl propionate	methyl propanoate
$C_6H_5-\overset{O}{\overset{\|}{C}}-O-C_6H_5$	$C_6H_5-\overset{O}{\overset{\|}{C}}-OH$	$HO-C_6H_5$		phenyl benzoate

Figure 11-12. *In naming an ester, its constituent acid and alcohol are identified and named; these names are combined, adding the ending* -ate.

Esters are neutral substances, insoluble in water and with generally delightful odors that give fruits their characteristic aromas. The names and structures of a number of esters from fruits are given in Fig. 11-13.

Polyesters

Suppose an esterification reaction is carried out between an acid that contains two carboxyl groups and an alcohol with two hydroxyl groups. The result will be a polymer, a *polyester*. Since each time an ester bond is formed, a molecule of water is produced, polymerization of this type is called *conden-*

	Name	Odor
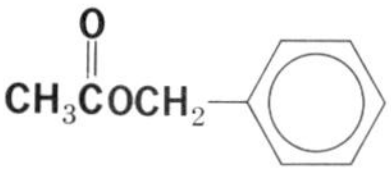 $CH_3C(=O)OCH_2-C_6H_5$	benzyl acetate	peach
$CH_3CH_2CH_2C(=O)OCH_2CH_3$	ethyl *n*-butyrate	apple
$CH_3C(=O)O(CH_2)_7CH_3$	*n*-octyl acetate	orange
$CH_3C(=O)OCH_2CH{=}C(CH_3)_2$	3-methyl-2-butenyl acetate	Juicy Fruit

Figure 11-13. *Some esters with familiar odors. Juicy Fruit does not occur naturally.*

sation polymerization to distinguish it from *addition polymerization* of alkenes, which we studied in Chapter 4. Dacron, the best known example of a polyester, is formed by the loss of water between a dicarboxylic acid, terephthalic acid, and a diol, ethylene glycol (Fig. 11-14).

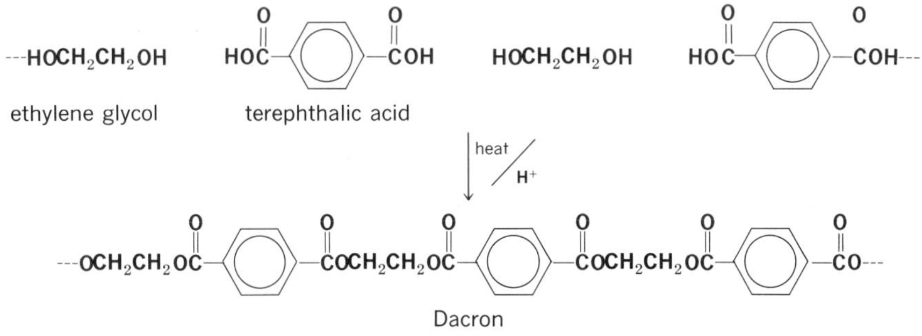

$$\text{---}HOCH_2CH_2OH \quad HOC(=O)-C_6H_4-C(=O)OH \quad HOCH_2CH_2OH \quad HOC(=O)-C_6H_4-C(=O)OH\text{---}$$

ethylene glycol   terephthalic acid

$$\downarrow \text{heat}/H^+$$

$$\text{---}OCH_2CH_2OC(=O)-C_6H_4-C(=O)OCH_2CH_2OC(=O)-C_6H_4-C(=O)OCH_2CH_2OC(=O)-C_6H_4-C(=O)O\text{---}$$

Dacron

Figure 11-14. *Dacron is a polyester, formed by the condensation polymerization of a diol and a diacid.*

Dacron is a linear polymer that usually contains about 80 units per chain, a solid which can be melted without decomposition. The molten polymer is forced through tiny holes in a spinneret and cools and solidifies in the form of thin fibers. These fibers are then stretched to about five times their original length. During the stretching process the individual polymer molecules become aligned so that they all lie parallel to the fiber axis; this allows the molecules to pack closely together, making the yarn stronger. Closer packing of the polymer molecules also makes it more difficult for other molecules to penetrate the fiber, and drawn Dacron is more chemically resistant and absorbs less water than the undrawn polymer. For the same reason, it is difficult to dye Dacron, since the dye molecules cannot find room between the polyester chains. In the form of a film, this same polyester, polyethylene terephthalate, is known commercially as Mylar. In Europe, this polyester is called Terylene, and the fibers Fortrel and Kodel made in the United States have similar structures. Still other polyesters are prepared from aliphatic diacids.

Formation of a polyester from a triol instead of a diol can give rise to a polymer that is not linear like Dacron but rather highly cross-linked and three-dimensional. Common members of this class are the *glyptals,* made from *gly*cerol and *phthal*ic acid, the isomer of terephthalic acid in which the carboxylic acid groups are *ortho* instead of *para* to one another.

Prob. 11-5. Draw a portion of the structure of a glyptal polymer from phthalic acid and glycerol.

Acid Halides and Anhydrides

An acid anhydride is formally derived from the loss of water between two carboxylic acid molecules. The most important acid anhydride is *acetic anhydride,* b.p. 140°; nearly half of the acetic acid manufactured yearly is converted into acetic anhydride. Acetic anhydride reacts readily and exothermically with water to give acetic acid; the equilibrium in this reaction lies so far to the right that very special industrial conditions are required to form acetic anhydride from acetic acid by loss of water (Fig. 11-15). This

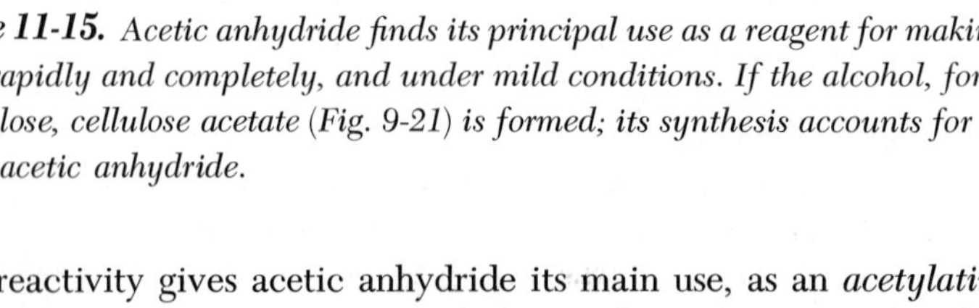

$$\underset{\text{acetic anhydride}}{CH_3{-}\overset{\overset{O}{\|}}{C}{-}O{-}\overset{\overset{O}{\|}}{C}{-}CH_3} + H_2O \xrightarrow[\text{required}]{\text{no catalyst}} CH_3{-}\overset{\overset{O}{\|}}{C}{-}OH + HO{-}\overset{\overset{O}{\|}}{C}{-}CH_3$$

$$+ \underset{\text{alcohol}}{ROH} \xrightarrow[\text{required}]{\text{no catalyst}} \underset{\text{ester}}{CH_3{-}\overset{\overset{O}{\|}}{C}{-}OR} + HO{-}\overset{\overset{O}{\|}}{C}{-}CH_3$$

Figure 11-15. *Acetic anhydride finds its principal use as a reagent for making acetate esters rapidly and completely, and under mild conditions. If the alcohol, for example, is cellulose, cellulose acetate (Fig. 9-21) is formed; its synthesis accounts for the major use of acetic anhydride.*

great reactivity gives acetic anhydride its main use, as an *acetylating agent* to make acetic acid esters of alcohols rapidly and completely under mild conditions. Thus while esterification of the hydroxyl groups of cellulose by acetic acid would require prolonged heating with acetic acid and a mineral

phenol + CO_2 $\xrightarrow[\text{heat, pressure}]{NaOH}$ salicylic acid

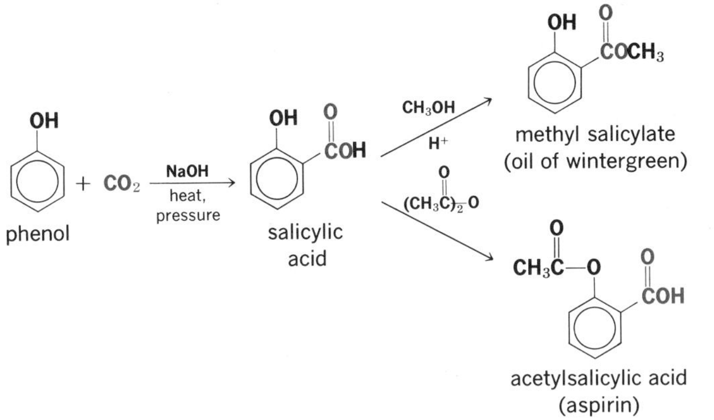

salicylic acid $\xrightarrow[H^+]{CH_3OH}$ methyl salicylate (oil of wintergreen)

salicylic acid $\xrightarrow{(CH_3CO)_2O}$ acetylsalicylic acid (aspirin)

Figure 11-16. *Salicylic acid, methyl salicylate, and acetylsalicylic acid are all important medicinal compounds. The parent acid is prepared from phenol and carbon dioxide.*

acid catalyst, conditions that cause extensive degradation of the cellulose molecule, acetic anhydride reacts quickly and in high yield at room temperature.

A great deal of acetic anhydride is also used in the production of *aspirin,* one of a number of useful derivatives of *salicylic acid,* *o*-hydroxybenzoic acid (Fig. 11-16). Recent research has suggested that aspirin functions by inhibiting the formation of prostaglandins, extremely potent hormonelike substances whose role in the body is not yet understood. The beneficial effects of esters of salicylic acid, the salicylates, have been known since antiquity: Leaves that contain sizable quantities of these pain- and fever-relieving compounds were prescribed by ancient physicians. Salicyclic acid is the active ingredient of the salicylates and is itself an analgesic or pain killer; it may be taken orally, but it is a strong enough acid to irritate the stomach. Aspirin, its ester with acetic acid, is less acidic and less irritating. The contents of the stomach itself are acidic, and aspirin passes through this organ unchanged; however, under the alkaline conditions in the intestines, it is hydrolyzed to sodium salicylate, which is absorbed through the intestinal walls. Methyl salicylate can also be taken internally or absorbed through the skin and thus finds much use in liniments and other topical applications.

Acid halides are the most reactive of the common derivatives of organic acids. They may be considered anhydrides between an organic acid and one of the hydrogen halides. The most common acid halide is *acetyl chloride;* like acetic anhydride, acetyl chloride reacts spontaneously with an alcohol to form an ester. Acid chlorides can be prepared rapidly and in nearly quantitative yield from an acid (Fig. 11-17). They are most commonly used in the laboratory to prepare esters or amides.

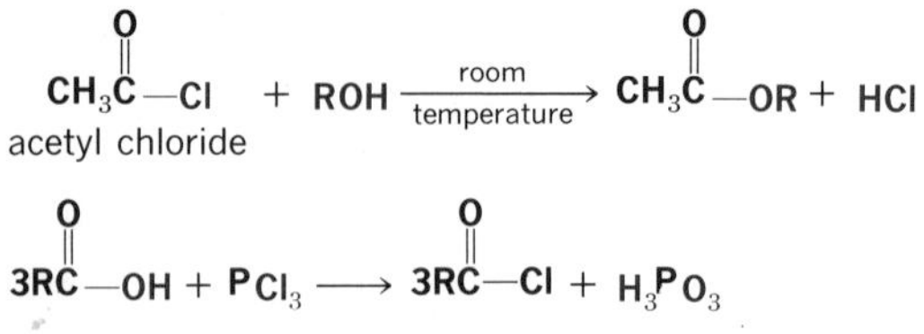

$$\underset{\text{acetyl chloride}}{CH_3\overset{O}{\overset{\|}{C}}-Cl} + ROH \xrightarrow[\text{temperature}]{\text{room}} CH_3\overset{O}{\overset{\|}{C}}-OR + HCl$$

$$3R\overset{O}{\overset{\|}{C}}-OH + PCl_3 \longrightarrow 3R\overset{O}{\overset{\|}{C}}-Cl + H_3PO_3$$

Figure 11-17. *Acid halides, especially acid chlorides, are widely used in the laboratory for the preparation of esters and other derivatives of acids. An acid may be converted in high yield to the acid chloride by treatment with phosphorus trichloride or thionyl chloride* ($SOCl_2$).

Amides

Because amides constitute the backbone of protein molecules, their chemistry is of extreme importance (see Chapter 13). Although formally derived from acids and ammonia or amines by loss of water, amides are most commonly made in the laboratory by the reaction of acid chlorides or anhydrides (Fig. 11-18). Esters will also react with amines to give amides, but higher temperatures are usually required. A carboxylic acid reacts with an amine to give a salt. If such an ammonium salt is heated above its melting point, water is eliminated and an amide results.

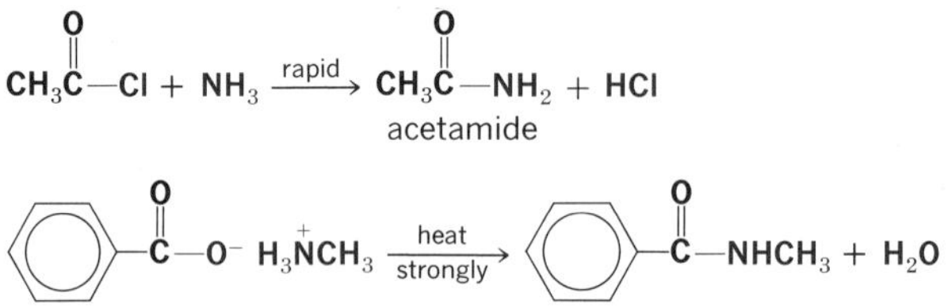

$$CH_3\overset{O}{\overset{\|}{C}}-Cl + NH_3 \xrightarrow{\text{rapid}} \underset{\text{acetamide}}{CH_3\overset{O}{\overset{\|}{C}}-NH_2} + HCl$$

$$C_6H_5-\overset{O}{\overset{\|}{C}}-O^- \ H_3\overset{+}{N}CH_3 \xrightarrow[\text{strongly}]{\text{heat}} C_6H_5-\overset{O}{\overset{\|}{C}}-NHCH_3 + H_2O$$

Figure 11-18. *Two methods for the synthesis of an amide. An acid halide (or anhydride) reacts rapidly with ammonia or an amine. An ammonium salt requires strong heating before it can be converted to an amide by loss of water.*

Prob. 11-6. Write an equation for the reaction of acetic anhydride with aniline to give acetanilide. Why will a tertiary amine, e.g., trimethyl amine, not form an amide?

Among condensation polymers, the polyamide nylon is undoubtedly the most important. A number of different nylons have become commercially

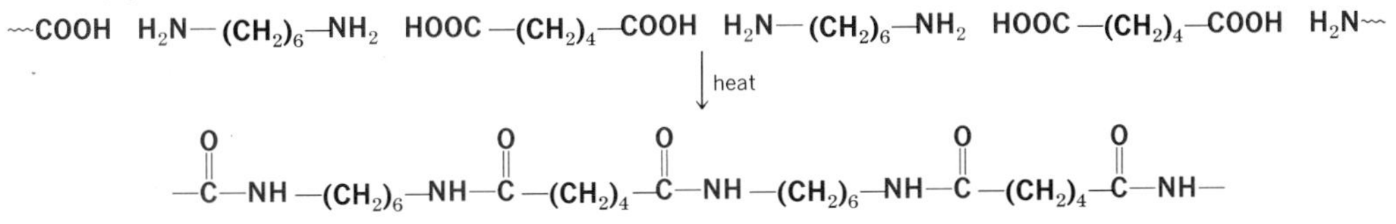

$$\sim COOH\ H_2N-(CH_2)_6-NH_2\ HOOC-(CH_2)_4-COOH\ H_2N-(CH_2)_6-NH_2\ HOOC-(CH_2)_4-COOH\ H_2N\sim$$

$$\downarrow \text{heat}$$

$$-\overset{O}{\overset{\|}{C}}-NH-(CH_2)_6-NH-\overset{O}{\overset{\|}{C}}-(CH_2)_4-\overset{O}{\overset{\|}{C}}-NH-(CH_2)_6-NH-\overset{O}{\overset{\|}{C}}-(CH_2)_4-\overset{O}{\overset{\|}{C}}-NH-$$

Figure 11-19. *Nylon 66 is a polyamide formed from a six-carbon diacid and a six-carbon diamine. Although other nylons are known, and several are produced commercially, Nylon 66 is especially cheap because both components can be produced starting with benzene.*

important, but Nylon 66 is produced in the largest amount. This polymer results from loss of water between the six-carbon diacid *adipic acid* and the six-carbon diamine *hexamethylenediamine* (Fig. 11-19); the number 66 is applied because there are six carbons in each component.

Reactions of Amides

In contrast to ammonia and the amines, the nitrogen atom in amides is not basic, and amides do not form salts in aqueous acid. As is so often the case, this modification of basicity finds its explanation in resonance. The basic properties of ammonia and its derivatives arise from the unshared electron pair on nitrogen, which can combine with a proton (Fig. 10-1). In an amide this electron pair is delocalized by resonance with the carbonyl group, which lowers its energy and makes it less available for reaction (Fig. 11-20). Amides

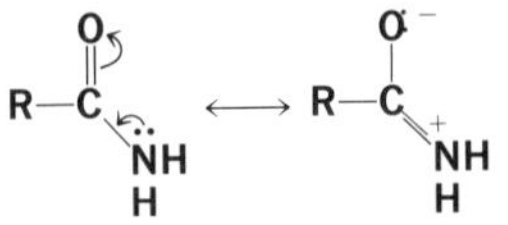

$$R-\overset{O}{\overset{\|}{C}}-\ddot{N}H-H \longleftrightarrow R-\overset{O^-}{\overset{|}{C}}=\overset{+}{N}H-H$$

Figure 11-20. *In an amide, the electron pair on nitrogen is shared by the oxygen of the carbonyl, as these resonance structures show. Amides are not basic and do not form salts in aqueous solution.*

are quite stable, compared to other acid derivatives, and vigorous conditions are required for their hydrolysis. The usual conditions are prolonged heating with concentrated hydrochloric acid. Amides may also be reduced, either catalytically with hydrogen (high temperatures and pressures are required) or with chemical reducing agents (lithium aluminum hydride is best) to give amines (Fig. 11-21).

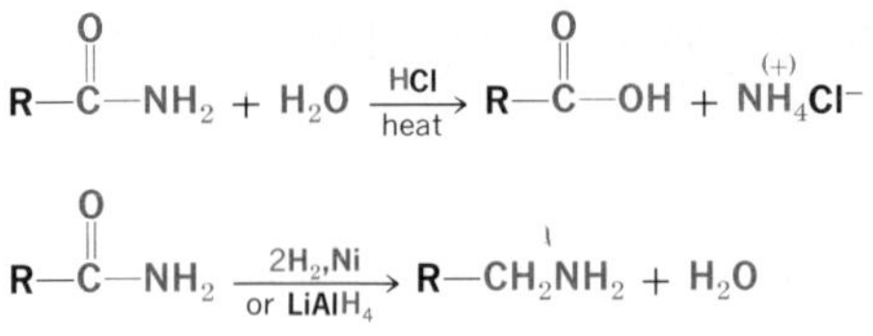

$$R-\overset{\overset{O}{\|}}{C}-NH_2 + H_2O \xrightarrow[heat]{HCl} R-\overset{\overset{O}{\|}}{C}-OH + \overset{(+)}{NH_4}Cl^-$$

$$R-\overset{\overset{O}{\|}}{C}-NH_2 \xrightarrow[\text{or } LiAlH_4]{2H_2,Ni} R-CH_2NH_2 + H_2O$$

Figure 11-21. *The usual way to hydrolyze an amide is to heat it with strong hydrochloric acid. Amides also serve as a useful source of amines by reduction.*

Urea and the Barbiturates

Urea, the diamide of carbonic acid, is an end product of the metabolism of proteins and is also widely used industrially in fertilizer and for the production of urea–formaldehyde resins. Urea is produced by heating ammonia and carbon dioxide together under pressure (Fig. 11-22). The reaction is slowly reversible so that urea makes a long-lasting fertilizer, gradually releasing its ammonia to the soil.

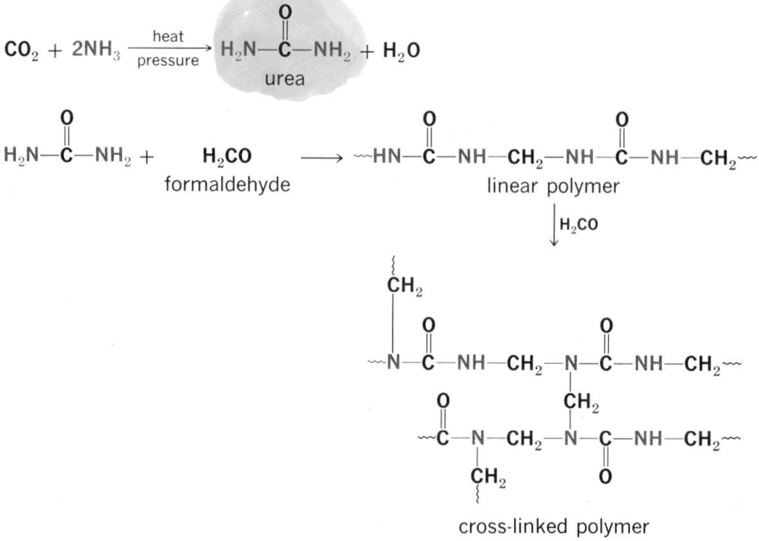

$$CO_2 + 2NH_3 \xrightarrow[pressure]{heat} \underset{\text{urea}}{H_2N-\overset{\overset{O}{\|}}{C}-NH_2} + H_2O$$

$$H_2N-\overset{\overset{O}{\|}}{C}-NH_2 + \underset{\text{formaldehyde}}{H_2CO} \longrightarrow \underset{\text{linear polymer}}{\sim HN-\overset{\overset{O}{\|}}{C}-NH-CH_2-NH-\overset{\overset{O}{\|}}{C}-NH-CH_2\sim}$$

$$\downarrow H_2CO$$

cross-linked polymer

Figure 11-22. *Carbon dioxide and ammonia react to form urea, which is useful as a fertilizer and which reacts with formaldehyde to form a hard, high molecular weight polymer.*

Barbiturates result from the reaction of urea with esters of substituted malonic acids, the dicarboxylic acid containing three carbons. Barbituric acid itself was first synthesized from diethyl malonate and urea in 1863, but it was not until 1903 that the barbiturates were introduced into medicine.

Hundreds of derivatives of barbituric acid have been prepared since that time; the structures of some which have gained the most medical importance are presented in Fig. 11-23. All of these drugs are habit-forming and available only on prescription.

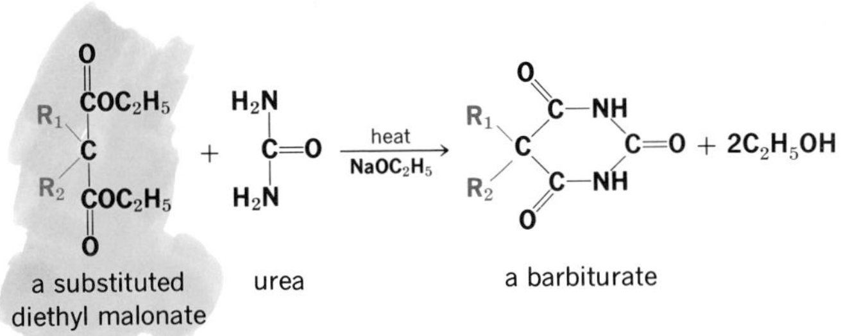

$$R_1R_2C(COOC_2H_5)_2 + (H_2N)_2C{=}O \xrightarrow[NaOC_2H_5]{heat} R_1R_2C(CONH)_2C{=}O + 2C_2H_5OH$$

a substituted diethyl malonate   urea   a barbiturate

R_1	R_2	*Barbiturate*	*Duration of Action*
CH_3CH_2-	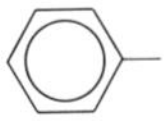 C_6H_5-	phenobarbital	long
CH_3CH_2-	$CH_3CH(CH_3)CH_2CH_2-$	amytal	intermediate
CH_3CH_2-	$CH_3CH_2CH_2CH(CH_3)-$	nembutal	intermediate
$CH_2{=}CHCH_2-$	$CH_3CH_2CH_2CH(CH_3)-$	seconal	short

Figure 11-23. *Barbiturates are derivatives of barbituric acid ($R_1 = R_2 = H$), prepared by reaction of urea with substituted diethyl malonates.*

Reactions of Acid Derivatives

Many of the reactions of acid derivatives involve their interconversion. In learning these interconversions, it is helpful to remember their order of reactivity, which is summarized in Fig. 11-24. Any derivative may be converted to any other below it by reaction with the appropriate reagent on the left. In order to go up the scale, the free acid must be prepared first. The most reactive derivative, an acid halide, may be converted into an anhydride, an ester, or an amide in one step, simply by reaction with an acid salt, an alcohol, or ammonia. An anhydride readily forms an ester or amide but cannot be converted directly into an acid halide in the laboratory. An ester gives an amide with ammonia but cannot be converted directly into an anhydride or acid halide; it must first be hydrolyzed to an acid.

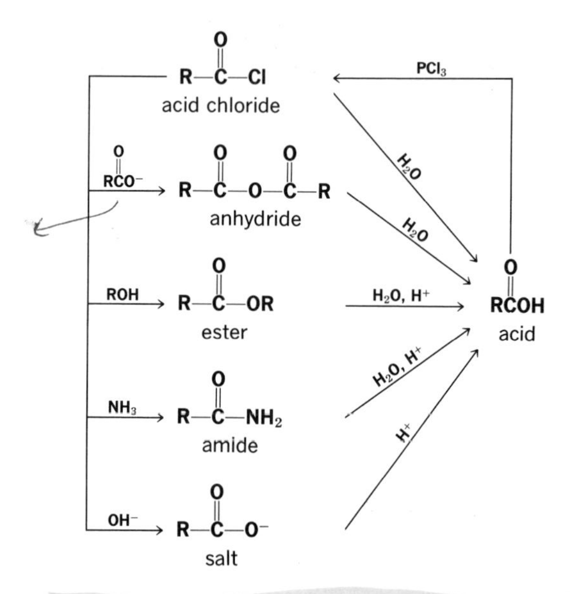

Figure 11-24. *In this chart the common derivatives of acids are arranged in order of decreasing reactivity.*

Prob. 11-7. Using Fig. 11-24, show how you would convert (a) benzoic anhydride to *N*-methylbenzamide and (b) ethyl propionate to propionic anhydride.

Reactions with Grignard Reagents

Esters react with two moles of a Grignard reagent to give tertiary alcohols (Fig. 11-25).

The addition of one mole of Grignard reagent to the carbon–oxygen double bond gives an unstable intermediate that breaks down to a ketone. A second mole of reagent then adds to the ketone, giving a tertiary alcohol,

$$\mathrm{R{-}\overset{\overset{O}{\|}}{C}{-}OR'} + 2\mathrm{R'MgX} \longrightarrow \xrightarrow{H^+} \mathrm{R{-}\underset{\underset{R''}{|}}{\overset{\overset{OH}{|}}{C}}{-}R''}$$

Figure 11-25. *Tertiary alcohols are formed from esters by reaction with a Grignard reagent.*

in which at least two of the groups attached to the carbinol carbon are the same. An exactly analogous series of reactions occurs with acid halides and anhydrides (Fig. 11-26).

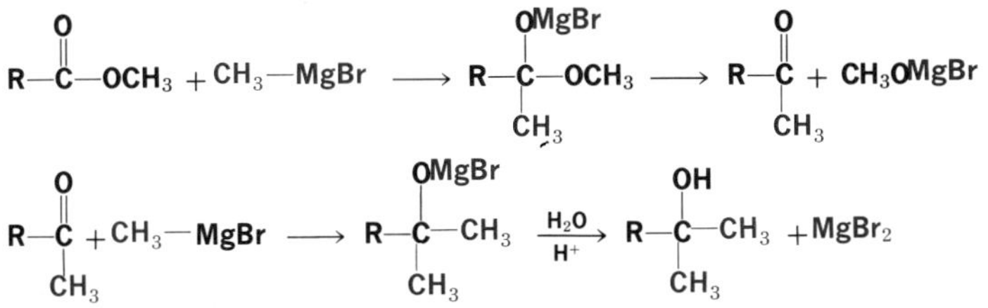

$$R-\overset{O}{\overset{\|}{C}}-OCH_3 + CH_3-MgBr \longrightarrow R-\underset{CH_3}{\underset{|}{\overset{OMgBr}{\overset{|}{C}}}}-OCH_3 \longrightarrow R-\underset{CH_3}{\underset{|}{\overset{O}{\overset{\|}{C}}}} + CH_3OMgBr$$

$$R-\underset{CH_3}{\underset{|}{\overset{O}{\overset{\|}{C}}}} + CH_3-MgBr \longrightarrow R-\underset{CH_3}{\underset{|}{\overset{OMgBr}{\overset{|}{C}}}}-CH_3 \xrightarrow[H^+]{H_2O} R-\underset{CH_3}{\underset{|}{\overset{OH}{\overset{|}{C}}}}-CH_3 + MgBr_2$$

Figure 11-26. *An ester (or an acid halide or an anhydride) reacts first with a Grignard reagent to form a ketone, which reacts further to give an alcohol.*

Ordinarily, it is not possible to make and isolate a ketone through reaction of an ester or other acid derivative with only one mole of Grignard reagent. However, other organometallic reagents can carry out this useful conversion (Fig. 11-27). An organocadmium compound, for instance, formed

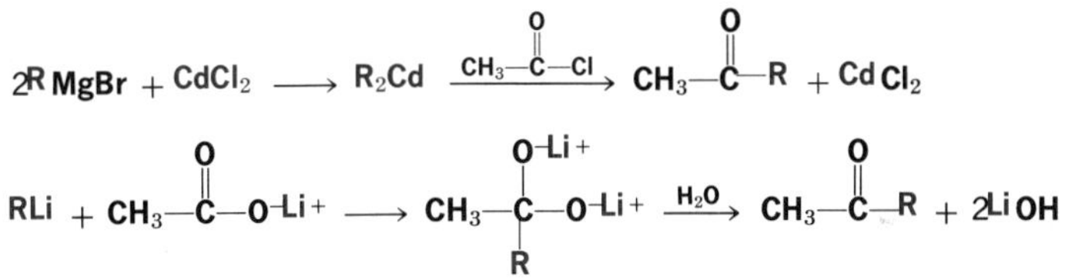

$$2RMgBr + CdCl_2 \longrightarrow R_2Cd \xrightarrow{CH_3-\overset{O}{\overset{\|}{C}}-Cl} CH_3-\overset{O}{\overset{\|}{C}}-R + CdCl_2$$

$$RLi + CH_3-\overset{O}{\overset{\|}{C}}-O^-Li^+ \longrightarrow CH_3-\underset{R}{\underset{|}{\overset{O^-Li^+}{\overset{|}{C}}}}-O^-Li^+ \xrightarrow{H_2O} CH_3-\overset{O}{\overset{\|}{C}}-R + 2LiOH$$

Figure 11-27. *Two methods for the synthesis of ketones from acid derivatives.*

from a Grignard reagent by reaction with cadmium chloride, yields a ketone when treated with an acid chloride. Furthermore, an organolithium reagent is able to react with the salt of an acid to form, after hydrolysis, a ketone. A Grignard reagent is not reactive enough to react under ordinary conditions with the already negatively charged carboxylate ion.

Nitriles

Nitriles do not contain carbonyl groups like the other acid derivatives of this chapter. Nevertheless, they can be prepared from acids, and it is convenient at this point to consider their reactions, which show many similarities to those of esters. Nitriles are most conveniently formed by the reaction of halides with cyanide ion or of diazonium salts with cuprous cyanide. Their hydrolysis

$CH_3C{\equiv}N$ acetonitrile

$CH_3CH_2C{\equiv}N$ propionitrile

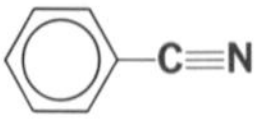

$C_6H_5{-}C{\equiv}N$ benzonitrile

$$RC{\equiv}N \underset{POCl_3}{\overset{H_2O,\ H^+}{\rightleftarrows}} R\overset{O}{\overset{\|}{C}}NH_2 \xrightarrow[H^+]{H_2O} R\overset{O}{\overset{\|}{C}}OH$$

$$R{-}C{\equiv}N + CH_3MgBr \longrightarrow R{-}\underset{CH_3}{\underset{|}{C}}{=}NMgBr \xrightarrow{H_2O} R{-}\underset{CH_3}{\underset{|}{C}}{=}NH \xrightarrow{H_2O} R{-}\overset{O}{\overset{\|}{C}}{-}CH_3$$

$$R{-}C{\equiv}N \xrightarrow[\text{or } LiAlH_4]{H_2/Pd} R{-}CH_2NH_2$$

Figure 11-28. *In addition to their usefulness as intermediates in the synthesis of acids and amides, nitriles give good yields of ketones when treated with Grignard reagents; they may also be reduced to primary amines.*

to acids, which has been discussed before, may be stopped at the amide stage after the addition of one molecule of water (Fig. 11-28).

PROBLEMS

1. Identify the alcohol and acid from which each of the following esters are made:

(**a**) ethyl butyrate
(**b**) methyl benzoate
(**c**) *n*-propyl propionate
(**d**) *t*-butyl *p*-nitrobenzoate
(**e**) $(CH_3)_2CHOOCCH_3$
(**f**) $CH_3(CH_2)_3COOCH_3$
(**g**) $HCOOCH_2CH_3$
(**h**) $C_6H_5COO(CH_2)_2CH_3$

2. Name the following compounds:

(**a**) Parts (e) to (h) of Prob. 1.
(**b**) $C_6H_5{-}\overset{O}{\overset{\|}{C}}{-}O{-}\overset{O}{\overset{\|}{C}}{-}C_6H_5$
(**c**) $C_6H_5{-}\overset{O}{\overset{\|}{C}}{-}NH_2$
(**d**) $(H_2N)_2C{=}O$
(**e**) $Cl{-}C_6H_4{-}C{\equiv}N$
(**f**) $CH_3COO^-Na^+$
(**g**) $CH_3CH_2\overset{O}{\overset{\|}{C}}Br$

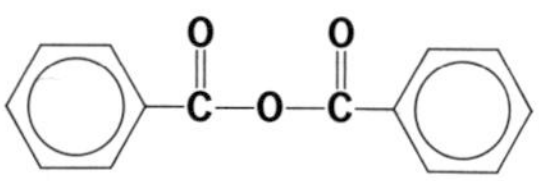

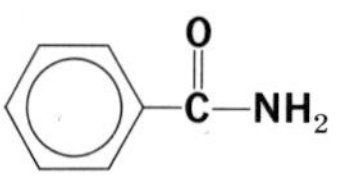

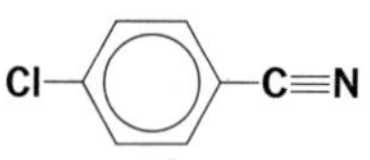

3. Complete the following reactions:

(**a**) $C_6H_5{-}COOCH_3 + NaOH \longrightarrow$

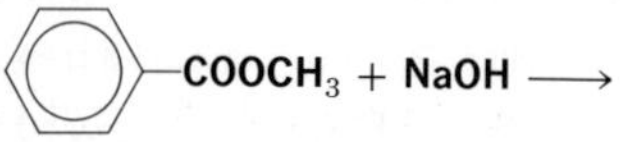

(b) $CH_3CH_2COOH + NaOH \longrightarrow$

(c) $C_6H_5-\overset{O}{\overset{\|}{C}}-O-\overset{O}{\overset{\|}{C}}-C_6H_5 + NH_3 \longrightarrow$

(d) $CH_3COOH + PCl_3 \longrightarrow$

(e) $(CH_3)_2CHCOCl + CH_3OH \longrightarrow$

(f) $CH_3-C_6H_4-CH_2CH_3 + KMnO_4 \longrightarrow$

(g) $Cl-\overset{O}{\overset{\|}{C}}(CH_2)_4\overset{O}{\overset{\|}{C}}-Cl + H_2N(CH_2)_6NH_2 \longrightarrow$

(h) $CH_3COOCH_3 + CH_3MgBr(2\ moles) \xrightarrow{H_3O^+}$

(i) $CH_3C{\equiv}N + CH_3MgBr(1\ mole) \xrightarrow{H_3O^+}$

(j) $N{\equiv}C-C_6H_4-C{\equiv}N + H_3O^+ \xrightarrow{heat}$

(k) $C_6H_5-MgBr + CO_2 \longrightarrow$

4. Explain why esters are usually hydrolyzed using sodium hydroxide, rather than hydrochloric acid, as a catalyst.
5. How could the following compounds be prepared starting from alcohols of three carbons or less, and benzene as the only organic reagent. (NaCN and CO_2 are considered inorganic.)
   (a) $CH_3CH_2CH_2COOH$

   (b) $CH_3COOCH_2CH_3$

   (c) $C_6H_5-\overset{O}{\overset{\|}{C}}-NH_2$

   (d) $(CH_3)_3CCOOH$

   (e) C_6H_5-CN

   (f) $HOOC-C_6H_4-COOH$
6. A student tried to make the Grignard reagent of *p*-bromobenzoic acid and to react it with CO_2 to get the dicarboxylic acid [Prob. 5(f)]. Instead he got benzoic acid. Explain. (No, the ether was not wet.)
7. Would you expect to get a useful fiber-forming polymer like nylon if you used adipic acid and 1,3,5-triaminopentane as your components? Explain.

8. As we shall see in the next chapter, long-chain aliphatic acids with an even number of carbon atoms occur plentifully in nature, but those with an odd number of carbons are rare. Devise a general method to increase the length of an aliphatic acid by one carbon: $R{-}COOH \longrightarrow R{-}CH_2COOH$. Several steps will be required.
9. How could you distinguish among the following compounds by simple visual chemical tests? Tell what you would do and see.
   (a) $CH_3COCH_2CH_2CH_3$, $CH_3CH_2CH_2CH_2CHO$, and $CH_3CH_2CH_2CH_2COOH$
   (b) cyclohexane, cyclohexanol, cyclohexylamine, and methyl cyclohexanecarboxylate
   (c) $C_6H_5{-}NH_2$, $C_6H_5{-}\overset{O}{\overset{\|}{C}}{-}NH_2$, $C_6H_5{-}\overset{O}{\overset{\|}{C}}OH$, and $C_6H_5{-}OH$
10. Show how the following compounds may be prepared using the Grignard synthesis.
    (a) $(C_6H_5)_3COH$ (b) $C_6H_5COCH_3$
    (c) *p*-$CH_3C_6H_4COOH$ (d) $C_6H_5\underset{OH}{\underset{|}{C}}(CH_3)C_2H_5$

---

short

12
fats, oils and detergents

Fats and oils are esters of long-chain carboxylic acids (the so-called *fatty acids*) with the triol glycerine, and are members of the general class of naturally occurring substances known as *lipids*, a class that also includes the waxes. They occur in both plants and animals. Large amounts of edible oils are obtained from cottonseeds and soybeans, as well as from milk and meats; even some fruits, especially olives and avocados, contain large amounts of oil.

Animal fats are generally obtained by *rendering*, a process involving slow heating, while seed oils are produced by pressing or solvent extraction. Oil presses are known to have been used in ancient Egypt, and the commercial production of oils has been going on for centuries. In addition to their importance as food, fats and oils are also used as lubricants, in paints and varnishes (as drying oils) and as raw materials for the manufacture of soap.

Fats and Oils

All naturally occurring fats and oils are *triglycerides*, i.e., triesters of glycerol (glycerine). A simple example is glyceryl tripalmitate, found in palm oil, the triester of glycerol and the 16-carbon, straight-chain, carboxylic acid palmitic acid (Fig. 12-1). Hydrolysis of the glyceride produces glycerol and the free fatty acid.

A large number of different fatty acids occur naturally as glyceryl esters in fats and oils. Naturally occurring fatty acids have a number of structural features in common. With rare exceptions their carbon skeletons are unbranched, with an even number of carbon atoms. We shall see why this is so when we consider their biosynthesis in Chapter 16. If they contain double bonds, and many of them do, these are always in the *cis* configuration. Structures for a number of common fatty acids are given in Figure 12-2.

The difference between oils and fats is merely one of melting point;

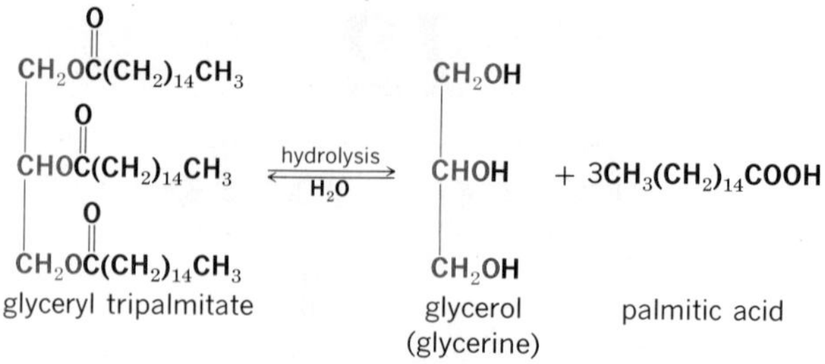

$$\begin{array}{l} \overset{O}{\underset{|}{CH_2OC(CH_2)_{14}CH_3}} \\ \overset{O}{\underset{|}{CHOC(CH_2)_{14}CH_3}} \\ \overset{O}{CH_2OC(CH_2)_{14}CH_3} \end{array} \underset{H_2O}{\overset{\text{hydrolysis}}{\rightleftharpoons}} \begin{array}{l} CH_2OH \\ | \\ CHOH \\ | \\ CH_2OH \end{array} + 3CH_3(CH_2)_{14}COOH$$

glyceryl tripalmitate — glycerol (glycerine) — palmitic acid

Figure 12-1. *Fats and oils are triesters of glycerol with long-chain carboxylic acids. Glyceryl tripalmitate is a* simple triglyceride *because all three of the ester groups are formed from the same acid. Most natural fats and oils are* mixed triglycerides, *containing two or three different acids.*

an oil is a liquid fat. Glycerides of the shorter fatty acids tend, naturally, to have lower melting points, but a more important observation is that oils usually contain a much greater proportion of glycerides of the unsaturated fatty acids than do fats. For example, corn and olive oil contain about 90% oleic and linoleic esters, while beef fat and butter contain nearly 50% myristic, palmitic, and stearic esters. Molecular models make this difference in melting point understandable, since the saturated chains are able to adopt a linear conformation that packs easily into a crystal, while the unsaturated esters, with their

Fatty acid	*Name*	*Source*
$CH_3(CH_2)_{10}COOH$	lauric acid	coconuts
$CH_3(CH_2)_{12}COOH$	myristic acid	widely found
$CH_3(CH_2)_{16}COOH$	stearic acid	animal fats
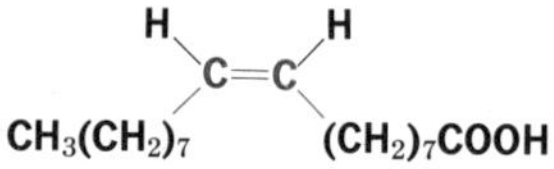 $CH_3(CH_2)_7CH{=}CH(CH_2)_7COOH$ (cis)	oleic acid	corn oil
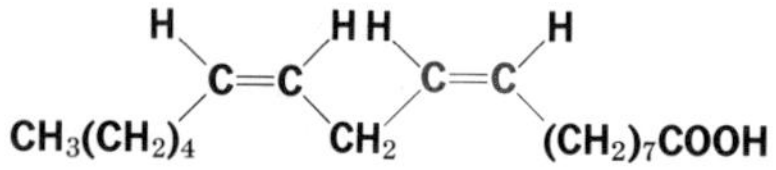 $CH_3(CH_2)_4CH{=}CHCH_2CH{=}CH(CH_2)_7COOH$ (cis, cis)	linoleic acid	cottonseed and soybean oils
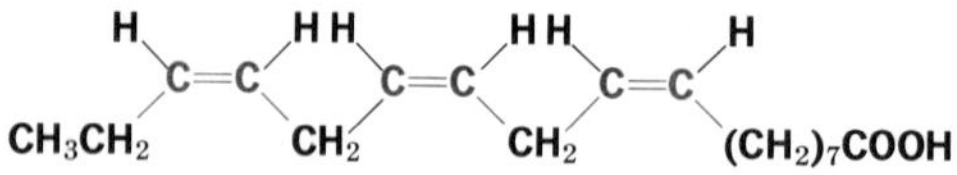 $CH_3CH_2CH{=}CHCH_2CH{=}CHCH_2CH{=}CH(CH_2)_7COOH$ (all cis)	linolenic acid	linseed oil
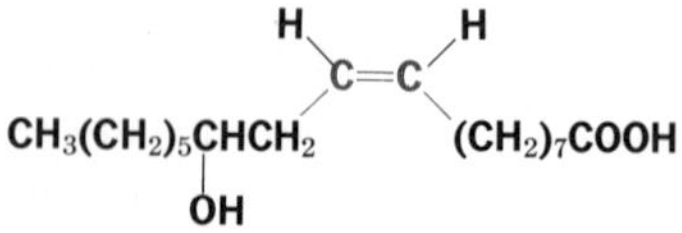 $CH_3(CH_2)_5CH(OH)CH_2CH{=}CH(CH_2)_7COOH$ (cis)	ricinoleic acid	castor oil

Figure 12-2. *The names and structures of some fatty acids. Note the* cis *configuration at each double bond, and the fact that each acid has an even number of carbon atoms.*

cis double bonds, are U-shaped and do not pack together easily. Since most people of Western cultures prefer fats to oils, a major industry has developed for the conversion of oils into fats by reduction of the double bonds with hydrogen (Fig. 12-3). Oleomargarine is produced in this way from cottonseed,

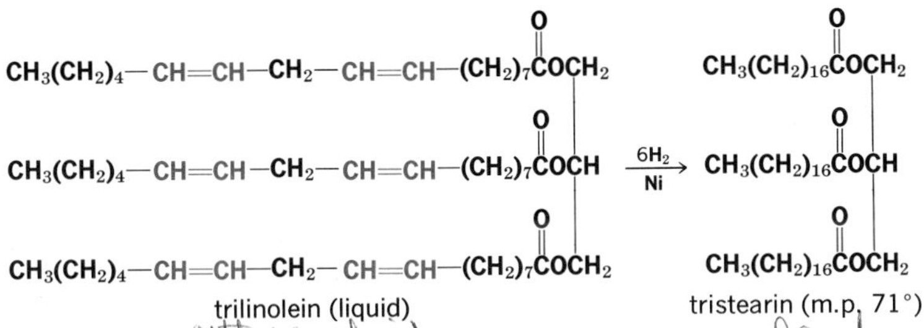

$$\begin{array}{lcl}
CH_3(CH_2)_4-CH{=}CH-CH_2-CH{=}CH-(CH_2)_7\overset{O}{\overset{\|}{C}}OCH_2 & & CH_3(CH_2)_{16}\overset{O}{\overset{\|}{C}}OCH_2 \\
CH_3(CH_2)_4-CH{=}CH-CH_2-CH{=}CH-(CH_2)_7\overset{O}{\overset{\|}{C}}OCH & \xrightarrow[Ni]{6H_2} & CH_3(CH_2)_{16}\overset{O}{\overset{\|}{C}}OCH \\
CH_3(CH_2)_4-CH{=}CH-CH_2-CH{=}CH-(CH_2)_7\overset{O}{\overset{\|}{C}}OCH_2 & & CH_3(CH_2)_{16}\overset{O}{\overset{\|}{C}}OCH_2 \\
\text{trilinolein (liquid)} & & \text{tristearin (m.p. 71°)}
\end{array}$$

Figure 12-3. *Hydrogenation of an oil often converts it to a fat. Linoleic acid is a major constituent of the triglycerides of cottonseed oil, while stearic acid is found in the triglycerides of lard.*

soybean, or corn oil; peanut oil is hydrogenated to give a peanut butter in which the oil does not separate. On the other hand, evidence has been presented that unsaturated triglycerides are metabolized more easily than saturated ones and lead to less cholesterol in the blood. Thus sales of highly unsaturated oils like safflower seed oil are increasing.

As might be inferred from their saturated character, fats are among the most stable of all organic compounds, rivaling petroleum in this regard. Centuries ago the Irish were in the habit of burying butter in large wooden kegs. The butter would turn rancid, but the butyric acid and other fluid or volatile products were eventually leached out or evaporated and the residue resembled cheese. Tubs of such "bog butter" buried in the eleventh to fourteenth centuries have been unearthed in modern times and found to be in edible condition.

> ***'Butter to eat with their hog***
> ***was seven years buried in a bog'***
> —Seventeenth-Century Rhyme

Drying Oils

Unsaturated molecules oxidize more easily than saturated ones. Therefore, oils turn rancid by air oxidation during storage more readily than fats do, though natural fats also contain some unsaturated esters and undergo slow air oxidation. This tendency of oils to react with oxygen can be put to good advantage in certain cases. Since oxygen has unpaired electrons, it is able to initiate a free-radical polymerization of highly unsaturated oils. These oils, known as drying oils, are used in paints, varnishes, and lacquers. When spread in a thin film by painting, the individual molecules are exposed to oxygen, polymerization ensues, and the whole surface is converted into gigantic molecules. Certain metallic salts, such as cobaltic stearate, are commonly added to catalyze the polymerization. Linoleum is produced by polymerizing unsaturated esters into a high molecular weight plastic, together with other materials that add body to the product (binders) or color it (pigments).

Fatty Alcohols

The carboxyl group of an acid or an ester may be reduced to a primary alcohol by hydrogenation at high temperatures and pressures (the usual commercial method) or by the use of chemical reducing agents like lithium aluminum hydride, the reagent of choice in the laboratory (Fig. 12-4). If either procedure

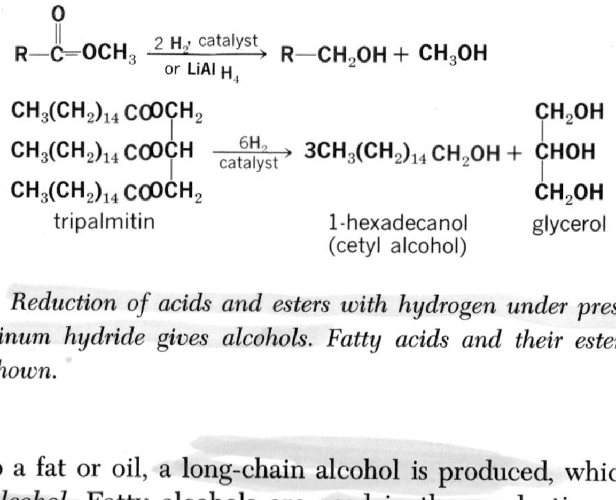

$$R{-}\overset{O}{\overset{\|}{C}}{-}OCH_3 \xrightarrow[\text{or LiAl}H_4]{2\ H_2,\ \text{catalyst}} R{-}CH_2OH + CH_3OH$$

$$\begin{matrix} CH_3(CH_2)_{14}COOCH_2 \\ CH_3(CH_2)_{14}COOCH \\ CH_3(CH_2)_{14}COOCH_2 \\ \text{tripalmitin} \end{matrix} \xrightarrow[\text{catalyst}]{6H_2} \begin{matrix} 3CH_3(CH_2)_{14}CH_2OH \\ \text{1-hexadecanol} \\ \text{(cetyl alcohol)} \end{matrix} + \begin{matrix} CH_2OH \\ CHOH \\ CH_2OH \\ \text{glycerol} \end{matrix}$$

Figure 12-4. *Reduction of acids and esters with hydrogen under pressure or with lithium aluminum hydride gives alcohols. Fatty acids and their esters give fatty alcohols, as shown.*

is applied to a fat or oil, a long-chain alcohol is produced, which is known as a *fatty alcohol.* Fatty alcohols are used in the production of synthetic detergents and also as components in polishes and cosmetics.

Waxes

Fatty alcohols are also found in nature as components of *waxes.* Plant waxes are complex mixtures of organic materials of high molecular weight that coat leaves and stems in hot or arid regions and reduce the evaporation of moisture. *Carnauba wax*, used as floor wax and in carbon paper, coats the leaves of the carnauba palm, which grows along river banks in Brazil. The leaves are cut up and beaten against saw horses in airtight buildings to collect the wax. *Beeswax* is secreted by eight wax glands of the bee. Approximately 1 lb of wax is produced for each 8 lb of honey. Since insects have a high surface to volume ratio, many produce a protective coating of wax to prevent dehydration.

Esters composed of long-chain fatty acids and long-chain fatty alcohols are always important constituents of waxes. Cetyl palmitate, an ester of cetyl alcohol, $CH_3(CH_2)_{14}CH_2OH$, and palmitic acid, is the wax *spermaceti* from the sperm whale.

Soap

Salts of long-chain fatty acids are known as *soaps*. Hydrolysis of a fat or oil with alkali produces a soap and glycerol (Fig. 12-5). The process of soap-

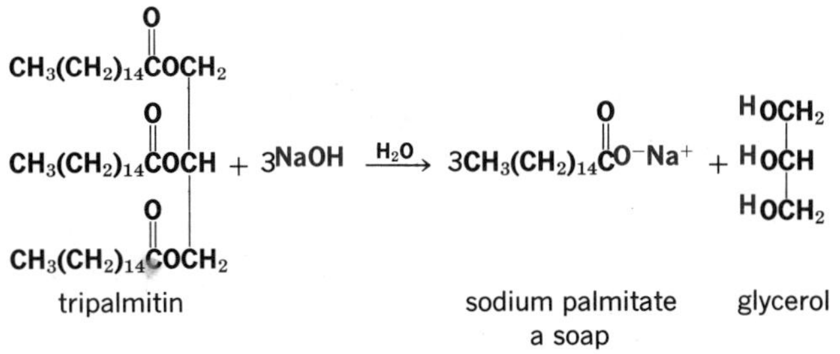

$$\underset{\text{tripalmitin}}{\begin{array}{l} CH_3(CH_2)_{14}\overset{O}{\overset{\|}{C}}OCH_2 \\ CH_3(CH_2)_{14}\overset{O}{\overset{\|}{C}}OCH \\ CH_3(CH_2)_{14}\overset{O}{\overset{\|}{C}}OCH_2 \end{array}} + 3NaOH \xrightarrow{H_2O} \underset{\substack{\text{sodium palmitate} \\ \text{a soap}}}{3CH_3(CH_2)_{14}\overset{O}{\overset{\|}{C}}O^-Na^+} + \underset{\text{glycerol}}{\begin{array}{l} HOCH_2 \\ HOCH \\ HOCH_2 \end{array}}$$

Figure 12-5. *Basic hydrolysis of a fat gives glycerol and the salts of the fatty acids, which are soaps.*

making has been known since antiquity, when the necessary lye was leached from the ashes of a wood fire. Even now, the home production of soap makes an interesting, economical, and instructive experiment (recipe in Fig. 12-6). Industrially, molten tallow (the fat of cattle and sheep) is treated with an excess of alkali in large vats. The mixture is heated and steam is bubbled through it. After hydrolysis is complete, the soap is precipitated ("salted out") by the addition of an inorganic salt like sodium chloride, filtered, and washed by successive solution in water and reprecipitation with salt. Glycerol is recovered from the aqueous solutions.

■ A RECIPE FOR THE SYNTHESIS OF SOAP: Purify bacon fat by mixing it with an equal amount of water and bringing it to a boil; remove from the fire, add 1 qt of cold water for each gallon of liquid, and let cool. Remove fat from top when firm. Slowly add 350 g of sodium hydroxide to 1200 ml of cold water. Melt 2700 g (6 lb) of fat, cool it to 30° (85°F), pour the lye solution at 24° (75°F) into the melted fat in a slow steady stream with slow even stirring. Continue stirring for 10 min until the material has the texture of thick honey. Pour into a wooden or heavy cardboard box lined with damp cotton cloth and cover with an old blanket or rug to retain heat. Let stand 24 hours, cut the soap into bars, and let age in a dry place for 2 weeks. Warning: Do not use glass, stoneware, or aluminum containers.

Figure 12-6. *A recipe for the synthesis of soap. Assuming that the consumption of bacon is not increased to get the fat, one can make soap in this way for only a few cents a cake.*

The usual toilet or household soap is the sodium salt of a fatty acid; these salts are relatively insoluble in water and can be obtained in dry, solid form. The potassium salts are more soluble and form gels; they are used in shaving creams and shampoos. Aluminum salts of acids form gels; mixed with gasoline, these gels are used in flame-throwers and napalm bombs. Calcium, magnesium, and iron salts of fatty acids are insoluble and precipitate, thus leading to the main disadvantage of soap. So-called "hard water" contains appreciable quantities of one or more of these three ions. When a soap is placed in hard water, the calcium, magnesium, or iron salt precipitates, forming a deposit on clothes or a ring around the bathtub. Hard water is "softened" by removing these three ions, usually by replacing them by sodium ion in a water softener. Another method to prevent the precipitation of these metal soaps is to tie up the metal ion extremely tightly as a *chelate* (from the Greek: "chele," for "claw"). Ethylenediaminetetraacetic acid (EDTA) (Fig. 12-7) is an example of a chelating agent that forms a stable, water-soluble salt with calcium and other metal ions.

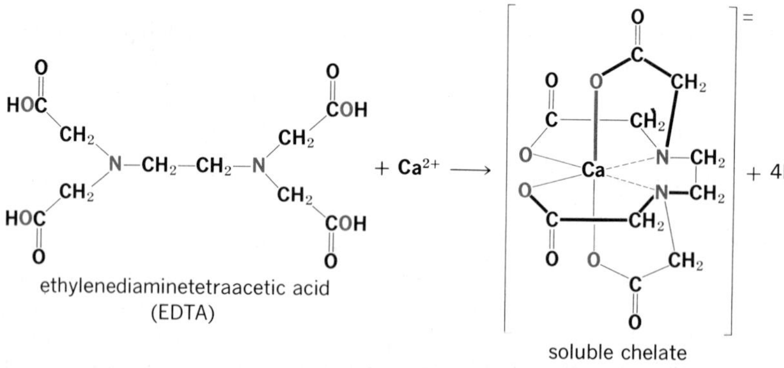

$(HOOC{-}CH_2)_2N{-}CH_2{-}CH_2{-}N(CH_2{-}COOH)_2 + Ca^{2+} \longrightarrow [\text{Ca-EDTA chelate}]^{=} + 4H^+$

ethylenediaminetetraacetic acid (EDTA)

soluble chelate

Figure 12-7. *Ethylenediaminetetraacetic acid forms stable, water-soluble salts with metals that cause soaps to precipitate, and so is used as a water softening agent.*

Soap cleans, primarily, because of its ability to emulsify fats, oils, greases, and other organic molecules. Most dirt, for instance, is held to clothes by a thin film of oil. If the oil is removed, the dirt can be rinsed away. A suspension of tiny oil droplets in water is an *emulsion.* If pure oil and water are shaken together briefly and allowed to stand, the two layers are regenerated after a short time. However, if a little soap is added to the mixture before shaking, a cloudy emulsion forms, with the oil dispersed in the form of tiny droplets throughout the water. This emulsion is indefinitely stable, showing no tendency to separate into layers of oil and water.

Soap owes its remarkable ability to act as an emulsifying agent to the curious combination of polar and nonpolar sites in its structure. Viewed from one end, a soap is a highly ionic salt, attracted to water (hydrophilic) and repelled by organic solvents (lipophobic). Seen from the other end, a soap is a long-chain hydrocarbon, repelling water (hydrophobic) and attracting organic solvents (lipophilic). If a little soap is added to water, the hydrophilic carboxylate ends dissolve in the water, but the hydrophobic hydrocarbon ends do not. As a consequence, soap makes a film on water one molecule thick, and accumulates on the surface of water. The surface tension of water is drastically reduced by this film. Therefore, water containing a little soap will wet glass, for instance, more readily than will pure water.

Soap also seeks out the interface between a layer of oil and water, with the carboxylate end dissolved in the water and the hydrocarbon end dissolved in the oil. If the oil is dispersed into tiny droplets throughout the water by shaking or stirring, soap molecules arrange themselves around the surface of each drop (Fig. 12-8). Since the surface of each drop is negatively charged,

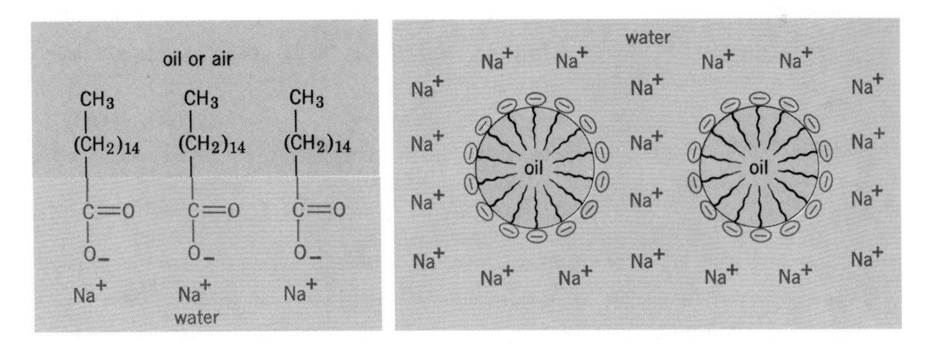

Figure 12-8. *The carboxylate end of a soap molecule dissolves in water; the hydrocarbon end dissolves in oil or is attracted to other molecules of the same composition.*

the drops repel one another and do not coalesce into a separate oil layer, and the oil is emulsified in the water.

To be effective, a soap must be only moderately soluble in water, and so the short-chain and the extremely long-chain fatty acids are ruled out; soaps containing 12 to 20 carbon atoms are the most useful. Ordinarily, soap has a higher specific gravity than water, but this can be reduced by blowing air into molten soap; in this way floating soaps are manufactured.

Synthetic Detergents

Many years ago it was recognized that the sodium salts of organic derivatives of sulfuric acid containing long alkyl chains resemble soap in their detergent properties but have the advantage that their metal salts are soluble so that they do not precipitate in hard water. At the present time approximately 80% of the detergents in use in the United States are of this synthetic type. Two different kinds of sulfuric acid derivatives are used, *sulfates,* which are esters of sulfuric acid, and *sulfonates,* in which the organic residue is attached directly to the sulfur atom.

A *sodium alkyl sulfate* detergent can be prepared by treating a long-chain alcohol with sulfuric acid and neutralizing the resulting ester with sodium hydroxide. As early as 1860 a commercial process was in use to produce a synthetic detergent from castor oil, which has a high percentage of the hydroxyl-containing fatty acid ricinoleic acid (Fig. 12-2).

With the availability of fatty alcohols by the hydrogenation of fats and oils, high quality synthetic detergents can be produced at moderate cost. One of the best is sodium lauryl sulfate (Fig. 12-9), derived from coconut and palm oil. It is used in toothpaste, cosmetics, and shampoos.

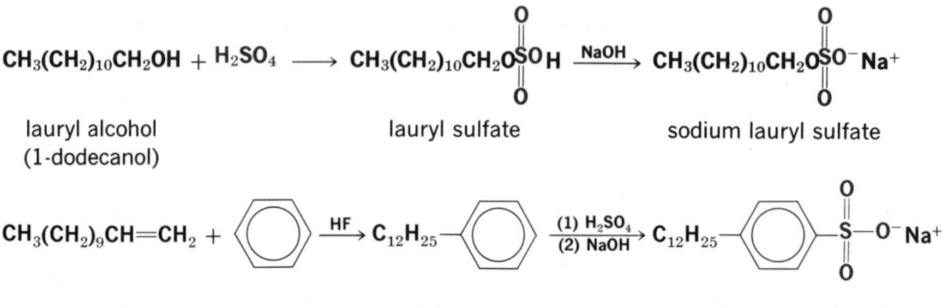

$$CH_3(CH_2)_{10}CH_2OH + H_2SO_4 \longrightarrow CH_3(CH_2)_{10}CH_2OSO_2OH \xrightarrow{NaOH} CH_3(CH_2)_{10}CH_2OSO_2O^- Na^+$$

lauryl alcohol (1-dodecanol) — lauryl sulfate — sodium lauryl sulfate

$$CH_3(CH_2)_9CH{=}CH_2 + C_6H_6 \xrightarrow{HF} C_{12}H_{25}{-}C_6H_5 \xrightarrow[(2)\ NaOH]{(1)\ H_2SO_4} C_{12}H_{25}{-}C_6H_4{-}SO_2{-}O^- Na^+$$

Figure 12-9. *Two important methods for the preparation of synthetic detergents. Reaction of an alcohol with sulfuric acid gives a sulfate, while aromatic substitution gives a sulfonate.*

The most common household detergents are *sulfonates,* salts of benzenesulfonic acids substituted with a long-chain alkyl group. These may be produced by Friedel–Crafts (Fig. 5-13) alkylation of benzene with a long-chain 1-alkene prepared from petroleum, followed by sulfonation (Fig. 5-12) and neutralization with sodium hydroxide. A typical synthesis is given in Fig. 12-9.

Synthetic detergents of this type, when used alone, are not the equal of soaps in cleaning power, but two important additives improve their cleaning power enormously. Sodium tripolyphosphate is used as a *builder;* it has the ability to break up and suspend certain clays, pigments, and other finely divided solids in aqueous solution and also to form chelates with numerous metal ions. Other phosphates also have detergent action of their own, and when combined with sulfonates the cleansing powers of both are greatly enhanced. Heavy duty household detergents also contain ½ to 1% of carboxymethylcellulose (CMC), prepared by the reaction of cellulose and chloroacetic acid in basic solution. This polymeric material has the ability to prevent redeposition of the dirt on the fabric once it has been removed by the detergents. Other common additives are whitening agents and sudsing enhancers or repressors; and, if the product is to be a soap powder, varying amounts of a granular salt like sodium sulfate are added to provide a satisfactory consistency.

Cationic and Nonionic Detergents

Sulfates and sulfonates, like soaps, are *anionic detergents,* since they carry a negative charge on the organic component. But from our discussion of the mode of action of a soap, we saw that what is needed is a long molecule with hydrophobic and hydrophilic ends. The nature of the charge, or even the presence of a charge at all, is not of primary importance. In a *cationic* or "invert" soap a trimethylammonium ion is attached to a long-chain alkyl group (Fig. 12-10). These soaps are not especially good detergents, but

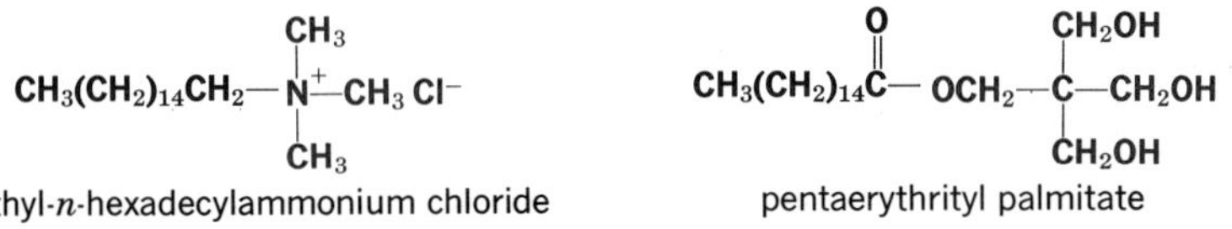

$$CH_3(CH_2)_{14}CH_2-\overset{+}{N}(CH_3)_3\ Cl^-$$

trimethyl-*n*-hexadecylammonium chloride

$$CH_3(CH_2)_{14}\overset{O}{\overset{\|}{C}}-OCH_2-C(CH_2OH)_3$$

pentaerythrityl palmitate

Figure 12-10. *Two synthetic detergents: a cationic soap and a nonionic soap. In each case the end of the molecule shown in red is soluble in water; the other end is oil-soluble.*

they do have bactericidal properties and so find wide use in medicinal soaps, salves, and ointments. They are also used as softening agents for clothes, since they are attracted to the negatively charged surface of textiles, where they act as lubricants. *Nonionic* soaps have one highly polar end, usually a polyalcohol, which is water-soluble. While not useful as detergents, they are incorporated into cosmetics and other products as emulsifying agents.

Phospholipids and Cell Membranes

In living systems the role of emulsifying agent is played by the phospholipids. An example is *phosphatidyl choline* or *lecithin* (Fig. 12-11), in which two of

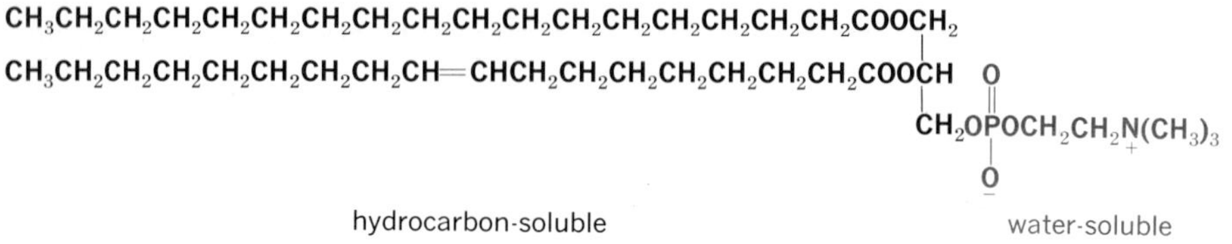

$$CH_3CH_2CH_2CH_2CH_2CH_2CH_2CH_2CH_2CH_2CH_2CH_2CH_2CH_2CH_2CH_2CH_2COOCH_2$$

$$CH_3CH_2CH_2CH_2CH_2CH_2CH_2CH_2CH{=}CHCH_2CH_2CH_2CH_2CH_2CH_2CH_2COOCH$$

$$CH_2O\overset{\overset{O}{\|}}{\underset{\underset{O^-}{|}}{P}}OCH_2CH_2\overset{+}{N}(CH_3)_3$$

hydrocarbon-soluble | water-soluble

Figure 12-11. *A phosphatidyl choline (lecithin) molecule, important as a constituent of cell membranes. Note the water-soluble polar end and the internal double bond in one of the fatty acids.*

the hydroxyl groups of glycerol are esterified by fatty acids and the third is joined to a choline ester of phosphoric acid. This latter group provides an ionic end of the molecule that is soluble in water, while the hydrocarbon chains of the fatty acids are, of course, not water soluble. Like soaps, phosphatidyl choline molecules, in contact with water, try to arrange themselves so that the hydrocarbon chains are in contact only with other, similar chains. One way they can do this is to form a *bilayer* two molecules thick, like that shown in Fig. 12-12. Such bilayers constitute the main framework of a cell membrane.

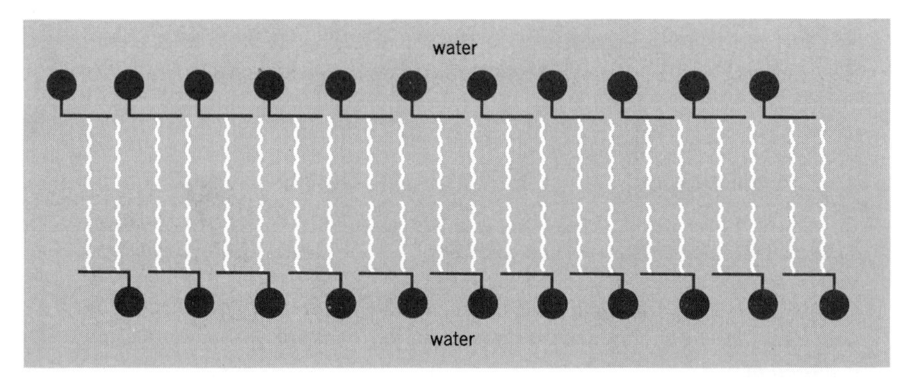

Figure 12-12. *In a cell membrane phosphatidyl choline molecules are arranged in a bilayer, whose hydrocarbon interior prevents the passage of polar molecules and ions. Protein molecules aid in the transport of these species through the membrane.*

Water can pass through such a bilayer, but larger, polar molecules and ions cannot. The membranes thus preserve concentration differences between different parts of a cell, or between a cell and its aqueous environment.

Fine differences in physical properties between cell membranes are controlled by the degree of unsaturation in the fatty acid side chains of the phosphatidyl choline molecules. We have already seen that glycerides containing *cis* double bonds are lower melting than saturated glycerides; therefore, membranes constructed of unsaturated phosphatidyl cholines are more liquid and allow more rapid movement of molecules through them than do similar membranes constructed from saturated fatty acid esters. It has also been noted that cells which must live and grow at a relatively low temperature have more unsaturated fatty acids in their membranes than those cells which live at higher temperatures.

In order for polar nutrients and ions to pass through the interior of these lipid membranes, their polar character must be disguised. Imbedded in the membranes are protein molecules whose job it is to enclose these hydrocarbon-insoluble nutrients and ions within their highly polar interior and to escort them through the membrane. By the proper expenditure of energy, and because proteins can be highly discriminatory in their ability to complex with and enclose ions and molecules, very large concentration differences can be built up between the inside and outside of a cell.

Related phospholipids (enol ether phosphatides or plasmalogens) occur in brain membranes. Instead of having glycerol esterified with two fatty acids, these molecules have one ester group and one long-chain, unsaturated ether group (Fig. 12-13). Recently, there has been great concern about poisoning by methylmercury, and it is thought that this pollutant interacts with the enol ether phosphatides. The metal ion catalyzes the addition of water across the reactive double bond in these molecules. The result is a hemiacetal, which readily hydrolyzes, breaking up the membrane of brain cells. The result is mental retardation and often death.

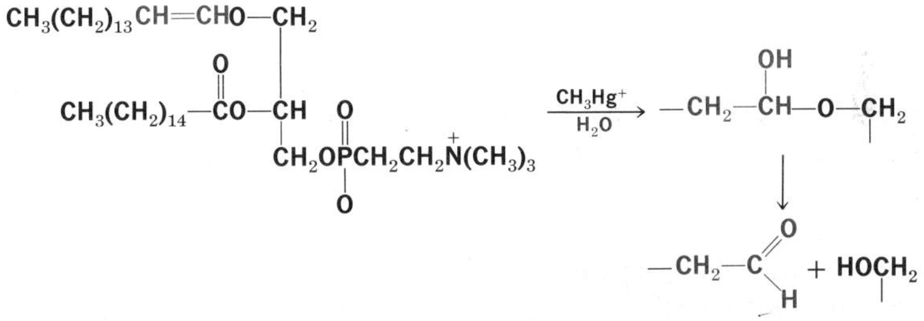

$CH_3(CH_2)_{13}CH{=}CHO{-}CH_2$

$CH_3(CH_2)_{14}{-}\overset{O}{\overset{\|}{C}}O{-}CH$

$CH_2O\overset{O}{\overset{\|}{P}}CH_2CH_2\overset{+}{N}(CH_3)_3$ (with O^- on P)

$\xrightarrow[H_2O]{CH_3Hg^+}$ $-CH_2{-}\overset{OH}{\overset{|}{C}}H{-}O{-}CH_2$

$\downarrow$

$-CH_2{-}C(=O)H \;+\; HOCH_2$

Figure 12-13. *Methylmercury poisoning is thought to arise from its catalytic action in hydrating the double bond of a brain cell phospholipid, thus disrupting the brain cell membranes.*

Still another group of phospholipids is that containing the sphinogolipids, in which glycerol and its derivatives are replaced by an amide derived from a fatty acid and sphingosine, a C_{18} aminodiol. Sphingomyelins such as that shown in Fig. 12-14 are found in nerve sheaths and the brain.

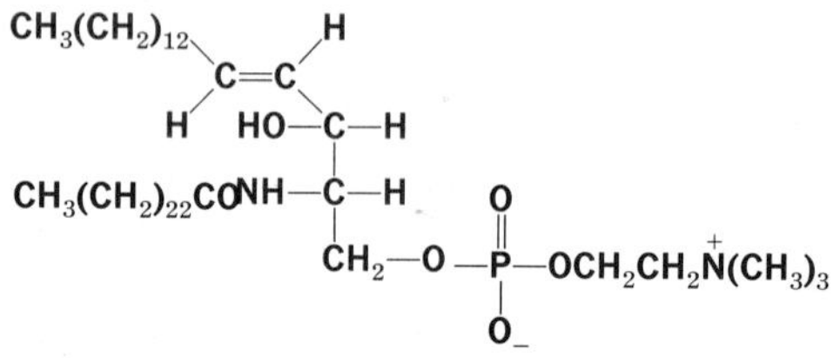

$CH_3(CH_2)_{12}$—CH=CH (trans, H on each carbon)—HO—C—H

$CH_3(CH_2)_{22}CONH$—C—H

CH_2—O—P(=O)(O$^-$)—$OCH_2CH_2\overset{+}{N}(CH_3)_3$

a sphingomyelin

Figure 12-14. *A sphingomyelin contains the unsaturated aminodiol sphingosine (in red) linked to a fatty acid by an amide bond and to choline phosphate by a phosphate bond. Note that the double bond in sphingosine is* trans *and that the fatty acid (lignoceric acid) has 24 carbon atoms.*

PROBLEMS

1. Draw the structure of each of the following:
   (a) glyceryl trilinolenate (b) stearyl alcohol
   (c) potassium oleate
   (d) trimethyl-*n*-octadecylammonium chloride
   (e) pentaerythrityl myristate
   (f) a phospholipid
   (g) lauryl stearate
2. What would be the product(s) of the reaction of trilinolein (Fig. 12-3) with each of the following reagents in excess?
   (a) H_2/Ni, heat (b) H_2/Pd, room temp.
   (c) NaOH/heat (d) $LiAlH_4$
   (e) O_3; then H_2O, Zn (f) Br_2
3. If the *cis* double bonds of an unsaturated triglyceride are isomerized to their *trans* isomers, the melting point increases nearly as much as it does upon reduction to a saturated triglyceride. Explain.
4. Suppose you wanted to try to develop a truly "natural" nonionic soap by combining the hydrocarbon chain from a fatty acid molecule to glucose. Show two different ways you might accomplish this, and explain why the product might have emulsifying properties.
5. Naturally occurring analogs to the product of Prob. 4 are the glycolipids. These molecules are analogous to phosphatidyl choline, but a sugar residue in the pyranose form is joined to carbon-3 of glycerol through an acetal linkage. If the sugar were α-D-glucose and the fatty acids both stearic acid, draw its structure.

13
amino acids and proteins

Three great classes of natural polymers, carbohydrates, proteins, and nucleic acids, are the building materials of life. Carbohydrates, as we saw in Chapter 9, are polyacetals constructed from glucose molecules as the monomeric units. Nucleic acids, the carriers of hereditary information in deoxyribonucleic acid (DNA) and ribonucleic acid (RNA), are, as we shall see in Chapter 14, polyesters. Proteins are polyamides, formed by the linking together of the amino group of one amino acid and the carboxyl group of another. Proteins are by far the most diverse of the three types of natural polymers. *Hemoglobin* is the protein that carries oxygen in our blood. Hair and feathers are made from the protein *keratin;* the connective tissue of flesh, muscle, and tendon is the protein *collagen. Enzymes,* organic molecules that catalyze the thousands of chemical reactions required to drive the machinery of life, are also proteins. And at the shorter end of the scale many small proteins, called peptides, are *hormones. Insulin,* which is excreted from the pancreas, is an example. In this chapter we shall examine the general structure of proteins, see how they are prepared in the laboratory, and examine how some of them act as enzymatic catalysts in biological systems. In the next chapter we shall see how they are synthesized in living systems under the control of the genes.

α-Amino Acids

The monomeric units for protein polymers are the 20 naturally occurring α-amino acids, carboxylic acids that are substituted in the α-position with an amino group. All but the simplest amino acid glycine, contain a chiral center at this α-carbon atom. Their stereochemistry at this chiral center is S, and they are said to belong to the L-series, in which the amino group is on the left when viewed as in Fig. 13-1 (compare the convention for D- and L-sugars). Since we know that carboxylic acids react with amines to form salts, individual amino acids exist as high-melting, water-soluble "inner salts" called *Zwitterions.*

The amino acids that occur in proteins differ from one another in the

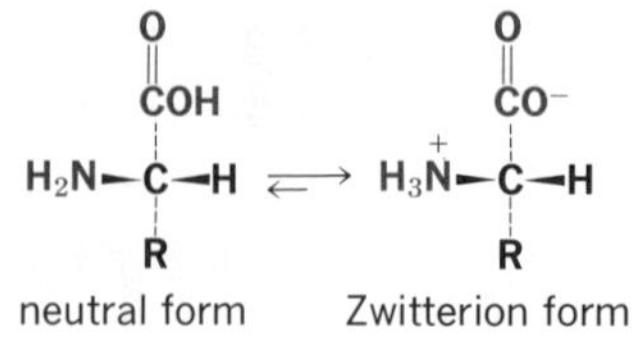

$$H_2N-\underset{R}{\overset{COH}{\overset{\|}{\overset{O}{}}C}}-H \rightleftharpoons H_3\overset{+}{N}-\underset{R}{\overset{CO^-}{C}}-H$$

neutral form Zwitterion form

Figure 13-1. *With one exception, all 20 α-amino acids found in proteins can be represented by the general formulas shown; their saltlike physical properties are in accord with their formulation as the Zwitterion form.*

structure of the substituent R. In the simplest amino acid glycine (which makes up a major portion of hair protein), R = H, and the molecule contains no chiral carbon atom. Since amides may be formed by strong heating of ammonium salts of carboxylic acids, a polymer with properties like a protein, *polyglycine,* may be formed on heating of glycine (Fig. 13-2; compare with

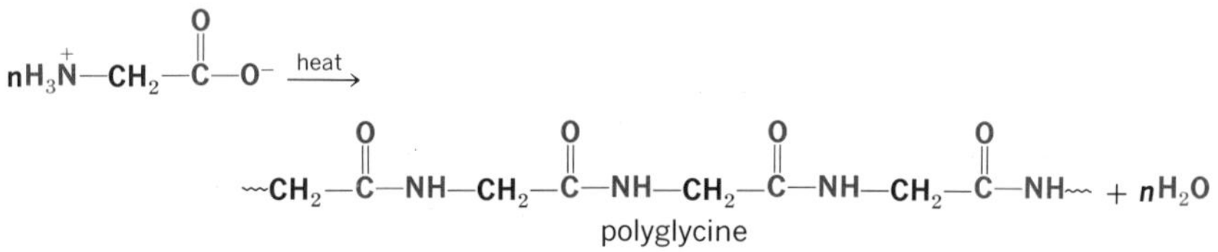

$$n H_3\overset{+}{N}-CH_2-\overset{O}{\overset{\|}{C}}-O^- \xrightarrow{\text{heat}}$$

$$\sim CH_2-\overset{O}{\overset{\|}{C}}-NH-CH_2-\overset{O}{\overset{\|}{C}}-NH-CH_2-\overset{O}{\overset{\|}{C}}-NH-CH_2-\overset{O}{\overset{\|}{C}}-NH\sim + nH_2O$$

polyglycine

Figure 13-2. *Polyglycine, a polyamide made by strong heating of glycine, resembles many proteins in physical and chemical properties.*

the structure of nylon, Fig. 11-19). Since the amide groups that hold together the amino acids in proteins are neutral and relatively nonreactive, the chemical and physical properties of proteins are mainly determined by the structure of the amino-acid side chains. It is usual to divide the amino acids into three groups: (1) those with *neutral, nonpolar* (hydrocarbon) side chains, (2) those with *neutral, polar* side chains, and (3) those with *acidic* or *basic* side chains.

Figure 13-3 gives the names and structures of the seven amino acids with hydrocarbon side chains. Tough, insoluble proteins like silk, wool, collagen, and elastin (found in ligaments) have a high proportion of one or more of these amino acids. Over 80% of the amino acids in silk, for example, are either alanine or glycine. Four of these nonpolar amino acids, valine, leucine, isoleucine and phenylalanine, are *essential amino acids,* which cannot be synthesized by thc human body, and must be obtained from proteins that are taken in and digested. *Proline* is the only amino acid that differs from the general structure given in Fig. 13-1; its side chain is part of a ring containing the amino group; thus proline is a secondary amine. This seemingly minor

glycine (Gly) alanine (Ala) valine (Val) leucine (Leu)

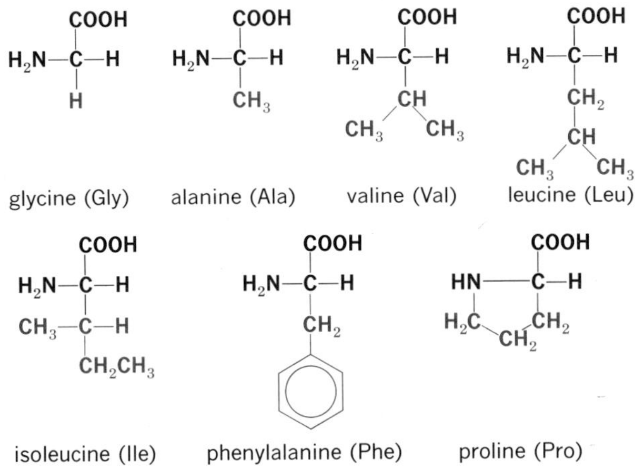

isoleucine (Ile) phenylalanine (Phe) proline (Pro)

Figure 13-3. *Seven amino acids with hydrocarbon side chains. Note especially the structure of proline, with its side chain attached to the amino group. The abbreviation for each amino acid is given in parenthesis.*

structural change has a profound effect on the structure of the polymer chain of proteins that contain proline units, as we shall see.

A second group of amino acids, shown in Fig. 13-4, has polar but

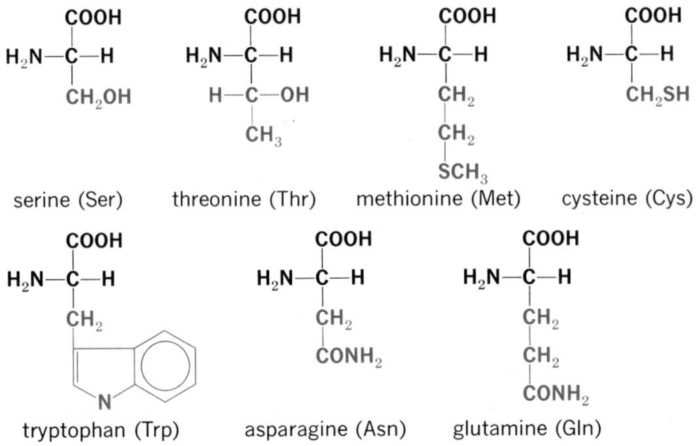

serine (Ser) threonine (Thr) methionine (Met) cysteine (Cys)

tryptophan (Trp) asparagine (Asn) glutamine (Gln)

Figure 13-4. *Seven amino acids that are polar but neutral.*

neutral side chains. Three of them, *threonine, methionine,* and *tryptophan,* are essential amino acids. The presence of a large number of polar amino acids increases the water solubility of a protein, while the functional groups of these molecules often play important roles in the action of enzymes and in other physiological properties of proteins. Cysteine, in particular, is crucial in securing the three-dimensional structure of proteins.

Finally, an important group of six amino acids has side chains containing acidic or basic groups (Fig. 13-5). These amino acids, in addition to promoting

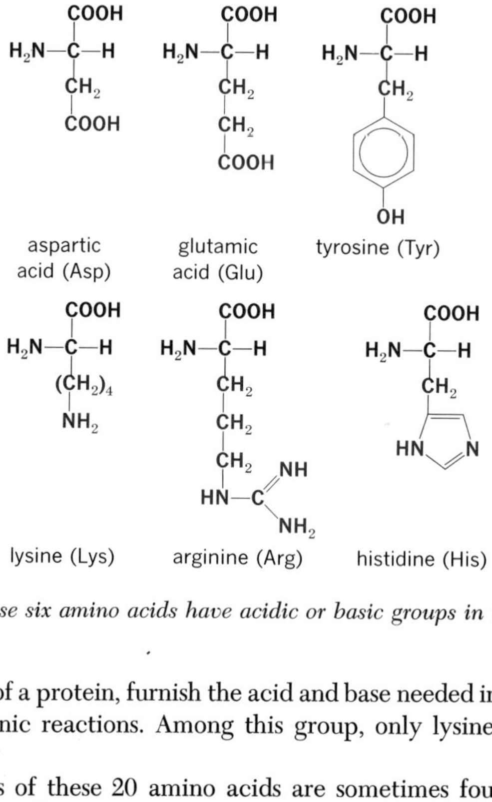

aspartic acid (Asp) glutamic acid (Glu) tyrosine (Tyr)

lysine (Lys) arginine (Arg) histidine (His)

Figure 13-5. *These six amino acids have acidic or basic groups in their side chains.*

water solubility of a protein, furnish the acid and base needed in the enzymatic catalysis of organic reactions. Among this group, only lysine is an essential amino acid.

Derivatives of these 20 amino acids are sometimes found in proteins. For example, after proline and lysine are incorporated into a protein, they are sometimes enzymatically oxidized by oxygen of the air to hydroxyproline (Hyp) and hydroxylysine (Hyl). Similarly, we shall see how two cysteine molecules may be oxidized to cystine (Cys-Cys). Other amino acids, including some of these twenty in the D- rather than the L- form, occur among lower plants, and microorganisms often utilize D-amino acids in forming antibiotics.

Reactions of Amino Acids

In general, the reactions of amino acids are those expected of the individual functional groups found in them. Of analytical importance are the reactions that serve as qualitative or quantitative tests. When amines are treated with nitrous acid, nitrogen gas evolves (see p. 198). In the Van Slyke procedure for analysis of amino acids, the nitrogen is collected and its volume measured under conditions of standard temperature and pressure. The chemist can determine the number of moles of amino acids present from the volume of gas evolved.

The ninhydrin test for amino acids (Fig. 13-6) is more complicated:

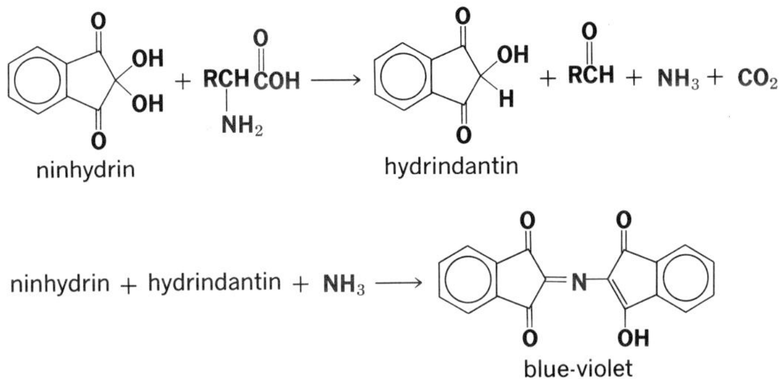

ninhydrin + $RCH(NH_2)COOH \longrightarrow$ hydrindantin + $RCHO + NH_3 + CO_2$

ninhydrin + hydrindantin + $NH_3 \longrightarrow$ blue-violet

Figure 13-6. *The reaction of ninhydrin with amino acids produces a blue-violet dye that can be used as a qualitative or quantitative test for amino acids.*

Ninhydrin oxidizes an amino acid to ammonia, carbon dioxide, and an aldehyde; a molecule of ninhydrin then reacts with both the ammonia produced and a molecule of reduced ninhydrin (called hydrindantin) to create a blue dye. The intensity of the blue color can be used as a measure of the amount of amino acid present. The reaction can also be used as a simple color test for the presence of amino acids.

In combination with paper chromatography, the ninhydrin reaction can be used to *identify* amino acids. For example, a protein can be hydrolyzed with strong acid to give a mixture of the amino acids present. A small amount of this mixture is then dissolved in water and applied (spotted) to a sheet of filter paper near the lower edge (at the origin). This is the area marked + at the bottom left of the paper in Fig. 13-7. A solvent (in this case, a solution containing phenol and water) is allowed to rise by capillary action through the paper, carrying the amino acids along with it at different rates; the less polar amino acids migrate faster than the more polar ones, which are more strongly attracted to the polar hydroxyl groups of the cellulose in the paper. The separation can be increased if the paper (a chromatogram) is dried and rotated by 90°, and a second solvent mixture (here, 2-butanol, formic acid, and water) is allowed to flow across the paper; this is called a two-dimensional paper chromatogram. To locate the separated amino acids, a solution of

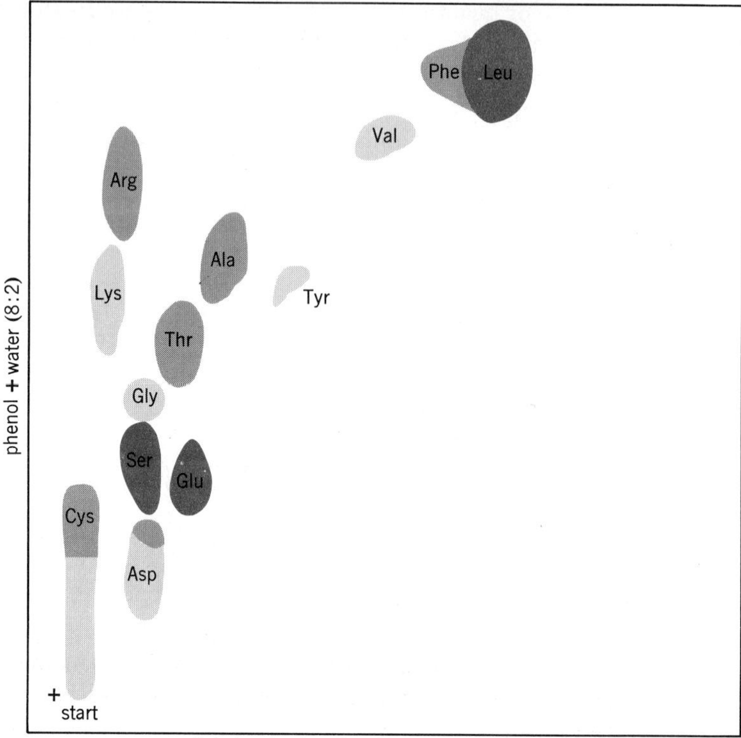

Figure 13-7. *A two-dimensional paper chromatogram of an amino-acid mixture, such as might result from the hydrolysis of a protein. (From F. Cramer,* Papierchromatographie, *4th ed., Verlag Chemie, Weinheim/Bergstr., 1958.)*

ninhydrin in alcohol is sprayed on the dried paper. Because of reactions like that shown in Fig. 13-6, a spot appears at those places where amino acids are present.

Peptides

Small polymers of amino acids are called *peptides*. A peptide from two amino acids is called a dipeptide; one from three amino acids, a tripeptide; and one with a few amino acids, an oligopeptide. Peptides are named by combining the names (or abbreviations) of the individual amino acids, starting by convention with the amino acid whose amino group is free (the so-called N-terminal amino acid) and ending with that whose carboxyl group is free (the C-terminal amino acid). Fig. 13-8 shows the structure of a tripeptide. The individual amino acids are held together by amide bonds (the *peptide*

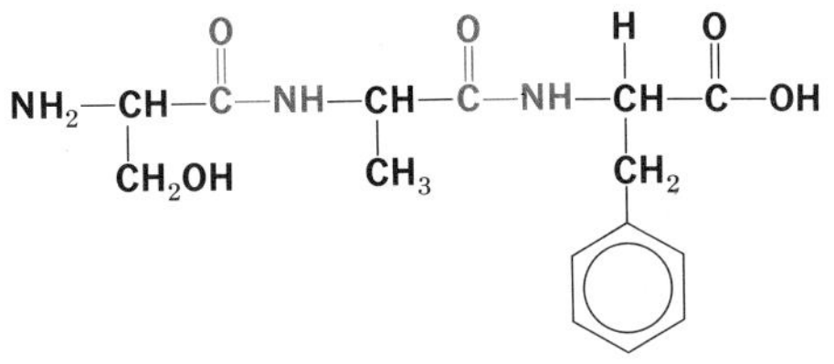

$$NH_2-CH(CH_2OH)-C(=O)-NH-CH(CH_3)-C(=O)-NH-CH(CH_2C_6H_5)-C(=O)-OH$$

serylalanylphenylalanine
(Ser-Ala-Phe)

Figure 13-8. *The structure of a tripeptide. Serine is the N-terminal amino acid, phenylalanine the C-terminal. Note the peptide bonds (in red).*

bonds), and the molecule is written with the N-terminal amino acid (serine) on the left and the C-terminal amino acid (phenylalanine) on the right. Tripeptides are usually high melting crystalline solids that exist in the Zwitterionic form, with the proton from the carboxyl group transferred to the terminal amino group. Note especially that the order in which the amino acids occur in a peptide (its *sequence* of amino acids) is an important part of its structure. The three amino acids serine, alanine, and phenylalanine can combine in six different ways to form a tripeptide (Ala-Ser-Phe, Phe-Ala-Ser, etc.). The possible arrangements for longer peptides and proteins with hundreds of amino acids chosen from among the 20 different possibilities, is virtually limitless and accounts for the great diversity of protein structures.

Sequencing of a Tripeptide

Suppose that a tripeptide were isolated from some natural source and we wished to know its structure. Complete hydrolysis and chromatography of the amino acids formed would tell us that it was composed of, for example, equimolar amounts of serine, alanine, and phenylalanine. Our next step is to *sequence* the peptide, i.e., to determine the order in which these amino acids occur. First, we might determine which of the three amino acids is N-terminal. Only the N-terminal amino acid has its α-amino group free; the other two α-amino groups are joined into peptide bonds. So we treat the original tripeptide with a reagent that combines specifically with free amino groups. One way of doing this is with 2,4-dinitrofluorobenzene; the amino nitrogen replaces the fluorine, and a yellow 2,4-dinitrophenyl derivative (DNP) of the tripeptide is formed. This is now hydrolyzed and chromatographed; two of the original amino acids (alanine and phenylalanine from the tripeptide of Fig. 13-8) will be unchanged, but the N-terminal amino acid will be missing. Instead, a new, yellow spot will appear corresponding in position to the 2,4-DNP derivative of serine (Fig. 13-9).

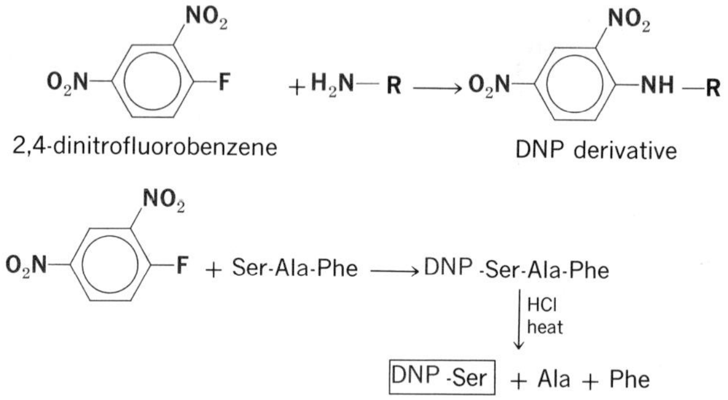

$$O_2N{-}C_6H_3(NO_2){-}F + H_2N{-}R \longrightarrow O_2N{-}C_6H_3(NO_2){-}NH{-}R$$

2,4-dinitrofluorobenzene — DNP derivative

$$O_2N{-}C_6H_3(NO_2){-}F + \text{Ser-Ala-Phe} \longrightarrow \text{DNP-Ser-Ala-Phe} \xrightarrow{\text{HCl, heat}} \boxed{\text{DNP-Ser}} + \text{Ala} + \text{Phe}$$

Figure 13-9. *To determine the N-terminal amino acid, the peptide is treated with 2,4-dinitrofluorobenzene and then hydrolyzed.*

The C-terminal end of a peptide may be identified in an analogous manner. For example, reaction with lithium borohydride (Fig. 13-10) will reduce a carboxylic acid group to an alcohol but will not reduce an amide. In this way the C-terminal amino acid will be reduced to an amino alcohol. Subsequent hydrolysis and chromatography will show phenylalanine to be missing from the hydrolyzate and will reveal the presence of the alcohol. With the N-terminal and C-terminal amino acids of a tripeptide known, the sequence is known.

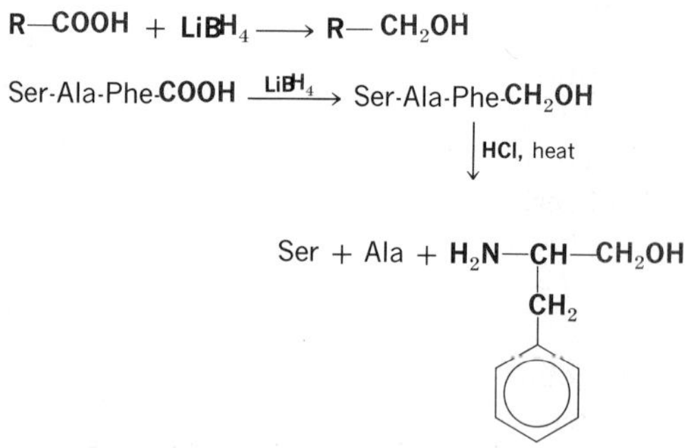

$$R{-}COOH + LiBH_4 \longrightarrow R{-}CH_2OH$$

$$\text{Ser-Ala-Phe-}\mathbf{COOH} \xrightarrow{LiBH_4} \text{Ser-Ala-Phe-}\mathbf{CH_2OH}$$

$$\xrightarrow{\text{HCl, heat}} \text{Ser} + \text{Ala} + H_2N{-}CH(CH_2C_6H_5){-}CH_2OH$$

Figure 13-10. *The C-terminal amino acid is revealed by reduction and hydrolysis.*

Degradation of Peptides

Larger peptides usually must be broken down into smaller fragments (*degraded*) for analysis. A number of quite specific enzymatic and chemical methods have been developed to degrade a peptide partially so that smaller peptides are formed without total hydrolysis into the individual amino acids. The details of the specific hydrolytic methods are beyond the scope of this book, but the logic by which the original structure is reconstructed from the structures of the smaller peptides is not. Suppose that we have isolated a hexapeptide whose hydrolysis shows that it contains one molecule each of alanine, tryptophan, leucine, aspartic acid, glycine, and phenylalanine. Reaction with 2,4-dinitrofluorobenzene, hydrolysis, and chromatography show glycine to be N-terminal, and reduction with $LiBH_4$, hydrolysis, and chromatography show aspartic acid to be C-terminal. From various methods of partial hydrolysis the three tripeptides shown in Fig. 13-11 are isolated and their

hexapeptide $\xrightarrow[\text{hydrolysis}]{\text{total}}$ **Gly, Ala, Tyr, Leu, Phe, Asp**

$\xrightarrow[\text{hydrolysis}]{\text{partial}}$ **Gly-Ala-Leu**

Ala-Leu-Phe

Phe-Tyr-Asp

Gly-Ala-Leu-Phe-Tyr-Asp

Figure 13-11. *Steps in the sequencing of a hexapeptide. See text for discussion.*

sequences determined by the methods of the last section. The sequence of amino acids in the original hexapeptide may be reconstructed by overlapping the amino acids in the tripeptide chains. Since glycine is known to be N-terminal, we begin with the tripeptide which contains that amino acid. We then see that two of the amino acids in this tripeptide overlap with two in another tripeptide, and there is a further overlap with the third tripeptide, which permits the complete sequence to be discovered. By exactly this method very long peptides can be sequenced, and by overlapping long peptide sequences the sequences of amino acids in proteins containing many hundreds of amino acids have been determined.

Cross-Linking by Cysteine

The amino acid cysteine contains the very reactive *thiol* or *sulfhydryl* group —SH, a group that plays an important role in many enzymes. Heavy metal ions like those from mercury form salts with the thiol group, thereby inactivating the enzymes and poisoning the living system containing them. Oxidation converts two cysteine molecules into *cystine* by joining them together through a sulfur–sulfur bond (Fig. 13-12). In peptides and proteins these disulfide

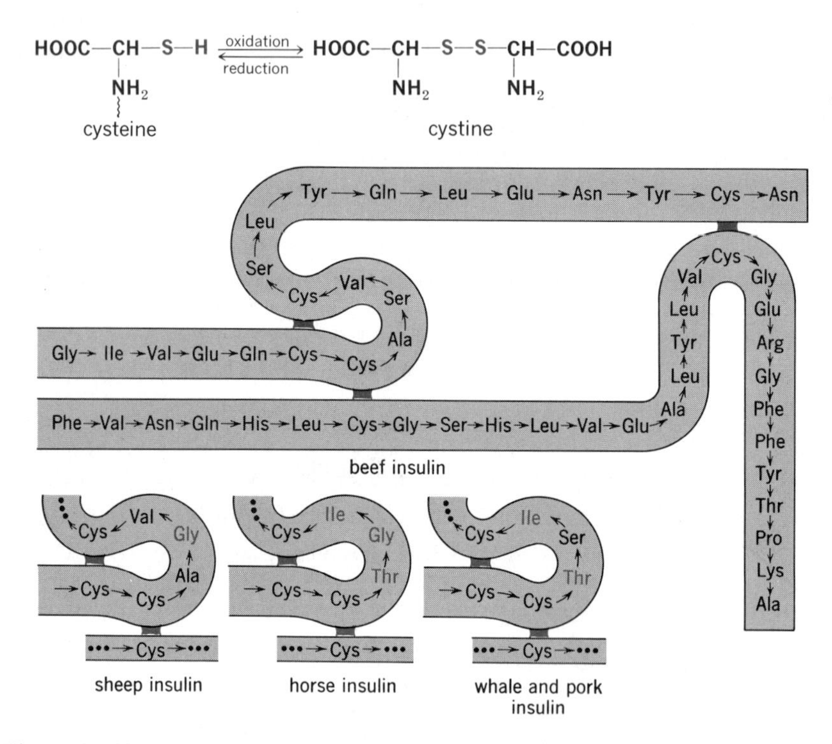

Figure 13-12. *By forming S—S linkages between cysteine molecules, separate peptide chains may be linked together or individual chains coiled into loops, as illustrated for insulin (S—S bonds in dark red). From* Introduction to Molecular Biology *by G. H. Haggis, et al., John Wiley & Sons, Inc., New York, 1964, p. 45.*

bridges (—S—S—) link various parts of the chain together. In *insulin*, the hormone that controls diabetes, disulfide linkages hold together two separate peptide chains, one of 30 and the other of 21 amino acids. The amino acid sequence in this molecule was determined by the English biochemist Sanger, who was awarded the Nobel Prize in 1958 for this work. Small differences, restricted to specific areas of the molecule, occur in the amino acids of insulin from different species of animals, but these do not alter the physiological function of the molecules. (Note that beef, horse, sheep, and pork insulins are slightly different, but that whale and pork insulins are identical). This is fortunate, for diabetics who become allergic to one type of insulin may begin using insulin from another source, and thereby avoid allergic reactions.

Hair protein gets a great deal of its body and shape from S—S bonds of cystine, and the chemistry involved in the home permanent wave takes advantage of this fact (Fig. 13-13). First, the hair is soaked in a chemical

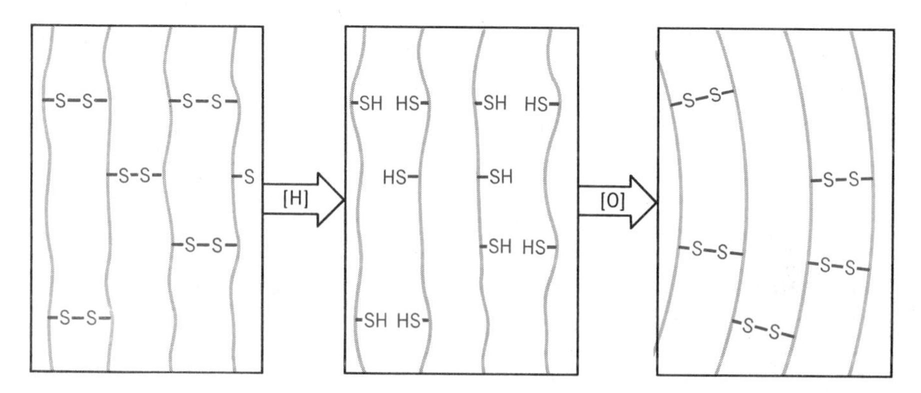

Figure 13-13. *In a home permanent the disulfide bonds in hair are first reduced to break the cross links. After the hair has been arranged in its desired shape, new disulfide bonds are formed by oxidation.*

reducing agent to break the S—S bonds and convert them to sulfhydryl groups. The hair proteins (*keratins*) are then no longer cross-linked, and can be arranged into curls. Reoxidation (by air or by the application of chemical oxidizing agents) reforms the disulfide bonds, and these help hold the hair in its new configuration.

Synthesis of Peptides

We know that an amide group, which comprises the peptide bond, forms readily from an activated carboxylic acid (e.g., an acid chloride or anhydride) and an amine (Fig. 11-18). Forming a peptide bond thus presents no special chemical problems. The problem in peptide and protein synthesis lies in the fact that amino acids contain other reactive groups which can combine with an activated carboxylic acid; therefore, these sites must be covered up with *protecting groups*. Suppose that we wished to synthesize glycylalanine; we might think to do so by reaction of the acid chloride of glycine with the amino group of alanine. But reaction could also occur with the amino group of another molecule of glycine or with the carboxyl group of alanine, both of which are reactive toward an acid chloride. So, before preparing the acid chloride, both of these potentially reactive sites must be "protected." This is illustrated schematically in Fig. 13-14.

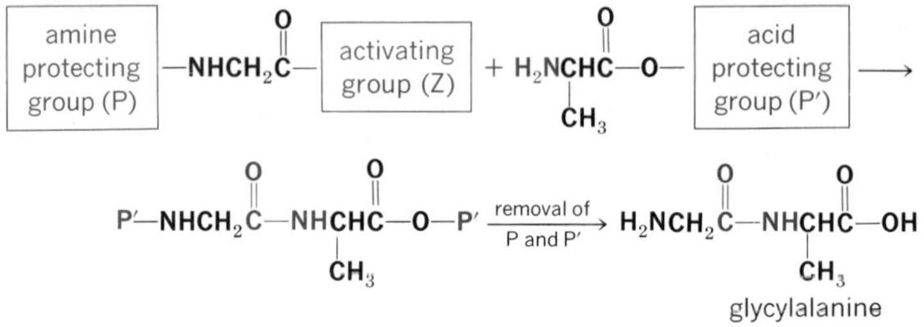

$$[\text{amine protecting group (P)}]\text{—NHCH}_2\overset{\overset{\large O}{\|}}{\text{C}}\text{—}[\text{activating group (Z)}] + \text{H}_2\text{N}\underset{\underset{\large CH_3}{|}}{\text{CH}}\overset{\overset{\large O}{\|}}{\text{C}}\text{—O—}[\text{acid protecting group (P}')] \longrightarrow$$

$$\text{P}'\text{—NHCH}_2\overset{\overset{\large O}{\|}}{\text{C}}\text{—NH}\underset{\underset{\large CH_3}{|}}{\text{CH}}\overset{\overset{\large O}{\|}}{\text{C}}\text{—O—P}' \xrightarrow[\text{P and P}']{\text{removal of}} \text{H}_2\text{NCH}_2\overset{\overset{\large O}{\|}}{\text{C}}\text{—NH}\underset{\underset{\large CH_3}{|}}{\text{CH}}\overset{\overset{\large O}{\|}}{\text{C}}\text{—OH}$$

glycylalanine

Figure 13-14. *In forming a peptide, all reactive sites except for the desired ones must be protected. To form a tripeptide either P or P′ might be removed from the protected dipeptide and the sequence repeated.*

Many different protecting and activating groups have been used in peptide synthesis. For the carboxyl end the protecting groups are usually esters, since they are easier to hydrolyze than amides and can be removed without hydrolyzing the peptide bond. For the amino protecting group special types of amides have been developed that may be removed under conditions which do not cleave peptide bonds or esters. If the amino acid side chains contain reactive sites (and except for the hydrocarbon side chains, most do), these too must be protected. Many other activating groups beside the chloride and anhydride have also been used. Obviously, the synthesis of a peptide is a tedious and time-consuming task. Nevertheless, a protein containing over 100 amino acids (ribonuclease, Fig. 13-19) has been synthesized by essentially these methods.

Automated Protein Synthesis

The laboratory synthesis of peptides and proteins has been greatly aided by the development of *solid phase protein synthesis,* a method susceptible to automation. The most tedious part of any organic synthesis is the isolation and purification of the products. One can buy protected amino acids, and their condensation into a dipeptide is relatively easy. Once the amide bond is formed, however, a whole series of extractions, washings, and crystallizations is required in order to purify the protected dipeptide, and these must be repeated for each subsequent step.

In the solid phase method, the C-terminal amino acid is attached, through an ester linkage, to a bead of insoluble polymer (Fig. 13-15). A polystyrene

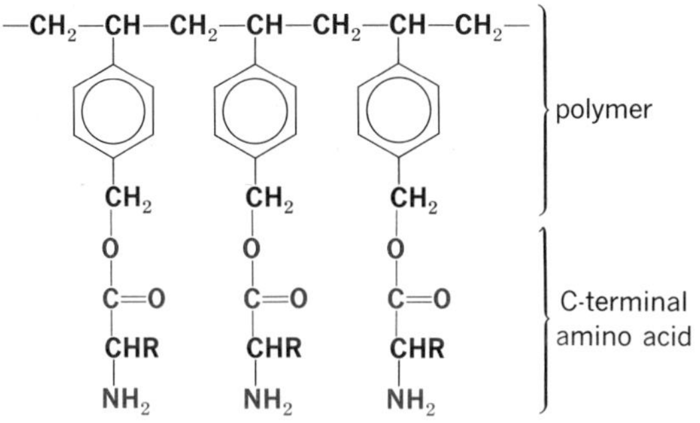

$-CH_2-CH-CH_2-CH-CH_2-CH-CH_2-$ } polymer

CH_2 / O / C=O / CHR / NH_2 } C-terminal amino acid

Figure 13-15. *In solid phase peptide synthesis the C-terminal amino acid is attached to an insoluble bead of polymer. After a reaction, solvent and impurities can be rinsed away from the product, which remains attached to the beads.*

derivative is used for this solid support. This polymer, with the amino acid bonded to it, is suspended in a solvent; then an activated amino acid, with its amino group protected, is added. Reaction takes place at the surface of the bead, and the chain is extended by one amino acid. Excess reagent is removed by filtration, and the polymer is washed with pure solvent and again filtered. Next the protecting group at the amino end of the chain is removed; again the product is purified merely by washing and filtration; and a third activated, protected amino acid is added in the same way. This is illustrated schematically in Fig. 13-16.

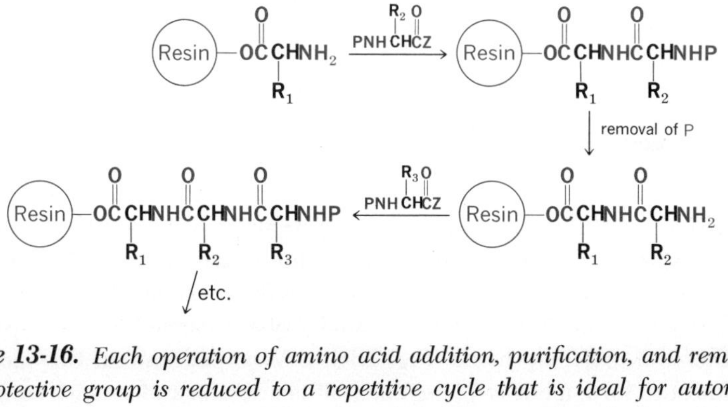

(Resin)—OCCHNH$_2$ (R$_1$) —PNHCHCZ (R$_2$)→ (Resin)—OCCHNHCCHNHP (R$_1$, R$_2$)

↓ removal of P

(Resin)—OCCHNHCCHNH$_2$ (R$_1$, R$_2$) —PNHCHCZ (R$_3$)→ (Resin)—OCCHNHCCHNHCCHNHP (R$_1$, R$_2$, R$_3$)

↓ etc.

Figure 13-16. *Each operation of amino acid addition, purification, and removal of the protective group is reduced to a repetitive cycle that is ideal for automation. Machines to carry out protein synthesis have been constructed, and ribonuclease (Fig. 13-19) has been synthesized using the solid phase method in such a machine.*

Three-Dimensional Structure of Proteins

In order to understand the physiological action of a protein, it is necessary to know more than just the sequence of amino acids it contains; we must also know how this extremely long chain is twisted and coiled in three dimensions. This is especially true of enzymes, for we shall see that their catalytic activity depends upon the close proximity in space of amino acid side chains which are far apart along the backbone of the molecule.

Suppose we prepare a polymer containing 100 units of a single amino acid. This molecule might be expected to assume an almost infinite variety of shapes, being randomly tangled like a long thread. On the contrary, most such polymers automatically acquire a regular structure in which the chain is coiled into a spiral known as a *helix*. Like a spiral staircase or a screw, a helix can be either right- or left-handed; for a polymer of L-amino acids, however, the right-handed helix seems to be a little more stable and is the helix found widely in nature. This is known as the secondary structure of a protein.

In the so-called α-helix suggested by Linus Pauling, the spiral is held together in part by hydrogen bonds. Each bond formed ties the N—H group of one peptide linkage to the oxygen of the C=O group above it on the next turn of the helix, four amino acid units farther along the chain. In Fig. 13-17, the drawing on the right shows how hydrogen bonding (C=O· · ·HN) stabilizes the α-helix. The medium red balls (R) represent amino acid side chains, the black balls carbon atoms, the dark red balls nitrogen atoms, and the light red balls oxygen atoms. There are 3.6 amino acid units for each turn of the helix, and after 18 amino acid units the chain returns exactly to its starting point. Notice that the substituent groups, which distinguish the amino acids from one another, project outward from this basic core. The drawing on the left shows the five turns required to return to the start of the helix. In this drawing a carbonyl group is represented by a dark red ball, an NH group by a light ball, and the remainder of the amino acid by a black ball.

Tertiary Structures of Proteins

If proteins were to exist exclusively in the form of an α-helix, all would be long rodlike molecules. The α-keratins (proteins of hair, skin, feathers, etc.) do have this form, but many other proteins are nearly spherical in shape and are known as *globular proteins*. Hemoglobin, which carries the oxygen in the blood stream, is an example. The actual oxygen carrier is not a protein, but rather the molecule heme (Fig. 13-18). Wrapped around this molecule so as almost to form a box is a long chain of nearly 150 amino acids composed of eight helical segments of protein. About 80% of this protein molecule is helical, and the helical segments are separated by short stretches of nonhelical

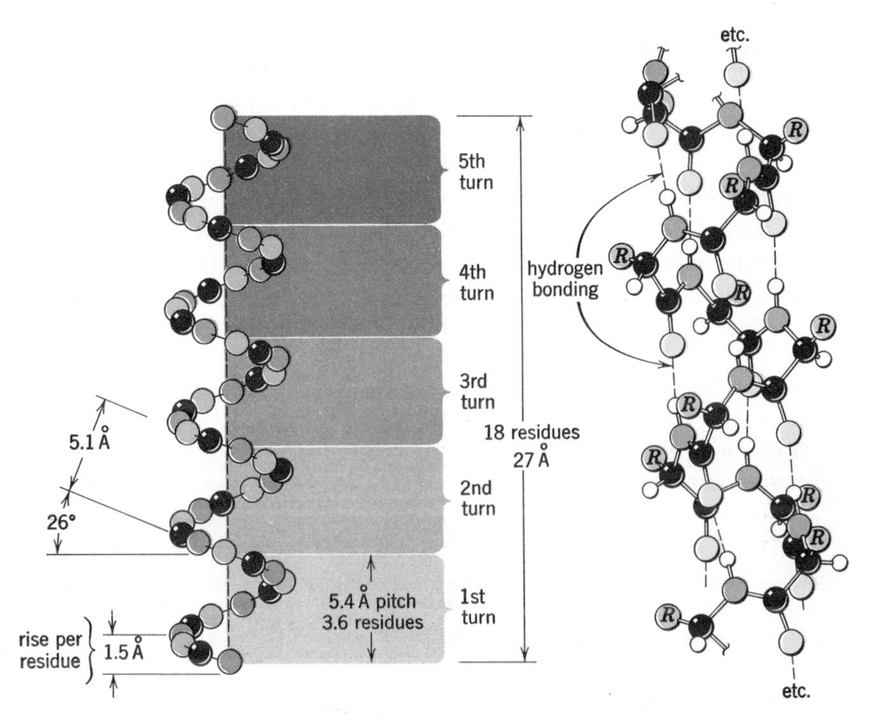

Figure 13-17. *A polypeptide coiled into an α-helix. The helix is held together by hydrogen bonds between N—H and C═O groups (shown as dashed lines in the drawing on the right). From "The Three-Dimensional Structure of a Protein Molecule" by J. C. Kendrew,* Scientific American, *December 1961. Used with permission.*

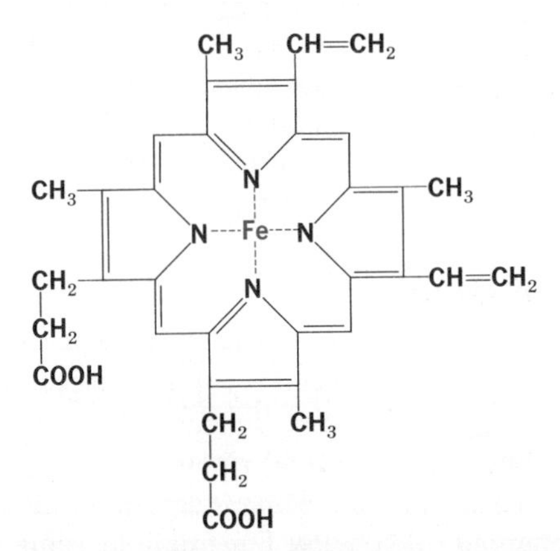

Figure 13-18. *The molecule* heme, *found in hemoglobin and other oxygen-carrying proteins. An oxygen molecule is bound to the iron in the lungs, transported through the bloodstream by the red blood cells, and released in the muscle.*

protein at the corners. What causes the bending of a protein chain out of the helical form? One factor that at least makes such bending easy is the presence of proline in the amino acid chain. This α-amino acid is unique in having its side chain attached to the amino group (Fig. 13-3). Proline in a polypeptide chain has no hydrogen atom on the amino nitrogen atom, and so cannot form a hydrogen bond to a carbonyl group, as do all the other amino acid in the α-helix. As a result the helix can bend in the vicinity of a proline unit. Significantly, human hemoglobin contains seven proline molecules, and a bend in the chain occurs in the vicinity of each.

In Fig. 13-19 the sequence of amino acids in the enzyme ribonuclease,

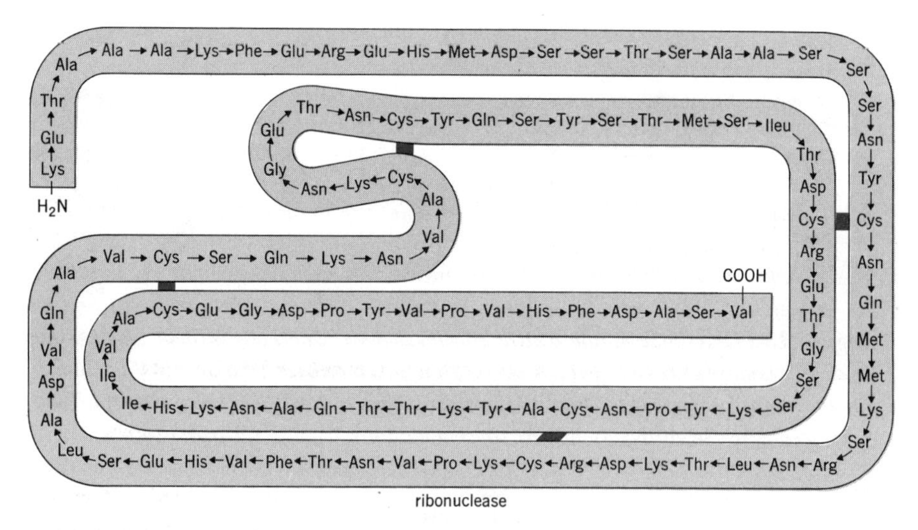

Figure 13-19. *The primary structure (amino acid sequence) of the enzyme ribonuclease. The S—S bonds, which help maintain the three dimensional shape of the molecule, are shown in red. Modified from* Introduction to Molecular Biology, *by G. H. Haggis, et al., Wiley, New York, 1964, p. 48.*

the first protein to be synthesized in the laboratory, is shown. This enzyme catalyzes the hydrolysis of the phosphate ester linkages in ribonucleic acid (RNA, Chapter 14). The four —S—S— linkages between cysteine sidechains are indicated. These bonds help to hold ribonuclease into its enzymatically active three dimensional shape. This shape is also determined by electrostatic interaction among charged side chains. For example, some amino acids (Glu, Asp) have side-chain carboxylate anions, while other side chains, e.g., Arg and Lys, may have positively charged ammonium ions. Naturally, the positive and negative charges of these groups will be attracted to each other, and they help in stabilizing the protein structure. The general side-chain polarity of the amino acids also plays an important role. Many proteins are water-soluble; in these structures the ionic and polar side chains point out into the water and the nonpolar side chains are gathered inside, forming a relatively water-free interior. In a sense the structure is maintained, at least in part, by the necessity of squeezing water out of the molecule. On the other hand, the proteins that are responsible for the conduction of ions and polar molecules through cell membranes (p. 235) have just the opposite polarity. Their outside is relatively nonpolar so as to be compatible with the hydrocarbon interior of the membrane, while their polar and ionic side chains are pointed inward to make a comfortable environment for the substances (polar *and* ionic) they will escort through the membrane.

Quaternary Structure of Proteins

By the *primary structure* of a protein we mean the sequence of amino acids in the molecule. Its *secondary structure* is the helical portion of the chain within the molecule. We say its *tertiary structure* is known when a complete, three-dimensional drawing can be made. Many proteins are known to have a *quaternary structure,* in which individual proteins are joined together by secondary valence forces (ionic attraction, hydrogen bonds) into larger units. Some enzymes, for example, contain two identical protein molecules that are inactive by themselves but which join together into an enzymatically active dimer. Hemoglobin is an example of a protein that contains four subunits, two molecules of α-hemoglobin and two of β-hemoglobin, each subunit with a molecule of heme. Each hemoglobin tetramer can be dissociated readily into the four subunits, since no covalent bonds are involved in its formation, but the tetramer reassembles itself spontaneously under physiological conditions. Each hemoglobin tetramer can therefore carry four molecules of oxygen. The interaction between the subunits in hemoglobin is such that uptake of the second, third, and fourth oxygen molecules occurs more readily than that of the first; this guarantees that each hemoglobin tetramer will enter the blood stream fully loaded with four molecules of oxygen.

Mechanism of Enzyme Action

An enzyme is a protein that catalyzes a chemical reaction. Several thousands of enzymes exist, and each catalyzes one or more of the several thousands of chemical reactions that occur in living systems. The tertiary structure of a number of enzymes is now known from X-ray studies, and chemists are beginning to be able to formulate reasonable theories about how they exert their catalytic activities. Each seems to have a "jaw" or cleft into which the *sub-*

strate, i.e., the molecule on which the enzyme acts, fits. Into this cleft, which is called the *active site,* project the side chains of the amino acids that actually carry out the catalysis. Ribonuclease, for example, catalyzes the hydrolysis of an ester between phosphoric acid and sugars in RNA (see Fig. 14-4). For an ester hydrolysis one needs acidic or basic groups, and the amino groups of lysine and the N—H bonds in histidine are utilized.

We mentioned earlier (p. 200) the enzyme acetylcholinesterase, which is important in the transmission of nerve impulses. Although the complete, three-dimensional structure for this enzyme has not yet been determined, we do know a great deal about how it reacts, and about how it is prevented from reacting by nerve gas and other nerve poisons.

Acetylcholinesterase is a protein of molecular weight about 250,000 with, most likely, four subunits. One enzyme molecule hydrolyzes approximately 800,000 molecules of acetylcholine per minute. The active site consists of two subsites, an anionic site that attracts the positively charged ammonium ion, and a hydrolytic site at which the actual hydrolysis occurs (Fig. 13-20).

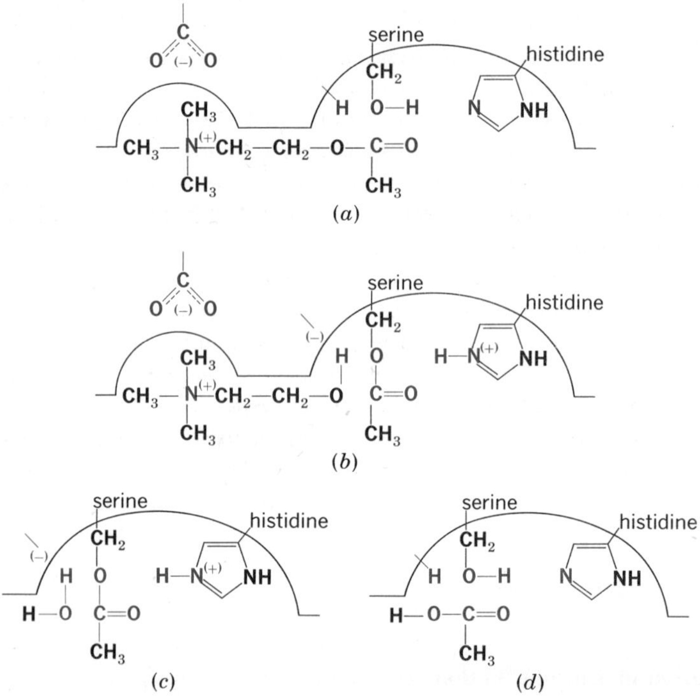

Figure 13-20. *Steps in the hydrolysis of acetylcholine by the enzyme acetylcholinesterase. In (a) the acetylcholine is drawn to the active site by electrostatic attraction. In (b) the acetyl group is transferred to the hydroxyl group of a serine molecule. In (c) the choline is replaced by a water molecule which in (d) receives the acetyl group and reforms the active enzyme.*

The hydrolysis occurs in two steps. First, the acetyl group is transferred to a serine molecule to form an *acetyl enzyme,* an intermediate that can actually be isolated. After this transfer of an acetyl group from acetylcholine to the enzyme, the choline molecule dissociates from the enzyme and a water molecule enters the active site, hydrolyzes the acetyl enzyme, forms acetic acid, and regenerates the active enzyme.

The mode of action on a molecular basis of the drugs and nerve gases whose structures are given in Fig. 10-10 can be understood in terms of Fig. 13-20. The ammonium salts, like decamethonium, simply occupy the anionic site and prevent access of acetylcholine. On the other hand, DFP and Parathion react to form phosphate esters of serine, analogous to, but more stable than, the acetyl enzyme (Fig. 13-21). The enzyme becomes *inactivated* and nerve transmission is disrupted, causing rapid death.

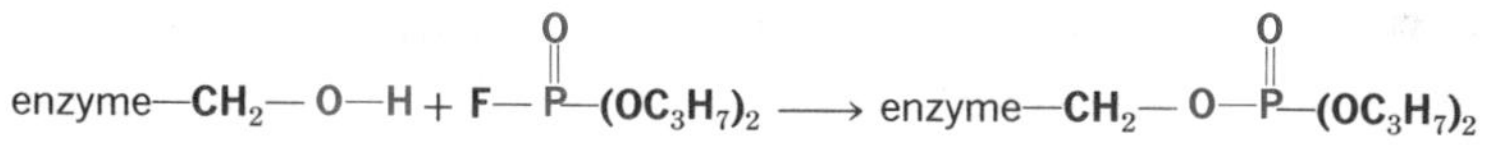

$$\text{enzyme—}CH_2\text{—O—H} + \text{F—}\overset{\overset{\displaystyle O}{\|}}{P}\text{—}(OC_3H_7)_2 \longrightarrow \text{enzyme—}CH_2\text{—O—}\overset{\overset{\displaystyle O}{\|}}{P}\text{—}(OC_3H_7)_2$$

Figure 13-21. *The nerve gas DFP forms a phosphate ester with the serine molecule of the enzyme active site, making it unavailable for reaction with acetylcholine.*

The modes of action of a number of enzymes are beginning to be understood; although they of course differ in detail from that for acetylcholinesterase, they all have in common the cooperative action of several amino-acid side chains which project into the active site and serve as donors or acceptors of protons or furnish the site for covalent bonding in an intermediate. As the mechanism of action of a particular enzyme becomes understood, it is sometimes possible to make medical use of the understanding. For example, based on the rational given above for the action of acetylcholinesterase, an antidote for nerve gas poisoning was designed that has been instrumental in saving many lives.

PROBLEMS

1. Give a structure for each of the following:
 (a) the Zwitterion form of an amino acid
 (b) the D-form of serine **(c)** polyalanine
 (d) an acidic amino acid **(e)** threonylmethionine
 (f) methionylthreonine **(g)** ninhydrin
 (h) DNP-glycine **(i)** cysteine

2. Tell what are meant by the primary, secondary, tertiary, and quaternary structures of a protein.

3. Describe the steps that would be needed to synthesize the tripeptide Gly-Ala-Val in the laboratory.
4. How many dipeptide structures would one have to establish to deduce the structure of a pentapeptide that gave the dipeptides on hydrolysis, assuming you knew which amino acid was N-terminane?
5. Give the IUPAC systematic names for serine, threonine, and alanine.
6. A common method for the preparation of amines is to treat an alkyl halide with ammonia (Fig. 10-3). Show how this reaction could be applied to the synthesis of glycine. Why should a large excess of ammonia be used in this reaction?
7. List three different chemical effects that might serve to maintain the three-dimensional structure of a protein.
8. A heptapeptide was isolated which, upon total hydrolysis, gave the following amino acids on chromatography: Met, Ser, Val, Gly, Phe, and Ile. When the heptapeptide was reduced with $LiBH_4$, and chromatographed, only the glycine spot was missing. Partial hydrolysis and sequencing gave the following peptides: Ile-Ser-Val, Met-Phe, Val-Ile, and Ser-Val-Met. What was the structure of the original heptapeptide?
9. Some peptides are cyclic, and so do not have N- or C-terminal amino acids. Such an octapeptide gave, on partial hydrolysis, the following tripeptides. What is its amino acid sequence?

   Gly-Cys-Ala　　Tyr-Cys-Phe
   Cys-Ala-Tyr　　Phe-His-Gly
10. A sample of 0.131 g of an amino acid was dissolved in water and treated with nitrous acid. Nitrogen gas was produced and collected. Under standard condition the gas totaled 22.4 ml. What was the molecular weight of the amino acid?
11. An isomer of threonine (Fig. 13-4) called allothreonine is a diastereoisomer belonging to the L-series of amino acids. What is its structure?

14 nucleic acids and the biosynthesis of proteins

The nucleic acids, DNA and RNA, are the third of the great classes of natural polymers. The carbohydrates and proteins furnish the material of construction of living systems and the machinery for carrying out the construction, while the nucleic acids contain the information, or the plans, that direct the work. In the nucleus of the cell we have the master information polymers, *d*eoxy-ribo*n*ucleic *a*cid, DNA, with its complementary strands twisted into a helix. The DNA is like a master tape, kept carefully stored. Segments of this master are copied as *r*ibo*n*ucleic *a*cid, RNA, which is actually used in the day to day operation of the cell.

The Structure of DNA

The nucleic acids have a *polyester* backbone, to which organic bases are attached; in that sense they may be compared to proteins with their polyamide backbone to which amino-acid side chains are attached (Fig. 14-1). Both DNA and RNA are linear polyesters formed from a triacid (phosphoric acid) and a di- or triol (a sugar derivative). Their structures are analogous to that of the polyester Dacron (Fig. 11-14).

Attached to each sugar of this backbone, DNA has one of four organic bases, adenine (A), cytosine (C), guanine (G), or thymine (T). These are complex nitrogen-containing molecules called *heterocycles* (because they contain within a ring an atom other than carbon, a hetero atom). They are attached to the sugar by an amino-ether linkage (Fig. 8-9), exactly analogous to the acetal linkage, except that an amine replaces one of the ethers (Fig. 14-2).

Deoxyribonucleic acid is then a polymer linking sugar-base components

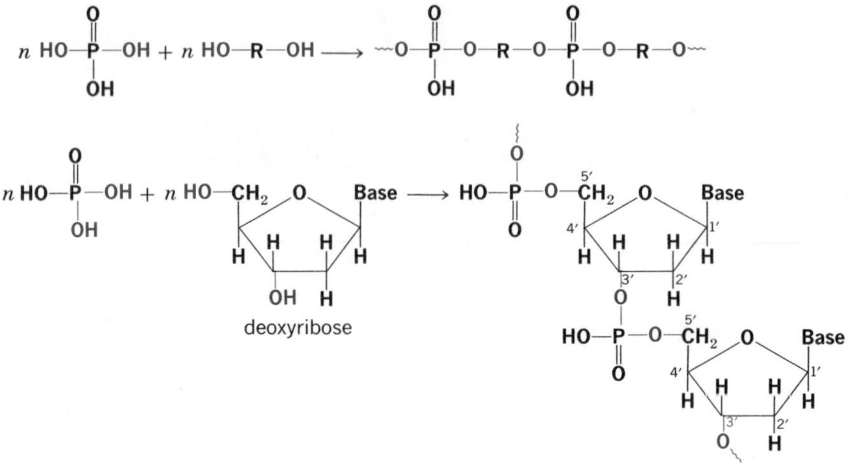

$$n\ \mathrm{HO{-}P(=O)(OH){-}OH} + n\ \mathrm{HO{-}R{-}OH} \longrightarrow \sim\mathrm{O{-}P(=O)(OH){-}O{-}R{-}O{-}P(=O)(OH){-}O{-}R{-}O}\sim$$

n HO—P(=O)(OH)—OH + n HO—CH_2 (deoxyribose ring: O, Base, H, H, H, H, OH, H) ⟶ HO—P(=O)(O~)—O—CH_2(5′) (ring: O, Base at 1′, 4′, 3′, 2′, H, H, H, H) 3′-O—P(=O)(HO)—O—CH_2(5′) (ring: O, Base at 1′, 4′, 3′, 2′, H, H, H, H, O~)

deoxyribose

Figure 14-1. *The backbone of DNA is a polyester between the sugar deoxyribose (with a base attached) and phosphoric acid. Note that such a polymer has a* direction, *with the 5′ end written on the left and the 3′ end on the right. The polymer is an acid, since one acid group of the phosphoric acid remains unesterified.*

(nucleosides), with phosphoric acid. An example of a possible segment of such a polymer is given in Fig. 14-3. Note the names of the four component nucleosides. Ribonucleic acid has an analogous structure, except that the sugar is *ribose*, with a hydroxyl group at the 2′ position, and the base uridine (U) replacing thymine (Fig. 14-4, on p. 260).

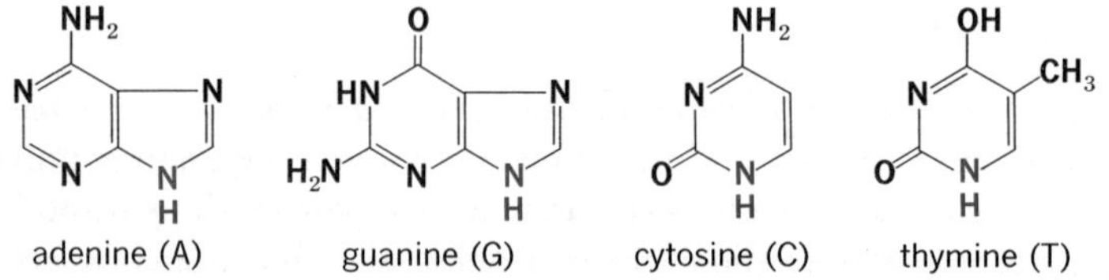

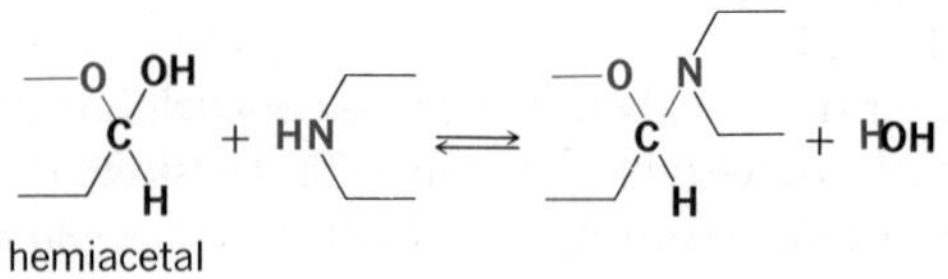

NH_2, N, N, N, N, H — adenine (A)

O, HN, N, H_2N, N, N, H — guanine (G)

NH_2, N, O, N, H — cytosine (C)

OH, CH_3, N, O, N, H — thymine (T)

—O OH C H + HN ⇌ —O N C H + HOH

hemiacetal

Figure 14-2. *The four heterocyclic bases found in DNA. Each is joined through the nitrogen atom in red to C-1 of deoxyribose in the polyester backbone. The bases A and G are* purines; *C and T are* pyrimidines.

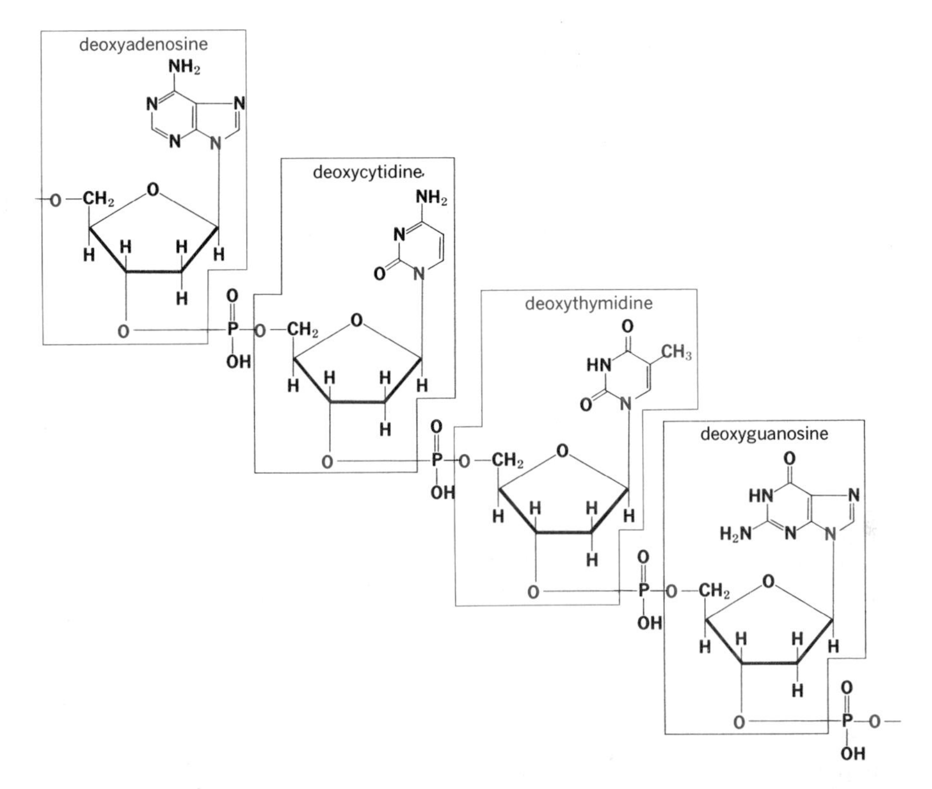

Figure 14-3. *Part of a DNA molecule. Note the way in which the bases are joined to the polyester backbone. Each box encloses a nucleoside.*

The Double Helix

The four bases in DNA form complementary *base-pairs;* the purine bases, adenine and guanine, pair with the pyrimidine bases, thymine and cytosine. Adenine pairs with thymine (A=T), and guanine pairs with cytosine (G≡C). Pairing occurs along the edge of the molecules, with hydrogen bonds forming to hold the complementary bases together (Fig. 14-5).

Deoxyribonucleic acid is a two-strand molecule with each strand the complement of the other. By this we mean that opposite each base in one strand is its hydrogen-bonded partner in the other. In addition to being complementary, the strands run in opposite directions, i.e., $3' \rightarrow 5'$ in one strand, $5' \rightarrow 3'$ in the other. The bases project from the backbone into the center and form hydrogen-bonds with their complements, giving a structure like a rope ladder (Fig. 14-6), coiled into a helix. In cell division, each strand of the double helix serves as a *template,* a model on which its complement

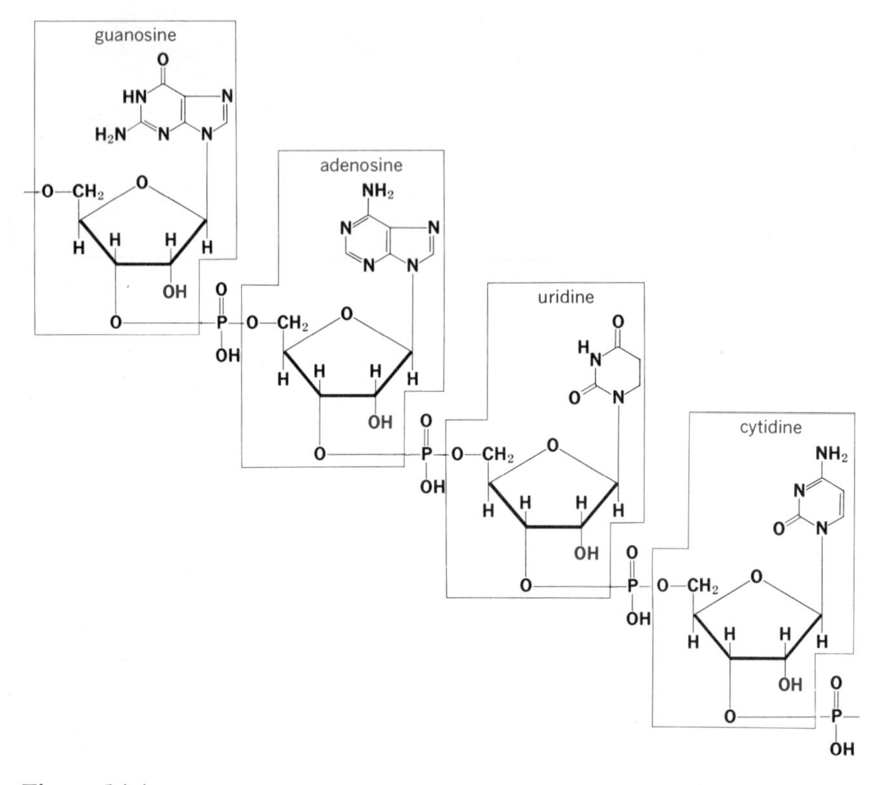

Figure 14-4. *Ribonucleic acid differs from deoxyribonucleic acid by the presence of a hydroxyl group on the sugar (in red) and by the absence of a methyl group from one base (uridine).*

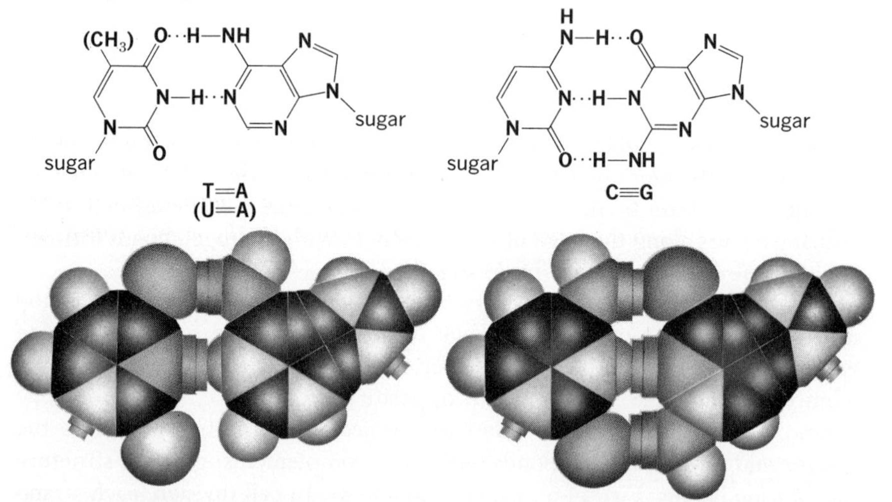

Figure 14-5. *Thymine-adenine (or uracil-adenine) and cytosine-guanine form base pairs that are held together by hydrogen bonding, as shown.*

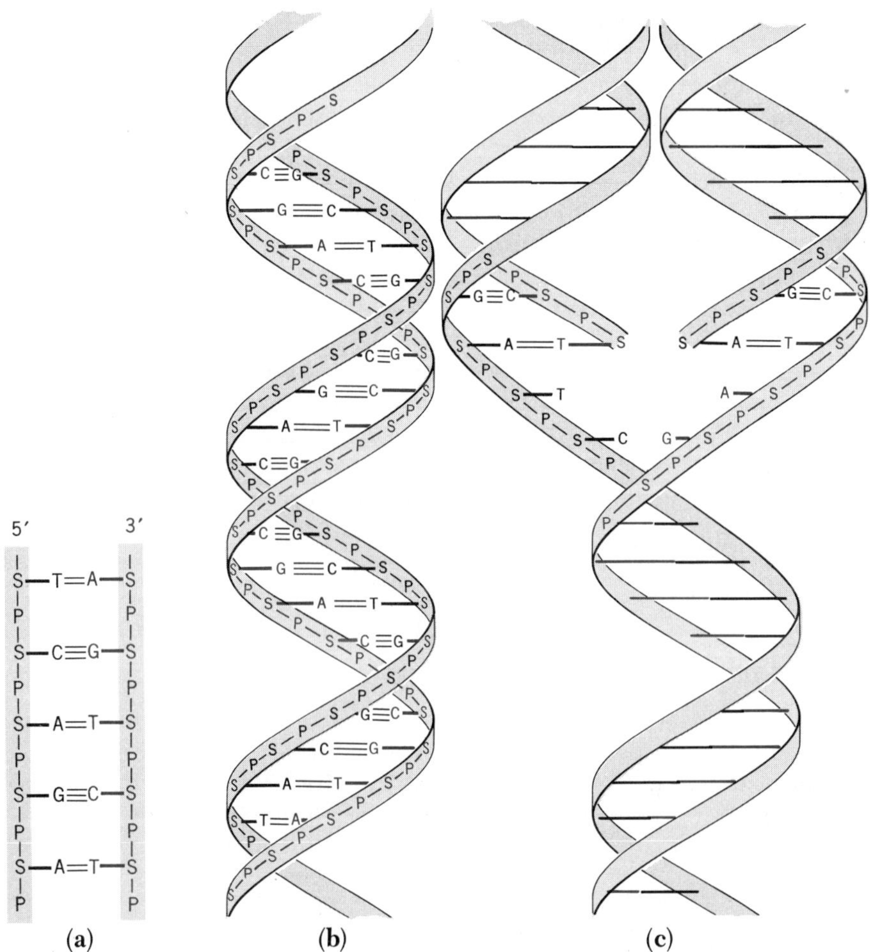

Figure 14-6. *In the double helix, two complementary DNA strands form (a) and wind into a helix (b). During gene replication each strand serves as a template for the reconstruction of the double helix. The drawing in (c) is meant to show schematically that replication of each complementary DNA strand will produce two identical DNA molecules. Replication does not take place simultaneously on each strand, and many details of the process are not understood.*

may be constructed. As a cell prepares to divide, its chromosomes, which contain the DNA, double so that each daughter cell contains a complete set of the genetic information.

RNA and the Genetic Code

A DNA molecule may have a molecular weight of many millions and contain the information needed for the syntheses of numerous protein molecules. A great deal, but far from everything, is known about how the information from the DNA is converted into the actual synthesis of a protein molecule. Here we can only give the main points and emphasize how the biological protein synthesis resembles that used in the laboratory solid phase method.

The first step in protein synthesis is the formation of messenger RNA (mRNA), which is a complement of a segment of the genetic message contained in the DNA and which is formed by a base pairing process of RNA on DNA like that described in the preceding section for the replication of DNA itself. The mRNA codes, by the sequence of its bases, for the sequence of amino acids in the protein to be synthesized. The genetic code is expressed in *base triplets* called *codons;* and the four bases U, C, A, and G can be arranged in triplets in 64 different ways. Since 64 code words are used to designate the 20 amino acids, the genetic code is *degenerate,* with several different words coding for the same amino acid. The genetic code is given in Fig. 14-7. Note that three triplets, UAA, UAG, and UGA code for termination of synthesis and one, AUG, codes for initiation as well as for methionine.

The Genetic Code

First position (5′ end)	Second Position				Third position (3′ end)
	U	C	A	G	
U	Phe	Ser	Tyr	Cys	U
	Phe	Ser	Tyr	Cys	C
	Leu	Ser	Term	Term	A
	Leu	Ser	Term	Trp	G
C	Leu	Pro	His	Arg	U
	Leu	Pro	His	Arg	C
	Leu	Pro	Gln	Arg	A
	Leu	Pro	Gln	Arg	G
A	Ile	Thr	Asn	Ser	U
	Ile	Thr	Asn	Ser	C
	Ile	Thr	Lys	Arg	A
	Met	Thr	Lys	Arg	G
G	Val	Ala	Asp	Gly	U
	Val	Ala	Asp	Gly	C
	Val	Ala	Glu	Gly	A
	Val	Ala	Glu	Gly	G

Figure 14-7. *The base triplets which code for the amino acids. The code word for methionine, AUG, also codes for initiation, while three words, UAA, UGA and UAG, all code for termination.*

The Ribosomes

The actual synthesis of proteins occurs on large particles found outside the cell nucleus which are called *ribosomes;* these play an analogous role to the polystyrene beads in the laboratory solid phase method of protein synthesis.

Ribosomes contain as many as 70 to 80 different proteins and a type of RNA known as ribosomal RNA (rRNA) whose function is as yet unknown. A bacterial cell may contain 10,000 ribosomes, an animal cell a million or more. All ribosomes consist of two subunits, one large and one small. Most is known about the ribosomes of the bacteria. *E. coli.* In E. coli cells the small subunit (called 30S ribosome because of its response in the ultracentrifuge) consists of 21 different proteins and a single molecule of rRNA with a molecular weight of about half a million. The large subunit of the *E. coli* ribosome (the 50S ribosome) consists of 35 different proteins and a much larger rRNA molecule (molecular weight about one million). The 30S and 50S subunits readily dissociate from one another but recombine during protein synthesis to a 70S ribosome with a particle weight of about three million. Analogous but larger subunits exist in animal ribosomes.

Transfer RNA (tRNA)

Before an individual amino acid can be recognized by a ribosome and incorporated into a protein, it must be esterified by its own characteristic transfer RNA (tRNA). Each cell contains about 60 different types of tRNA, more than one for each amino acid because of the degeneracy of the genetic code. Transfer RNAs are single-chain polymers of about 80 nucleic acids. Many have had their sequence of nucleic acids determined, and all tRNA can be written in a generalized cloverleaf pattern so as to maximize base pairing within the molecule. When written in this way, the amino acid is found attached to a ribose molecule at the 3′ end and the *anticodon* (i.e., the complement of the *codon* or triplet which codes for that amino acid) is found in the bottom leaf of the cloverleaf (Fig. 14-8).

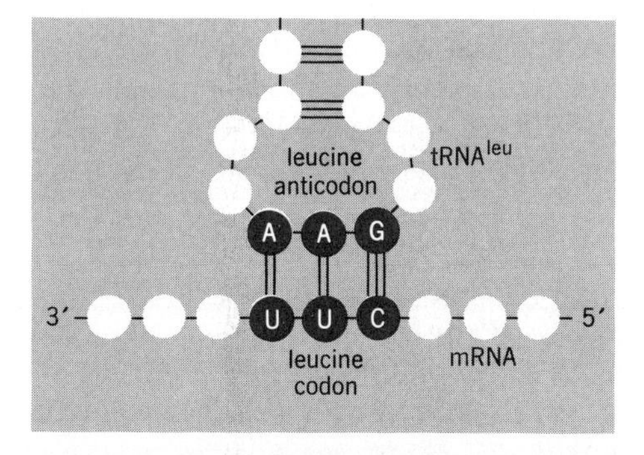

Figure 14-8. *Transfer RNAs are single-stranded RNA molecules to which amino acids become attached. In the bottom loop of their cloverleaf structure is the anticodon, the complement of the code word for the amino acid they contain. Note how the anticodon in tRNA^{leu} is the complement of the codon for leucine in mRNA.*

If an acid is to be esterified, it should first be activated; in biological systems this is usually accomplished by the formation of an anhydride with a substituted phosphoric acid (Fig. 14-9). More will be said about this activation in Chapter 15. Each of the 20 amino acids found in proteins has its own enzyme system that recognizes it, activates it, and attaches it to the correct tRNA.

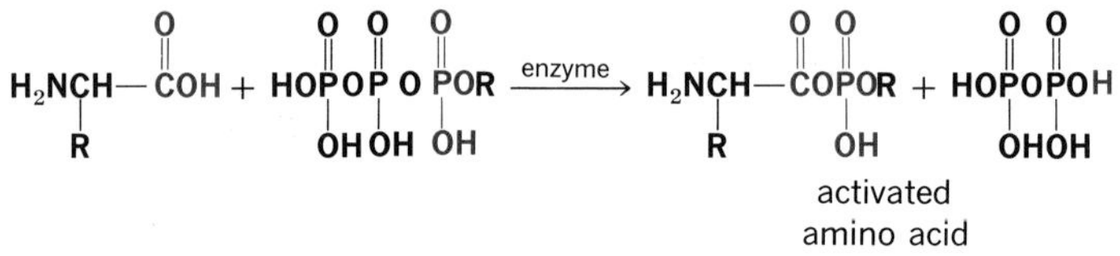

$$H_2NCH(R)-C(=O)OH + HOP(=O)(OH)OP(=O)(OH)OP(=O)(OH)OR \xrightarrow{\text{enzyme}} \underset{\text{activated amino acid}}{H_2NCH(R)-C(=O)OP(=O)(OH)OR} + HOP(=O)(OH)OP(=O)(OH)OH$$

Figure 14-9. *In biological systems amino acids are activated for reaction by the formation of an anhydride with phosphoric acid.*

Biosynthesis of Proteins

As with other polymerizations we can divide the steps in protein biosynthesis into initiation, propagation, and termination. As the genetic code shows, there is only one initiation codon, AUG, coding for methionine. When it occurs as the initial codon on mRNA, AUG codes for *N*-formylmethionine, esterified to a special initiation $tRNA_f$, which carries the anticodon. Note that the amino group of methionine, which will become the *N*-terminal end of the protein, is protected as an amide, just as we might protect it in the laboratory.

Initiation thus involves assemblage of mRNA, *N*-formylmethionyl-$tRNA_f$ and a ribosome as in II, Fig. 14-10. Several steps involving ribosome subunits, proteins, and Mg^{2+} are required for this assemblage, but the chemical details are not understood. Next, a second amino acyl–tRNA ester, one with the proper anticodon as specified by the next three bases on the mRNA (in our example that from alanine) enters an adjacent site on the ribosome, to give III. In this alanyl ester the amino group is not protected and is free to react with the adjacent methionine ester group to form an amide. We have seen

Figure 14-10. *Steps in the biosynthesis of a protein. I. The mRNA, which contains the code for the amino acid sequence, becomes associated with the 30S ribosome subunit. II. The first triplet is always AUG, and in response* N-*formylmethionine, as its tRNA ester, enters the P site. III. Directed by the second triplet, the appropriate tRNA ester enters the A site. IV. The free amine on the amino acid of the A site reacts with the ester on the P site to form a peptide bond. V.* Translocation *occurs, with mRNA moving so as to bring a new codon into site A. VI. The cycle repeats with a new amino acid.*

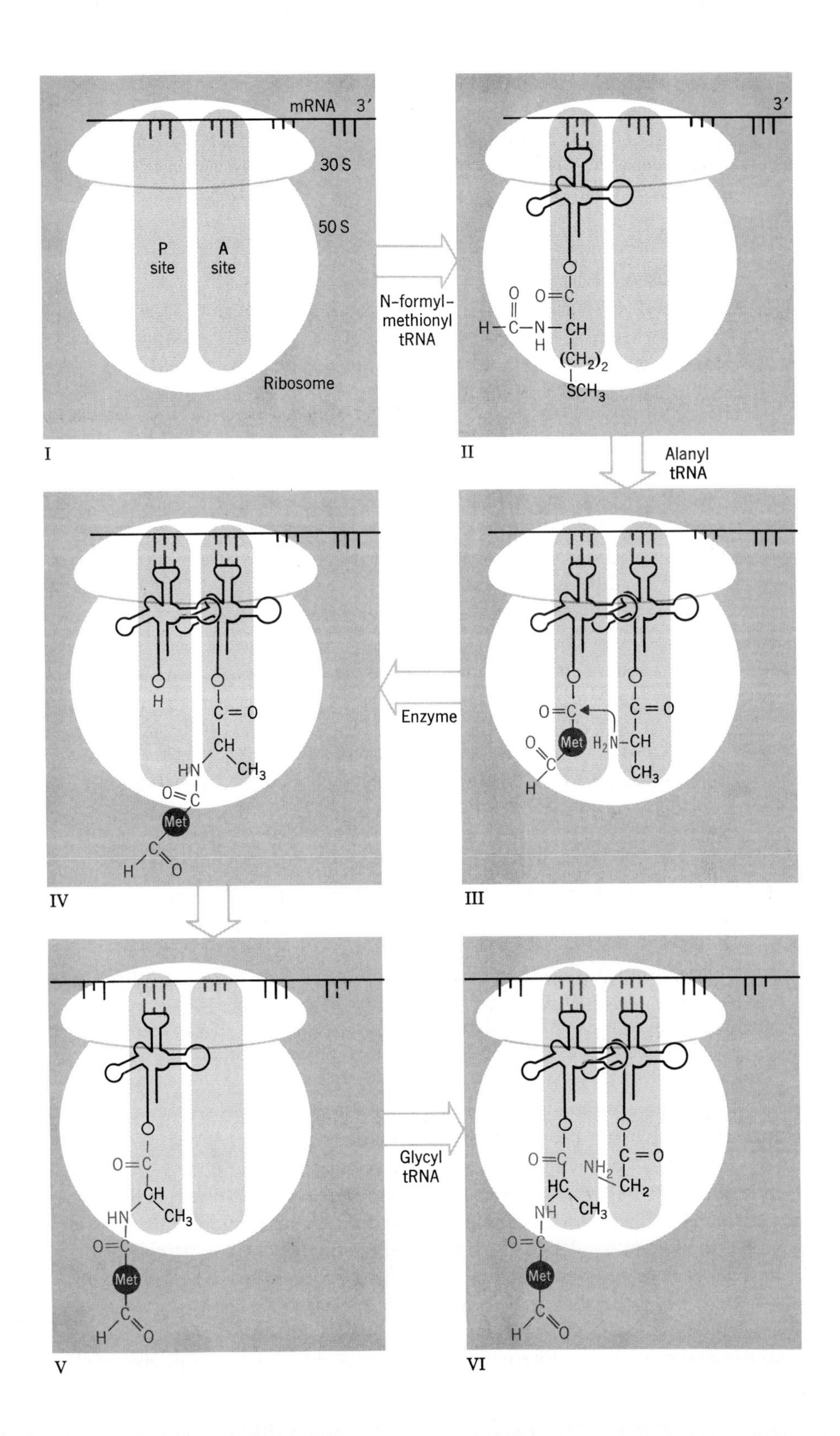
mRNA 3′
30 S
50 S
P site
A site
Ribosome
I
N-formyl-methionyl tRNA
3′
II
Alanyl tRNA
Met
III
Enzyme
IV
V
Glycyl tRNA
VI

that such a reaction is energetically favorable but slow. Here the reaction is speeded by enzymatic catalysis. By this means a peptide bond is formed, and the *N*-formylmethionyl group is transferred from its $tRNA_f$ to the adjacent A site, as in IV.

Finally, the elongation cycle is completed by *translocation* in which the mRNA moves one codon along through the ribosome, freeing $tRNA_f$, advancing the second tRNA, and bringing a new codon into position, as in V. A new tRNA is now free to enter, provided its anticodon is complementary (VI). Note that each new amino acid enters the chain at the C-terminal end; this is the reverse of the solid phase laboratory method of protein synthesis, where the initial amino acid remains attached to the bead and the chain is elongated at the N-terminal end.

Termination occurs when one of the three termination codons is encountered (Fig. 14-11). The tRNA ester, which is now a long protein chain, is hydrolyzed to an acid and the initiating *N*-formyl methionine is removed by hydrolysis. The ribosomes, mRNA, and tRNA dissociate and are ready to reassemble for additional syntheses.

Mutations

We can summarize the flow of information in protein biosynthesis as follows. First, one part of a DNA chain is copied to give mRNA. The sequence of bases in mRNA is thus complementary to those in the DNA; it is this sequence that determines the amino acid sequence in the protein. A change in the base sequence on the DNA will lead to a change in the complementary base sequence of mRNA, and this could lead to a change in the amino acids that are incorporated into the protein. Such a change in DNA is called a *mutation*. Here the degeneracy of the genetic code helps; many of the most common mutations that occur in DNA still code for the same amino acid or one with a similar side chain in terms of polarity, acidity, etc. However, the change of a single base in DNA may have serious consequences. An example is the disease *sickle cell anemia.* Persons with this disease produce an abnormal hemoglobin that is less soluble than ordinary hemoglobin and precipitates in tiny blood vessels, causing intense pain. In addition the abnormal hemoglobin is destroyed by the spleen, causing anemia.

Sickle cell anemia is caused by the change of a single base in DNA so that the mRNA which codes for the β-hemoglobin molecule specifies valine (GUA or GUG), rather than glutamic acid (GAA or GAG) for the seventh amino acid from the N-terminal end. Each of the two β-hemoglobins in the hemoglobin molecule has this one amino acid change; the two α-hemoglobins are the same in normal and in sickle cell hemoglobin. The change from an acidic to a hydrocarbon side chain on just 2 out of 600 amino acids is sufficient to cause the observed difference in properties (Fig. 14-12 on p. 268).

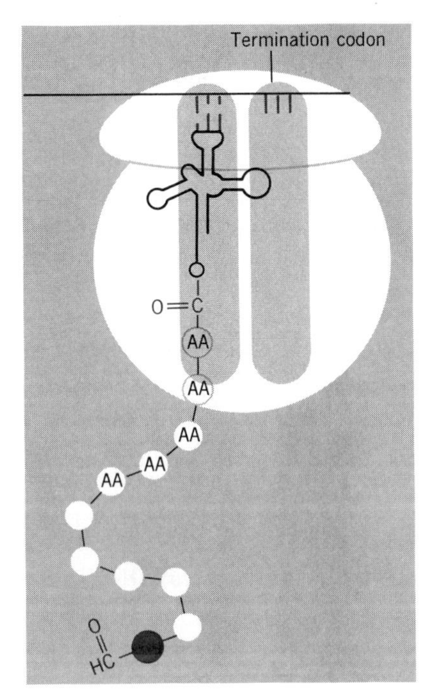

$$\xrightarrow[H_2O]{enzymes} \mathrm{H\overset{O}{\overset{\|}{C}}NH{-}\underset{\underset{\underset{SCH_3}{|}}{\underset{CH_2}{|}}}{\underset{CH_2}{\underset{|}{C}H}}{-}\overset{O}{\overset{\|}{C}}{-}NH{-}\underset{CH_3}{\underset{|}{C}H}{-}\overset{O}{\overset{\|}{C}}{-}NH\cdots\cdots\overset{O}{\overset{\|}{C}}{-}NH{-}\underset{R}{\underset{|}{C}H}{-}\overset{O}{\overset{\|}{C}}OH}$$

$$\Big\downarrow \, H_2O,\ enzymes$$

$$\mathrm{H\overset{O}{\overset{\|}{C}}OH + CH_3SCH_2CH_2\overset{NH_2}{\overset{|}{C}H}{-}\overset{O}{\overset{\|}{C}}{-}OH + HNH{-}\underset{CH_3}{\underset{|}{C}H}{-}\overset{O}{\overset{\|}{C}}{-}NH\cdots\cdots\overset{O}{\overset{\|}{C}}{-}NH{-}\underset{R}{\underset{|}{C}H}{-}\overset{O}{\overset{\|}{C}}OH}$$

Figure 14-11. *Termination of the protein biosynthesis begun in Fig. 14-10. At the termination signal the protein ester is hydrolyzed and, in subsequent steps, the initial* N-*formylmethionine is removed. The abbreviation* AA *stands for an unspecified amino acid.*

Mutations can be caused by accidental errors in copying DNA in preparation for cell division, but it can also be caused by ultraviolet light and by chemicals. Nitrous acid is a chemical that causes mutations, a *mutagen*. We know that it converts aromatic amines, by way of their diazonium salts, into phenols; it reacts analogously with nucleic acid bases, converting adenine, for example, into a base hypoxanthine, that resembles guanine in hydrogen-bonding capacity (Fig. 14-13).

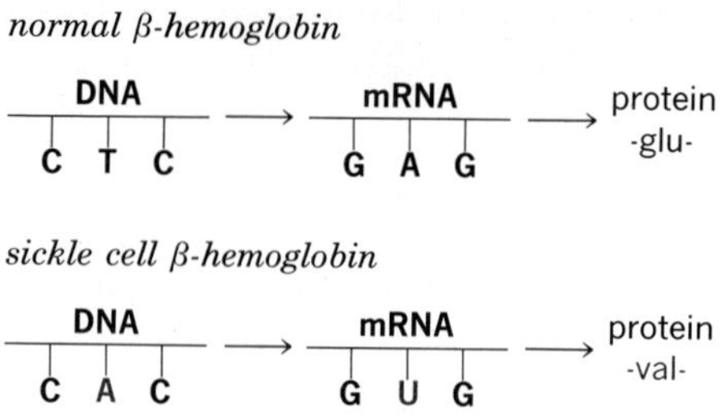

normal β-hemoglobin

DNA C T C ⟶ mRNA G A G ⟶ protein -glu-

sickle cell β-hemoglobin

DNA C A C ⟶ mRNA G U G ⟶ protein -val-

Figure 14-12. *In a point mutation a single base in DNA becomes changed, leading to the incorporation of the wrong amino acid in a protein. If the change affects the function of the protein, it can lead to disease or death.*

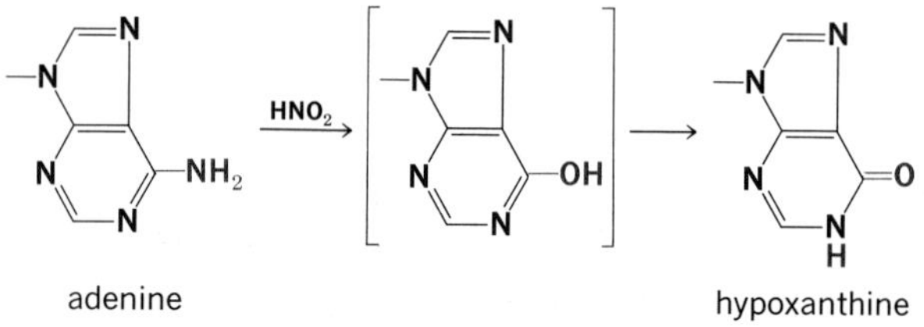

adenine $\xrightarrow{HNO_2}$ [ ] ⟶ hypoxanthine

Figure 14-13. *Nitrous acid causes mutations by converting adenine into hypoxanthine. The latter base pairs with cytosine, rather than with thymine, the base partner of adenine.*

Antibiotics and Their Mode of Action

Antibiotics are organic substances produced by microorganisms that are able to inhibit the growth of other microorganisms, especially bacteria or fungi. Antibiotics encompass a large number of structural types: peptides, sugars, lipids, and various other compounds, simple and complex. In the last two decades the mode of action of a number of antibiotics has been elucidated; many of them have been found to interfere with various aspects of the machinery that produces proteins. In this section we shall note the structure of several such antibiotics and discuss how they exert their activity.

Penicillin

Undoubtedly, the most famous and most widely used antibiotics are the penicillins. Although at first sight penicillin seems to have a completely unfamiliar structure, close examination reveals its origin; note that incorporated in the ring structures are two common amino acids, cysteine and valine

(Fig. 14-14). In penicillin G the cysteine amino group is joined to phenyl-acetic acid through an amide linkage. Other penicillins are amides of various other acids.

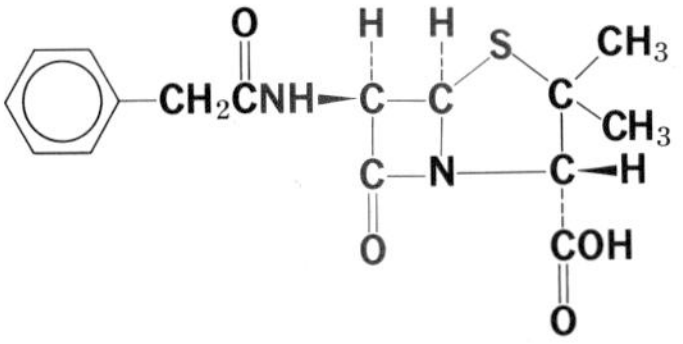

penicillin G

Figure 14-14. *Penicillin is synthesized by a mold and is clearly derived from the amino acids cysteine (in red) and valine.*

Penicillin inhibits the growth of bacteria by interfering with reactions that knit the bacterial cell walls together. These cell walls consist of carbohydrate and protein chains tied together with short peptide chains. The final stage of cell wall synthesis is the cross-linking of the peptides; it is this reaction with which penicillin interferes, probably because the enzyme responsible for the cross-linking mistakes penicillin for the peptide and is somehow inactivitated. A number of other antibiotics also interfere with this or other steps in bacterial cell wall manufacture, thus preventing bacterial growth.

Other Antibiotics

Puromycin has been of great aid in elucidating the steps in the biosynthesis of proteins. As its structure shows (Fig. 14-15), it closely resembles a nucleoside derivative of an amino acid but is an amide rather than an ester. It binds with the ribosome in place of an aminoacyl tRNA and becomes incorporated in the protein chain. At that stage protein synthesis stops because the next amino acid is unable to react with the amide bond in puromycin. Many other antibiotics, including tetracycline (Fig. 16-5) also interfere with various aspects of protein synthesis.

Two quite complicated antibiotics are the saccharide streptomycin and the modified peptide actinomycin. The former (Fig. 9-16) binds to the ribosome, causes misreading of mRNA (and hence forms proteins with wrong amino acid sequences), and interferes with initiation. The latter is a bright red compound that binds, by hydrogen bonding, to the DNA double helix and prevents copying (Fig. 14-16). In the presence of actinomycin no new mRNA can be formed, so no new proteins are produced.

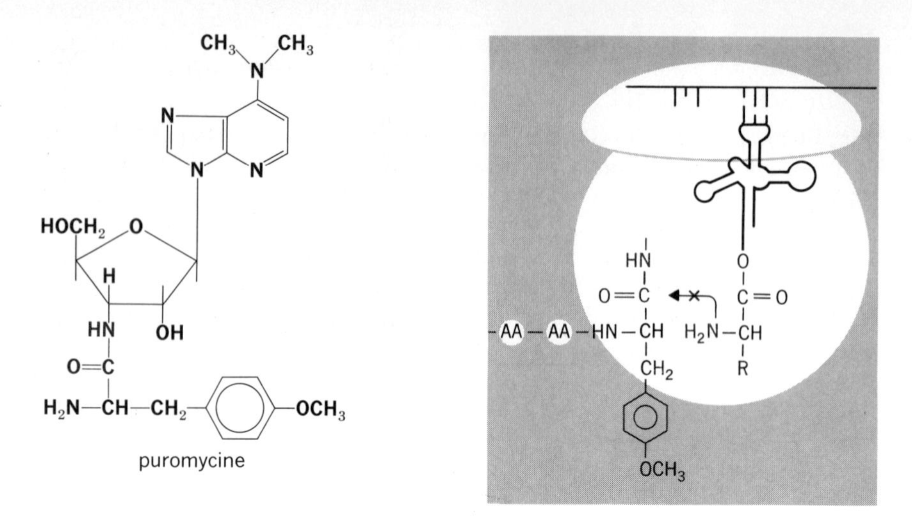

Figure 14-15. Puromycin has an amino acid attached to an amino sugar. When puromycin is incorporated into a protein chain, protein synthesis stops because the amino group of the next amino acid is unable to react with an amide group, which is much less reactive than the ester group usually present (compare Fig. 14-10).

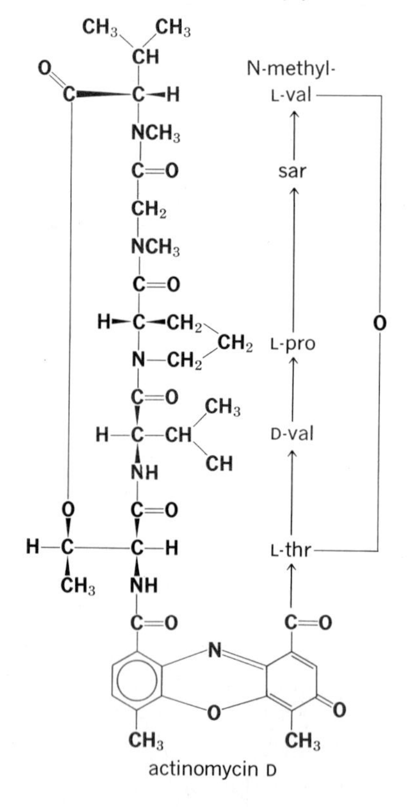

Figure 14-16. A modified peptide antibiotic. Note the presence of D-amino acids. The two peptide chains are identical; one is drawn out, the other abbreviated. Sar stands for sarcosine.

Drug Resistance

Given time, microorganisms develop resistance to antibiotics, usually by synthesizing an enzyme that modifies or destroys them. For example, penicillin-resistant bacteria have developed enzymes that hydrolyze the amide bond (four membered ring) of penicillin; the resulting hydrolyzed compounds are inactive as antibiotics (Fig. 14-17). Streptomycin is sometimes inactivated when a bacteria esterifies it with phosphoric acid.

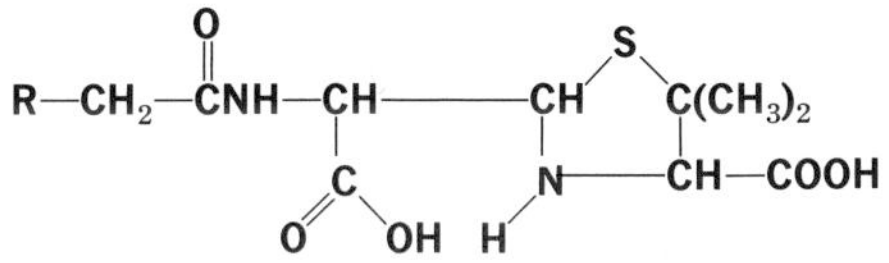

```
                 O                       S
                 ‖                     /   \
R—CH₂—CNH—CH—————————CH      C(CH₃)₂
                  |                   |       |
                  C                  N——————CH—COOH
                //  \               /
               O     OH            H
```

Figure 14-17. *Penicillin-resistant bacteria contain an enzyme that hydrolyzes penicillin, thus inactivating it.*

PROBLEMS

1. Give the structure of each of the following:
   (a) deoxyribose **(b)** a pyrimidine
   (c) a purine **(d)** a base pair
   (e) two base triplets for cysteine
   (f) the anticodon for initiation
   (g) *N*-formylmethionine
   (h) ribose-5-phosphate
   (i) the three termination codons

2. Suppose a mutation occurred so as to replace U by C. Which would be *least* harmful, for this substitution to occur in the first, second, or third position of a triplet? Why?

3. Give the anticodon for each of the following codons: CUU, AAG, and GCG.

4. If one section of a strand of DNA read as follows: UUUGCAACCCGA, what amino acid sequence would be generated? Suppose that during replication the first base (U) was accidentally omitted. What sequence would the remaining segment code for? How long would this misreading continue down the DNA chain? Do you think a base change (say U $\rightarrow$ A) or a base omission would probably have a more serious effect on an organism?

5. Compare by means of chemical equations the solid phase laboratory synthesis and the biosynthesis of peptides. Make a detailed comparison in terms of activating steps, blocking groups, etc.

15
metabolism and enolate reactions

By *metabolism* we mean the series of reactions by which biological nutrients, especially carbohydrates, lipids (fats and oils), and proteins are broken down chemically to supply energy for living systems. These chemical reactions must ultimately involve oxidations, and oxygen from the air is the eventual oxidant. In this chapter we shall see how some kinds of metabolism occur and how the energy released in the process is stored for later use in biosynthesis, the topic of Chapter 16. But before we can discuss metabolism or biosynthesis, we must introduce a new class of organic reactions that are especially useful for forming and breaking carbon–carbon bonds, the so-called *enolate condensation reactions*. After all, the formation of large organic molecules from small ones is at the very heart of organic chemistry, and yet the reactions we have discussed by which this may be accomplished, most notably the Grignard synthesis (Fig. 7-19), are hardly suitable for the aqueous environment of living systems.

Enolate Ions

Although we cannot use a Grignard reagent in biological systems, the principle by which carbon–carbon bonds are formed with these reagents remains crucial. Remember that we considered the reactive carbon in Grignard reagents to be negatively charged, i.e., a carbanion, and we found it to add to carbon–oxygen double bonds. It was also pointed out that the cyanide ion (itself a carbanion) is one of the few anions to add to a carbonyl group to give a stable product. Both of these reactions are reviewed in Fig. 15-1.

We must now recognize a new source of carbanions, one by which they can be formed reversibly in aqueous or alcoholic solutions. These carbanions arise by removal of a proton from the carbon adjacent to a carbonyl group. These *α-hydrogens* of aldehydes, ketones, esters, and their derivatives are only

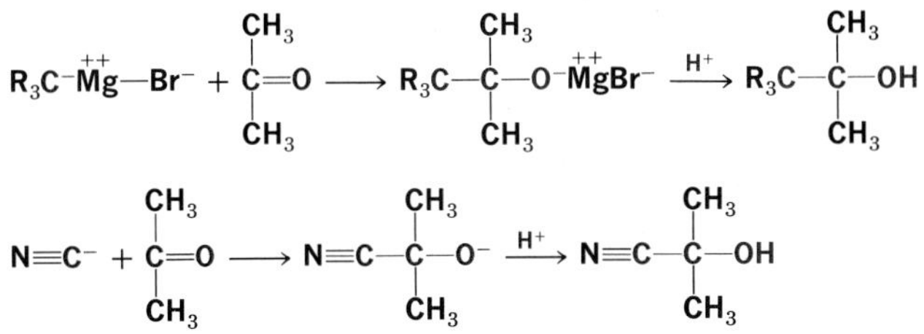

$$R_3C^{-}\overset{++}{Mg}-Br^{-} + \underset{CH_3}{\overset{CH_3}{C}}=O \longrightarrow R_3C-\underset{CH_3}{\overset{CH_3}{C}}-O^{-}\overset{++}{Mg}Br^{-} \xrightarrow{H^+} R_3C-\underset{CH_3}{\overset{CH_3}{C}}-OH$$

$$N\equiv C^{-} + \underset{CH_3}{\overset{CH_3}{C}}=O \longrightarrow N\equiv C-\underset{CH_3}{\overset{CH_3}{C}}-O^{-} \xrightarrow{H^+} N\equiv C-\underset{CH_3}{\overset{CH_3}{C}}-OH$$

Figure 15-1. *A new carbon–carbon bond is formed when a carbanion adds to a carbonyl group.*

slightly acidic, but acidic enough so that small but chemically significant amounts of carbanions are produced simply by dissolving carbonyl compounds in basic solution. Examples of the equilibria established are given in Fig. 15-2.

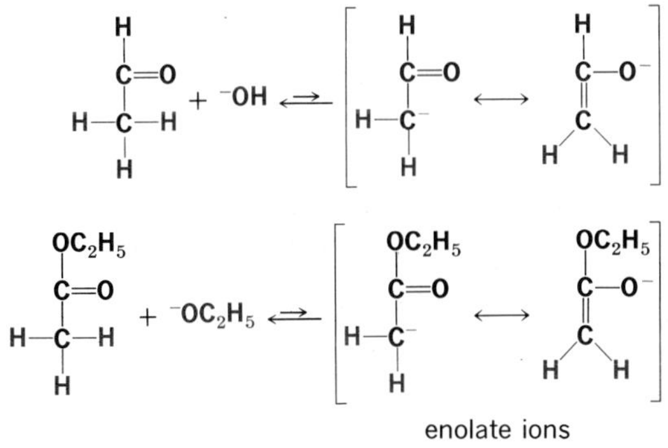

$$H-\underset{H}{\overset{\overset{H}{|}C=O}{C}}-H + {}^{-}OH \rightleftharpoons \left[H-\underset{H}{\overset{\overset{H}{|}C=O}{C^{-}}} \longleftrightarrow \overset{\overset{H}{|}C-O^{-}}{\underset{H\quad H}{C}} \right]$$

$$H-\underset{H}{\overset{\overset{OC_2H_5}{|}C=O}{C}}-H + {}^{-}OC_2H_5 \rightleftharpoons \left[H-\underset{H}{\overset{\overset{OC_2H_5}{|}C=O}{C^{-}}} \longleftrightarrow \overset{\overset{OC_2H_5}{|}C-O^{-}}{\underset{H\quad H}{C}} \right]$$

enolate ions

Figure 15-2. *Hydrogens on carbons adjacent to carbonyl groups (α-hydrogens, shown in red) are slightly acidic because the carbanion formed by their removal is delocalized, shared by oxygen of the carbonyl group.*

The acidity of these α-hydrogens is much enhanced over those aliphatic hydrogens which are not adjacent to a carbonyl group because of resonance in the carbanion, by which the negative charge is shared by the carbonyl oxygen as well as by the α-carbon atom. This is shown in resonance terminology in Fig. 15-2. There is an exact analogy between the enhanced acidity of the α-hydrogens of a carbonyl compound and the enhanced acidity of the hydroxyl group in a carboxylic acid; both types of protons occupy a similar position with respect to the carbon–oxygen double bond (Fig. 15-3).

When a carbonyl compound is brought into contact with base, an equilibrium with its anion is established. For simple aldehydes, ketones, and

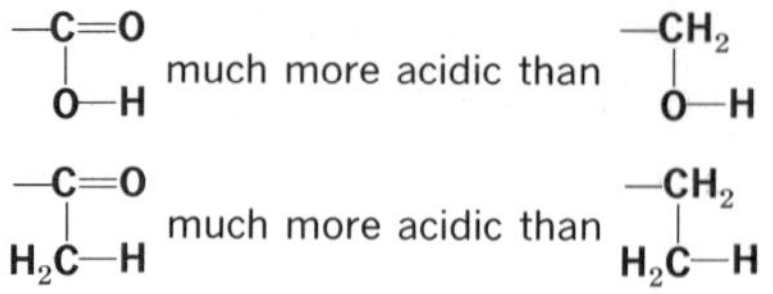

—C=O (O—H) much more acidic than —CH_2 (O—H)

—C=O (H_2C—H) much more acidic than —CH_2 (H_2C—H)

Figure 15-3. *The α-hydrogens are more acidic than normal aliphatic hydrogens, just as carboxylic hydrogens are more acidic than normal hydroxylic hydrogens.*

esters, there is usually only an extremely small amount of the carbanion present at equilibrium. Nevertheless, reactions can be carried out utilizing this anion; as it is used up by reaction, more carbanion is generated to reestablish the equilibrium. The situation is exactly analogous to that which prevails for reducing sugars, where despite the fact that only a trace of free aldehyde is present, the sugar gives all aldehyde reactions by way of the equilibrium between hemiacetal and aldehyde.

Carbanions formed by the removal of an α-hydrogen are known as *enolates,* since they may also be considered to be salts of an *enol,* compounds with a hydroxyl group substituted on a double bond (*ene*-alkene, *ol*-alcohol). See Fig. 15-4.

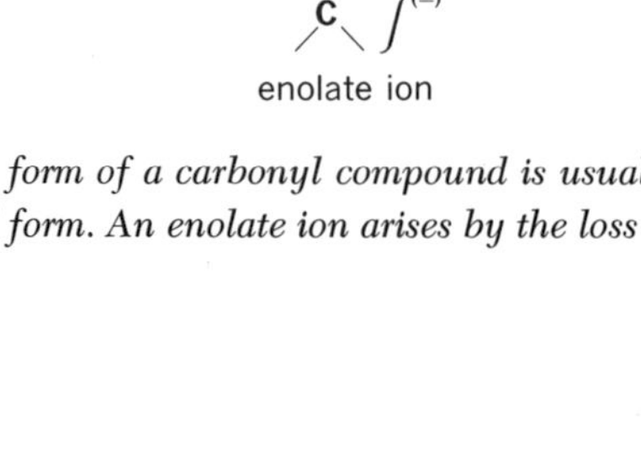

C=O, C—H ⇌ C—OH, C

carbonyl form

enol form
(unstable)

C=O, C (−)

enolate ion

Figure 15-4. *The enol form of a carbonyl compound is usually unstable and* tautomerizes *to the carbonyl form. An enolate ion arises by the loss of a proton from either tautomer.*

The Aldol Condensation

Since an enolate ion is a carbanion, it can add to the carbon of a carbonyl group. Suppose that we add acetaldehyde to an aqueous base. A small amount of the carbanion is formed, but most of the aldehyde is still in the carbonyl form. We then have a carbanion and a carbonyl compound in the same solution, and addition can occur. The result is *aldol,* a product that is both an *ald*ehyde and an alcoh*ol.* This *condensation reaction* (so-called because two molecules have been condensed into one) is reversible, but the equilibrium lies on the side of aldol (Fig. 15-5).

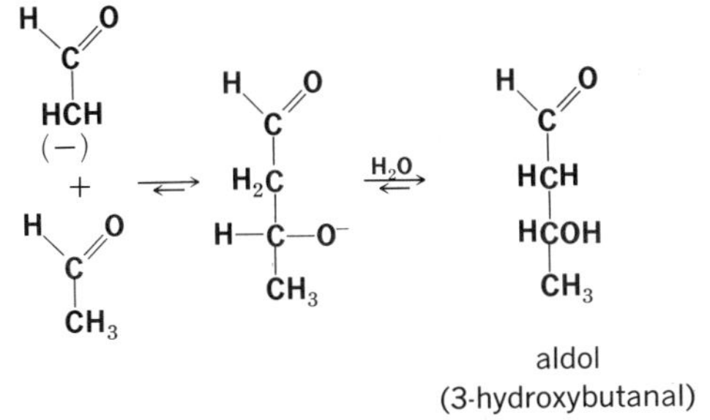

H O
C
HCH
(−)
+
H O
C
CH_3

$\rightleftharpoons$

H O
C
H_2C
H—C—O^-
CH_3

$\xrightarrow{H_2O}$

H O
C
HCH
HCOH
CH_3

aldol
(3-hydroxybutanal)

Figure 15-5. *The treatment of acetaldehyde with base gives aldol, in which a carbanion derived from one aldehyde molecule adds to the carbonyl group of another. All reactions are reversible; starting with aldol, some dissociation to acetaldehyde occurs in base.*

The Claisen Condensation

A similar reaction occurs when an ester is dissolved in base, although since an ester is hydrolyzed by sodium hydroxide, an alkoxide must be used as the base. Addition across the carbonyl group leads to a hemiketal; such compounds are unstable and quickly break down by loss of alcohol. The result is a β-ketoester (Fig. 15-6).

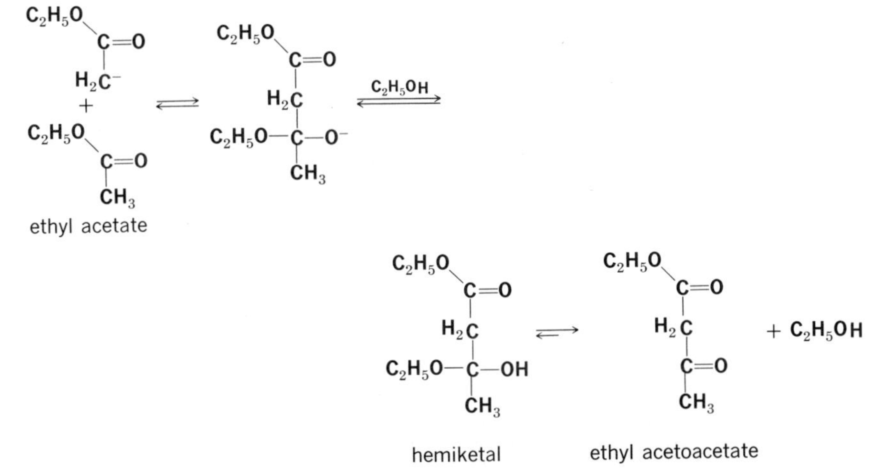

C_2H_5O
C=O
H_2C^-
+
C_2H_5O
C=O
CH_3
ethyl acetate

$\rightleftharpoons$

C_2H_5O
C=O
H_2C
C_2H_5O—C—O^-
CH_3

$\xrightarrow{C_2H_5OH}$

C_2H_5O
C=O
H_2C
C_2H_5O—C—OH
CH_3
hemiketal

$\rightleftharpoons$

C_2H_5O
C=O
H_2C
C=O
CH_3
ethyl acetoacetate

+ C_2H_5OH

Figure 15-6. *In the* Claisen condensation *the carbanion from an ester adds to the carbonyl group of another ester molecule; the result is a β-hydroxy-β-ethoxyester which, being a hemiketal, is unstable and yields a β-ketoester.*

β-Dicarbonyl Compounds

Naturally, if a proton is alpha to two carbonyl groups, as are for instance either of the two hydrogens on the methylene group of ethyl acetoacetate (Fig. 15-6), it is still more acidic. Such hydrogens are about as acidic as those of phenol; compounds of this type are completely converted to their salts by hydroxide or alkoxide ions (Fig. 15-7). The β-dicarbonyl compounds are

a β-dicarbonyl compound

enolate anion

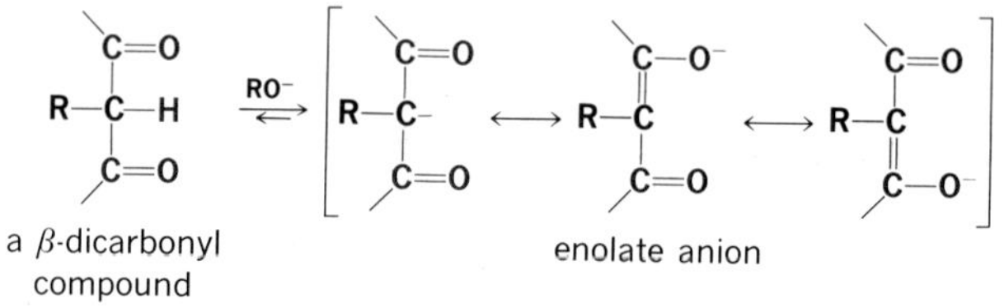

Figure 15-7. In the enolate anion of a β-dicarbonyl compound the negative charge is shared by two oxygens and a carbon; such an anion is quite stable, and β-dicarbonyl compounds can be converted completely to their anions in base.

readily cleaved into two carbonyl compounds by the reverse of the Claisen condensation. For much the same reason a β-keto acid readily loses carbon dioxide (it is easily *decarboxylated*). These various properties of β-dicarbonyl compounds are used for synthetic purposes in the *acetoacetic ester synthesis* (Fig. 15-8). Ethyl acetoacetate (acetoacetic ester) is easily converted to its

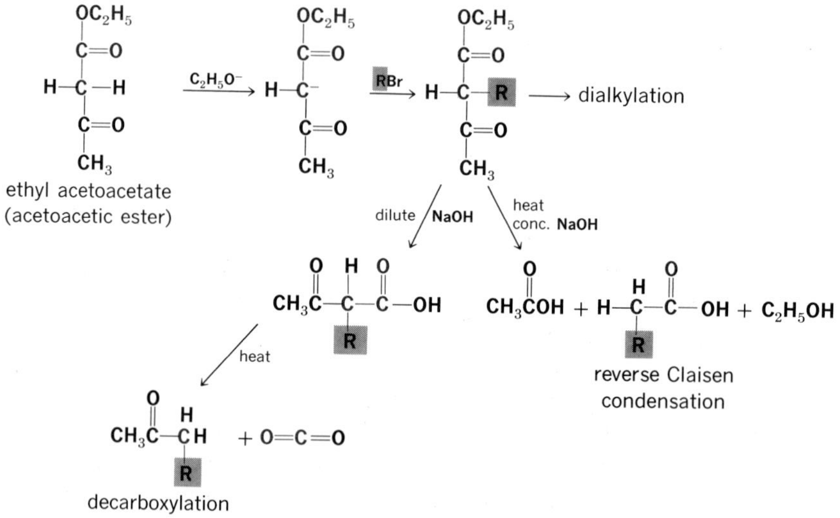

ethyl acetoacetate (acetoacetic ester)

$\xrightarrow{C_2H_5O^-}$ $\xrightarrow{RBr}$ ⟶ dialkylation

dilute NaOH; heat conc. NaOH

$CH_3C(=O)-CH(R)-C(=O)-OH$

$CH_3COOH + H-CH(R)-C(=O)-OH + C_2H_5OH$

reverse Claisen condensation

heat

$CH_3C(=O)-CH_2R + O{=}C{=}O$

decarboxylation

Figure 15-8. In the acetoacetic ester synthesis one or both acidic α-hydrogens are replaced by alkyl groups, and the resultant alkylated β-keto ester is converted into either an acid or a ketone.

enolate anion by sodium ethoxide, and, like other anions. this salt will carry out displacement reactions with suitable halides. Hydrolysis of the resultant ester and decarboxylation by heating gives a ketone. Alternatively, treatment with very concentrated sodium hydroxide causes a reversal of the Claisen condensation and the formation of a substituted acid.

Energy Production in Metabolism

Carbohydrates remain the major source of energy for most peoples of the world, so it is appropriate to begin our discussion of biological transformations with the steps in the metabolism of glucose in the body. The overall reaction for the complete combustion of glucose to carbon dioxide and water is given in Fig. 15-9; note that 690 kcal of energy is released per mole of glucose in

$$C_6H_{12}O_6 + 6O_2 \longrightarrow 6CO_2 + 6H_2O + 690 \text{ kcal/mole}$$

Figure 15-9. *Complete combustion of glucose releases energy.*

this conversion. This energy is immediately available to maintain body temperature, but biological systems must do more than remain warm, they must also store energy in forms that will be useful in carrying out chemical reactions. For instance, energy is required to synthesize a protein from its constituent amino acids or DNA from its nucleic acids.

ADP-ATP

The most common storehouse for chemical energy in biological systems is *adenosine triphosphate* (ATP), a nucleoside derivative in which energy is stored in the anhydride bonds. Energy is required to convert *adenosine diphosphate* (*ADP*) and phosphoric acid to ATP and water (Fig. 15-10). If we could somehow couple the breakdown of glucose to the buildup of ATP, energy could be stored. This energy could be used later either to generate heat or to cause some chemical reaction to occur that would be energetically unfavorable otherwise. A simple analogy might help here. Suppose that we have a hot furnace whose energy is going to waste. We might store some of this energy by passing acetic acid through the furnace and converting some of it to acetic anhydride and water, a reaction that is *endothermic*. If we got cold later on, we could get the energy back by adding acetic anhydride to water, since heat is given off in this reaction (*exothermic*). We could also use

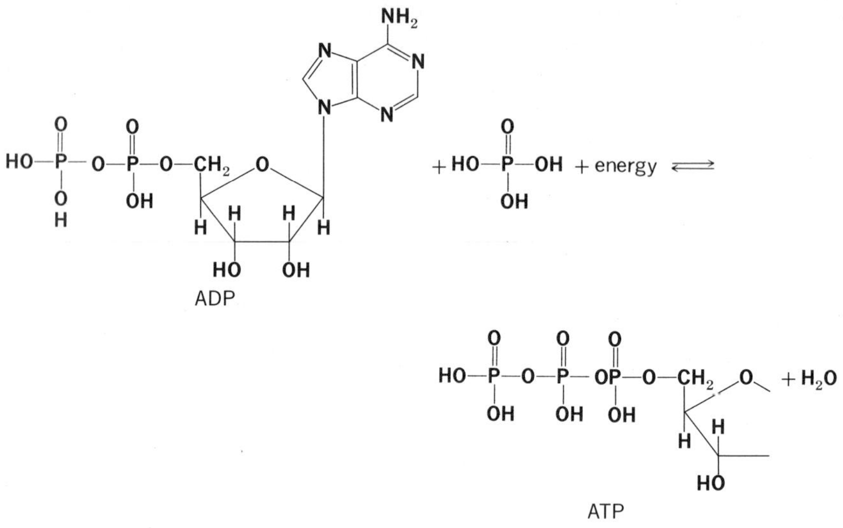

Figure 15-10. Energy from metabolism is stored for later use by converting adenosine diphosphate to adenosine triphosphate.

the anhydride in a chemical reaction that would not occur with the acid itself, e.g., to form an amide. As we discuss the pathway of carbohydrate metabolism, we shall see that the breakdown of glucose is coupled with the synthesis of ATP from ADP.

Glucose Metabolism

Most carbohydrates used nutritionally are consumed in the form of starch or sucrose, and the digestive tract contains specific enzymes for the hydrolysis of these di- and polysaccharides to glucose. For simplicity, let us assume that pure glucose has been eaten and follow the way it is metabolized.

In the first step glucose reacts with ATP to form glucose 6-phosphate and ADP; this reaction, like all the other metabolic reactions we shall discuss, is catalyzed by an enzyme. Since ATP is an acid anhydride, the reaction occurs readily and goes to completion. However it uses stored energy in the form of ATP, and in that sense is counterproductive (Fig. 15-11).

Next glucose 6-phosphate is isomerized to fructose 6-phosphate, and the latter is esterified again by ATP to fructose 1,6-diphosphate (Fig. 15-12).[1]

[1] This isomerization is easier to see in Fig. 15-12 if we remember that glucose (or its 6-phosphate) can exist in a free aldehyde form with an acidic α-hydrogen. That aldehyde can equilibrate with its enol form, which is also the enol for the keto form of fructose (or its 6-phosphate). The entire enzymatic isomerization glucose $\rightleftarrows$ fructose can be simulated in the laboratory in basic solution.

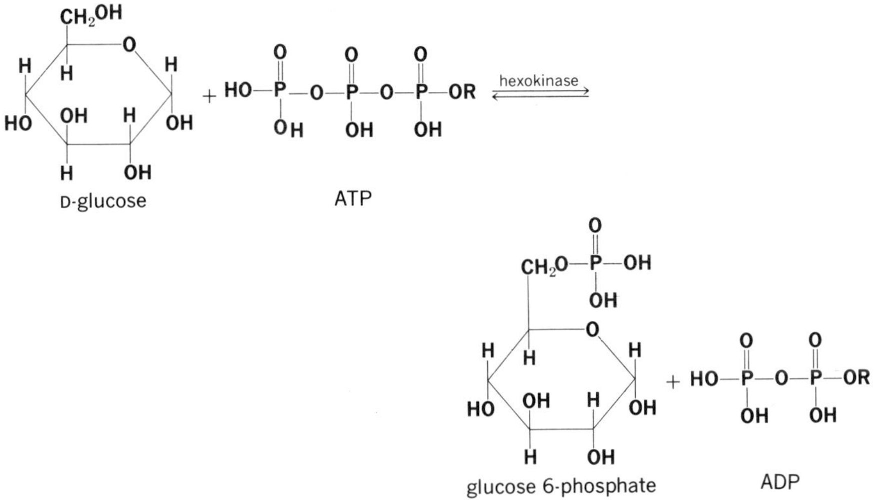

Figure 15-11. Glucose metabolism begins with the formation of glucose-6-phosphate. A molecule of ATP is used up in this step.

Fructose 1,6-diphosphate is then cleaved into two three-carbon fragments. Note that fructose, in its open-chain form, is a β-hydroxyketone, a *ketol,* exactly analogous to aldol, a β-hydroxyaldehyde (Fig. 15-5). A reversal of an

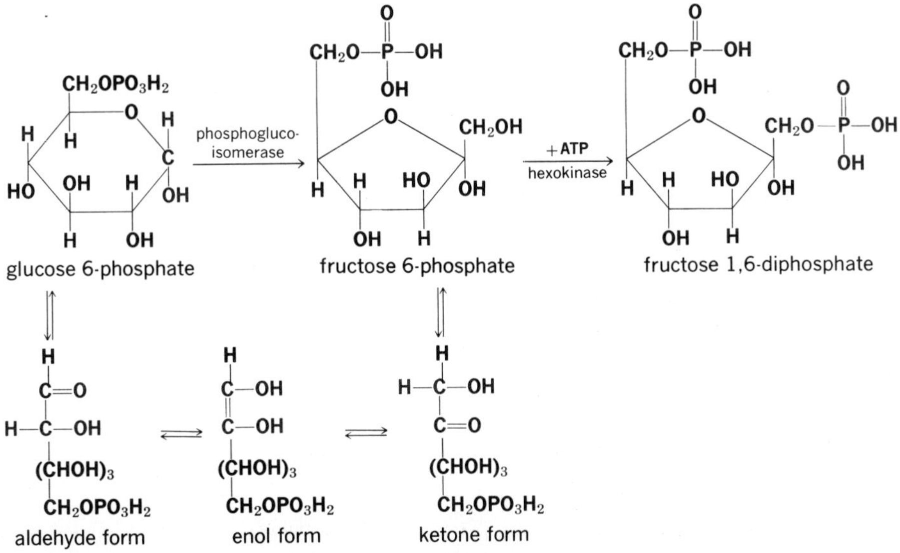

Figure 15-12. Glucose and fructose have the same enol form, through which they can be readily interconverted. Glucose metabolism proceeds through fructose 1,6-diphosphate.

aldol condensation may then occur, to give an aldehyde and a ketone, as shown in Fig. 15-13.

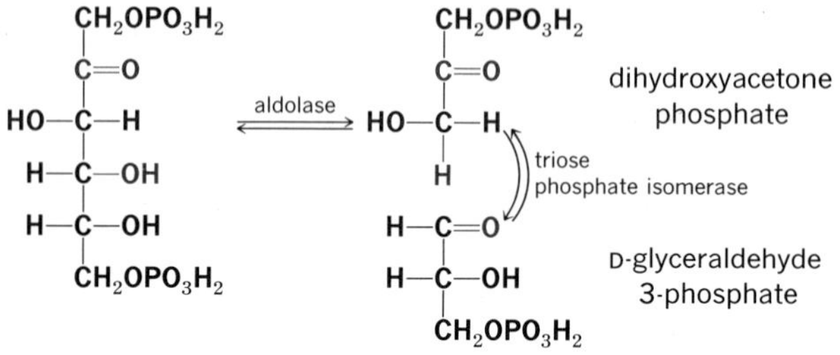

$CH_2OPO_3H_2$
C=O
HO—C—H
H—C—OH
H—C—OH
$CH_2OPO_3H_2$

aldolase ⇌

$CH_2OPO_3H_2$
C=O
HO—C—H
H
dihydroxyacetone phosphate

triose phosphate isomerase

H—C=O
H—C—OH
$CH_2OPO_3H_2$
D-glyceraldehyde 3-phosphate

Figure 15-13. *Fructose 1,6-diphosphate is cleaved into 2 three-carbon fragments by a reverse aldol reaction.*

Further metabolism occurs only through glyceraldehyde 3-phosphate, and another enzyme catalyzes the interconversion of the phosphates of glyceraldehyde and dihydroxyacetone, two compounds which, like glucose and fructose, have a common enol form. To give an idea of the catalytic efficiency of some enzymes, one molecule of triose phosphate isomerase can catalyze the interconversion of nearly one million molecules per minute.

Up to this point, the reactions in the metabolism of glucose have involved only phosphate ester formation, tautomerization, or reverse aldol condensations, all of which require only acidic or basic catalysis and no oxidation or reduction reactions. In the next step, glyceraldehyde 3-phosphate is oxidized to glyceric acid 1,3-diphosphate. Obviously, an oxidizing agent is needed, and the one provided by nature is nicotinamide adenine dinucleotide, NAD^+ (Fig. 15-14).

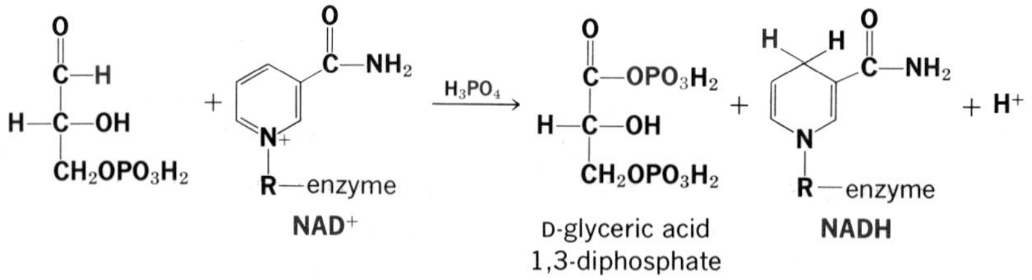

O
C—H
H—C—OH
$CH_2OPO_3H_2$

\+

O
C—NH_2
N^+
R—enzyme
NAD^+

H_3PO_4 →

O
C—OPO_3H_2
H—C—OH
$CH_2OPO_3H_2$
D-glyceric acid 1,3-diphosphate

\+

H H O
C—NH_2
N
R—enzyme
NADH

\+ H^+

Figure 15-14. *Oxidation of glyceraldehyde 3-phosphate to glyceric acid 1,3-diphosphate involves the reduction of NAD^+, whose reactive part is the vitamin niacin.*

Several facts are worthy of note about this reaction. In the first place the oxidizing agent, NAD^+, contains as its reactive part the vitamin *niacin* (nicotinamide). Vitamins are compounds that are required in small amounts by man, yet cannot be synthesized by the body and so must be included in the diet. Vitamin deficiency leads to disease; a deficiency of niacin causes *pellagra* in man.

Note also that the oxidation of the aldehyde has led, not to the carboxylic acid, but rather to an anhydride between the organic acid and phosphoric acid. An aldehyde is easily oxidized, and a large amount of heat would be liberated if the carboxylic acid were formed. By forming the anhydride instead, energy is stored in a highly reactive chemical species. In the next step this species is used to form the more generally useful ATP from ADP (Fig. 15-15).

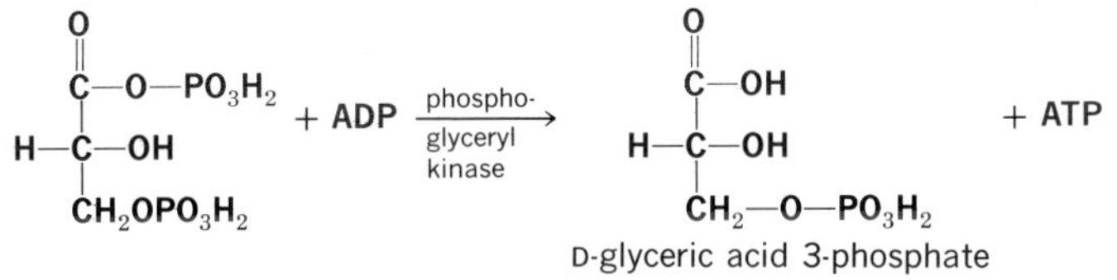

$$\begin{array}{c} O \\ \| \\ C{-}O{-}PO_3H_2 \\ | \\ H{-}C{-}OH \\ | \\ CH_2OPO_3H_2 \end{array} + ADP \xrightarrow[\text{glyceryl kinase}]{\text{phospho-}} \begin{array}{c} O \\ \| \\ C{-}OH \\ | \\ H{-}C{-}OH \\ | \\ CH_2{-}O{-}PO_3H_2 \\ \text{D-glyceric acid 3-phosphate} \end{array} + ATP$$

Figure 15-15. *In the formation of glyceric acid phosphate some of the energy released in glucose metabolism is stored in ATP.*

Since two molecules of glyceraldehyde 3-phosphate have been produced from each molecule of glucose, two ATP molecules are formed here from each glucose, and two have been used previously in forming fructose diphosphate. At this point, then, we have formed as many ATP molecules as we have used.

Metabolism continues by the transfer of the phosphate group from the 3- to the 2-position, loss of water to form an ester of the enol form of a ketone, and production of the important metabolic intermediate pyruvic acid and another molecule of ATP from ADP (two more moles of ATP based on glucose), as shown in Fig. 15-16.

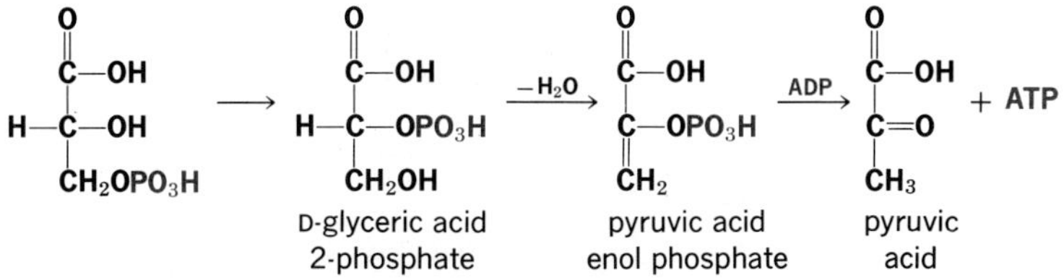

$$\begin{array}{c} O \\ \| \\ C{-}OH \\ | \\ H{-}C{-}OH \\ | \\ CH_2OPO_3H \end{array} \longrightarrow \begin{array}{c} O \\ \| \\ C{-}OH \\ | \\ H{-}C{-}OPO_3H \\ | \\ CH_2OH \\ \text{D-glyceric acid} \\ \text{2-phosphate} \end{array} \xrightarrow{-H_2O} \begin{array}{c} O \\ \| \\ C{-}OH \\ | \\ C{-}OPO_3H \\ \| \\ CH_2 \\ \text{pyruvic acid} \\ \text{enol phosphate} \end{array} \xrightarrow{ADP} \begin{array}{c} O \\ \| \\ C{-}OH \\ | \\ C{=}O \\ | \\ CH_3 \\ \text{pyruvic} \\ \text{acid} \end{array} + ATP$$

Figure 15-16. *In a series of simple steps, glyceric acid 3-phosphate is converted into pyruvic acid with the formation of another molecule of ATP from ADP.*

Anaerobic Glucose Metabolism

Thus far, glucose metabolism is the same, no matter whether the process takes place in human muscle tissue, in the liver, or in a bottle of wine fermenting in the cellar of a French chateau. The subsequent steps diverge; the conversion of glucose to pyruvic acid is an oxidation, but in an anaerobic process (that which occurs in the absence of air) no oxygen, the ultimate oxidizing agent, is available, and, as a consequence, a reduction step must be coupled to the oxidation to keep things in balance. In muscle tissue, an anaerobic process takes place and the pyruvic acid is reduced to lactic acid (Fig. 15-17). The

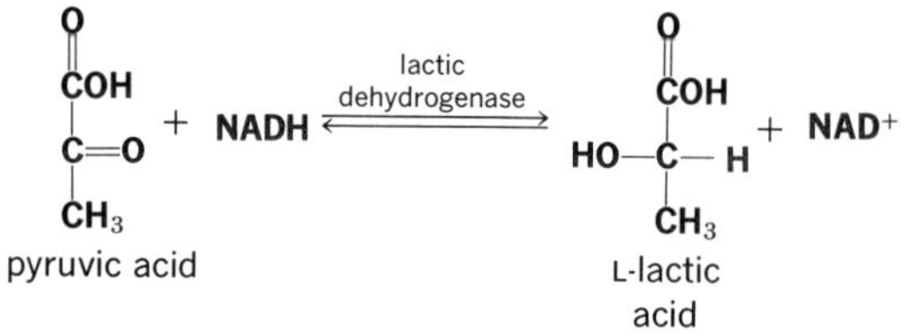

$CH_3COCOOH$ (pyruvic acid) + NADH $\xrightleftharpoons{\text{lactic dehydrogenase}}$ $CH_3CH(OH)COOH$ (L-lactic acid) + NAD^+

Figure 15-17. *Pyruvic acid produced in muscle is reduced to L-lactic acid, which is then transported by the blood to the liver.*

reducing agent is the same NADH formed in the previous oxidation step; so a small amount of NAD^+ can catalyze the metabolism of a great deal of glucose in the muscle by being oxidized and reduced many times.

A variant on the metabolism of glucose in muscle is found in alcoholic fermentation. Here the chemical reaction sequence leads from grain or grapes to ethanol. This is also an anaerobic process, but pyruvic acid is decarboxylated before reduction (Fig. 15-18). Acetaldehyde is an intermediate and is reduced to ethanol. Again the reduction is brought about by NADH so that oxidation and reduction are kept in balance.

Before discussing how lactic and pyruvic acid are converted ultimately to carbon dioxide we shall look briefly at the metabolism of fats.

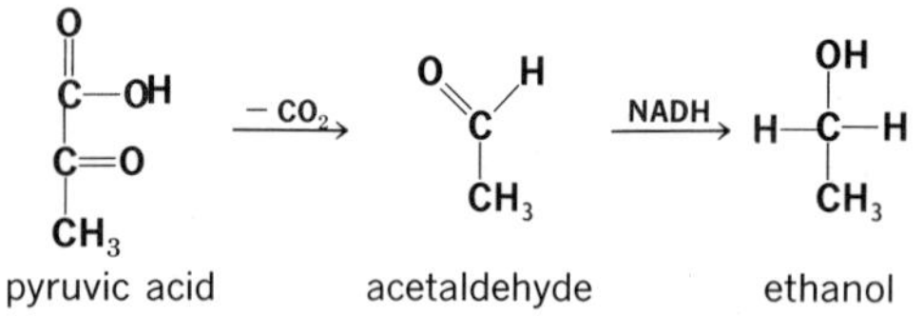

$CH_3COCOOH$ (pyruvic acid) $\xrightarrow{-CO_2}$ CH_3CHO (acetaldehyde) $\xrightarrow{\text{NADH}}$ CH_3CH_2OH (ethanol)

Figure 15-18. *In fermentation, pyruvic acid loses CO_2 to form acetaldehyde, which in turn is reduced to alcohol.*

Metabolism of Lipids

In higher animals ingested lipids pass through the stomach relatively unaffected by the acidic environment. Once in the alkaline conditions of the small intestine, however, the triglycerides are largely hydrolyzed with the help of enzymes known as *lipases*. From the intestines the hydrolyzed lipids enter the bloodstream and are transported to organs for further metabolism. Passage through the intestinal walls is a complex process, and far from completely understood. In part, the fatty acids themselves are involved, in part the mono-, di-, and tri-glycerides. Some of these molecules pass directly into the bloodstream; some arrive there through the lymphatic system.

Once in the bloodstream the fatty acids are reesterified with glycerol, and the triglycerides are complexed with blood protein and transported either to storage sites for deposition as "depot fat" or to body organs, largely the liver, for metabolism. Depot fat is itself available for transport to the liver; lipids are generally transferred back and forth among the various sites. In the liver the triglycerides are rehydrolyzed to fatty acids and oxidized, ultimately to carbon dioxide and water. A large amount of energy is released in this oxidation, considerably more per unit of weight than in the metabolism of glucose (Fig. 15-19).

$$\underset{\text{palmitic acid}}{CH_3(CH_2)_{14}\overset{\overset{O}{\|}}{C}OH} + 23O_2 \longrightarrow 16CO_2 + 16H_2O + 2338 \text{ kcal/mol}$$

Figure 15-19. *Fatty acids are oxidized in the body to carbon dioxide and water. A great deal of energy is released, considerably more per gram than from oxidation of carbohydrates.*

As with glucose, a large number of steps are involved in this conversion. From many observations it has long been recognized that fatty acids are broken down (and synthesized) in units of two carbon atoms. For instance, almost without exception, the fatty acids found in fats of higher plants and animals contain even numbers of carbon atoms. Another important observation was that of Knoop who, in 1904, prepared fatty acids in which one of the hydrogens of the terminal methyl group had been replaced by a phenyl group. When these fatty acids were fed to dogs, the side chains were metabolized, but the aromatic ring was resistant to metabolism and was excreted unchanged. When the aliphatic chain had an even number of carbon atoms, phenylacetic acid was the excretion product, while benzoic acid was formed from fatty acids with an odd number of carbon atoms in the side chain (Fig. 15-20). Both observations are consistent with the view that two-carbon units are important in lipid metabolism.

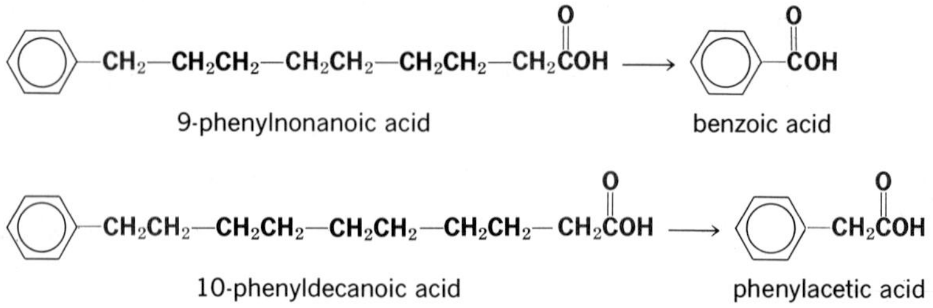

$C_6H_5-CH_2-CH_2CH_2-CH_2CH_2-CH_2CH_2-CH_2\overset{O}{\overset{\|}{C}}OH \longrightarrow C_6H_5-\overset{O}{\overset{\|}{C}}OH$

9-phenylnonanoic acid — benzoic acid

$C_6H_5-CH_2CH_2-CH_2CH_2-CH_2CH_2-CH_2CH_2-CH_2\overset{O}{\overset{\|}{C}}OH \longrightarrow C_6H_5-CH_2\overset{O}{\overset{\|}{C}}OH$

10-phenyldecanoic acid — phenylacetic acid

Figure 15-20. *When phenyl-substituted fatty acids are metabolized, the end product is benzoic or phenylacetic acid, depending on whether the side chain has an odd or an even number of carbon atoms.*

The multistep pathway by which fatty acids are broken down into two-carbon units is now well understood. A key role in this process, as well as in many others, is played by *coenzyme A* (CoA), which contains the vitamin *pantothenic* acid (Fig. 15-21), a member of the B-vitamin group.

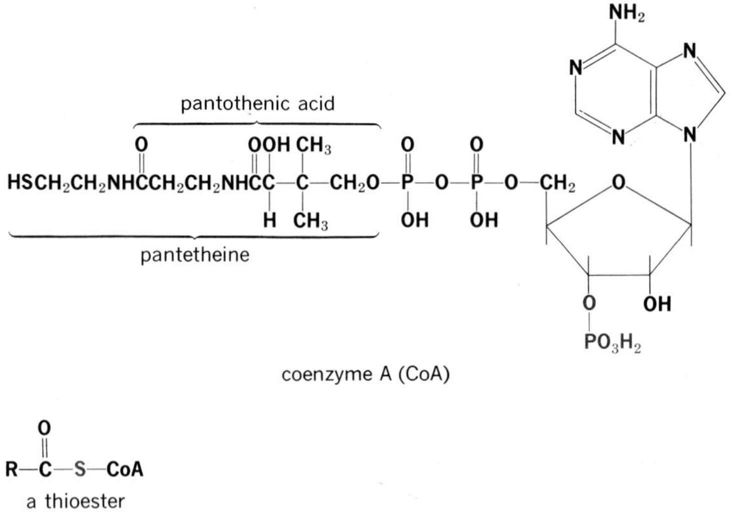

pantothenic acid

$HSCH_2CH_2NH\overset{O}{\overset{\|}{C}}CH_2CH_2NH\overset{O}{\overset{\|}{C}}\overset{OH}{\underset{H}{C}}-\overset{CH_3}{\underset{CH_3}{C}}-CH_2O-\overset{O}{\underset{OH}{\overset{\|}{P}}}-O-\overset{O}{\underset{OH}{\overset{\|}{P}}}-O-CH_2$ (ribose, with adenine (NH_2, N, N, N, N) attached; ring O; ring substituents $O-PO_3H_2$ and OH)

pantetheine

coenzyme A (CoA)

$R-\overset{O}{\overset{\|}{C}}-S-CoA$

a thioester

Figure 15-21. *Coenzyme A, through its thioesters, plays an important role in many enzymatic systems. Thioesters are nearly as reactive as anhydrides.*

In fatty acid metabolism the acid is first converted to its thioester with coenzyme A and then oxidized (vitamin B_2, riboflavin, is involved in this oxidation) to an unsaturated acid (Fig. 15-22). Next water is added to the

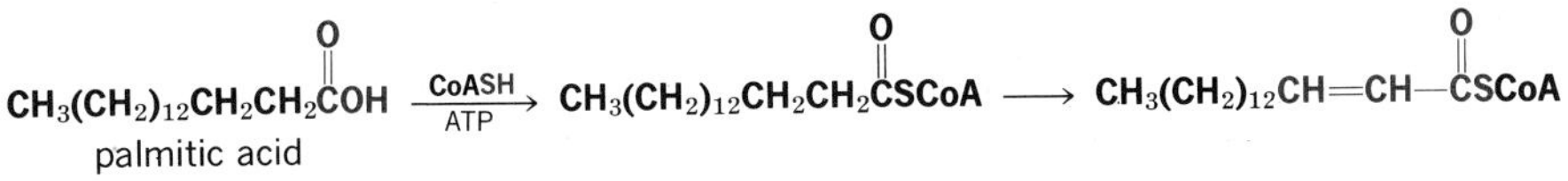

$$CH_3(CH_2)_{12}CH_2CH_2\overset{O}{\overset{\|}{C}}OH \xrightarrow[ATP]{CoASH} CH_3(CH_2)_{12}CH_2CH_2\overset{O}{\overset{\|}{C}}SCoA \longrightarrow CH_3(CH_2)_{12}CH{=}CH{-}\overset{O}{\overset{\|}{C}}SCoA$$

palmitic acid

Figure 15-22. *In fatty acid metabolism the acid is esterified with an SH group of coenzyme A and then oxidized to a conjugated unsaturated derivative. All steps are catalyzed by enzymes.*

double bond, giving the β-hydroxy thioester, which in turn is oxidized by NAD^+ to the β-keto thioester (Fig. 15-23). Finally, this molecule is cleaved in a reverse Claisen condensation reaction (see Fig. 15-6) to give the thioester

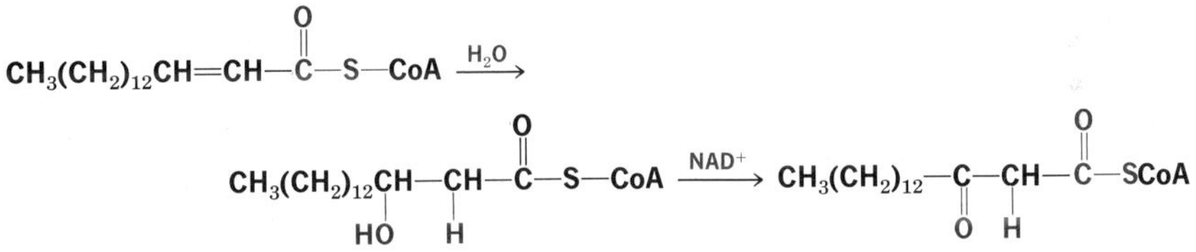

$$CH_3(CH_2)_{12}CH{=}CH{-}\overset{O}{\overset{\|}{C}}{-}S{-}CoA \xrightarrow{H_2O}$$

$$CH_3(CH_2)_{12}\underset{HO}{\underset{|}{C}}H{-}\underset{H}{\underset{|}{C}}H{-}\overset{O}{\overset{\|}{C}}{-}S{-}CoA \xrightarrow{NAD^+} CH_3(CH_2)_{12}{-}\underset{O}{\underset{\|}{C}}{-}\underset{H}{\underset{|}{C}}H{-}\overset{O}{\overset{\|}{C}}{-}SCoA$$

Figure 15-23. *The unsaturated thioester is hydrated and oxidized to a β-keto thioester.*

of acetic acid, acetyl coenzyme A (acetyl CoA), as shown in Fig. 15-24, and the CoA thioester of a fatty acid with two carbon atoms fewer than the original acid. The metabolic steps then repeat, chopping the fatty acid down by two carbons at each cycle until it is completely converted to acetyl CoA. In this sequence only one molecule of ATP is required, that needed to make the original fatty acid–CoA thioester.

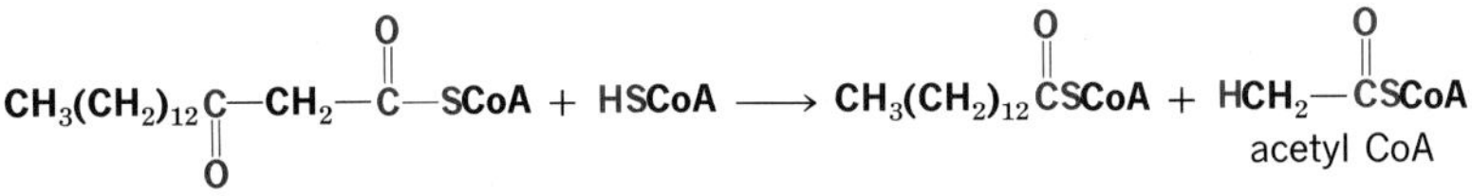

$$CH_3(CH_2)_{12}\underset{O}{\underset{\|}{C}}{-}CH_2{-}\overset{O}{\overset{\|}{C}}{-}SCoA + HSCoA \longrightarrow CH_3(CH_2)_{12}\overset{O}{\overset{\|}{C}}SCoA + HCH_2{-}\overset{O}{\overset{\|}{C}}SCoA$$

acetyl CoA

Figure 15-24. *In a reverse Claisen reaction the β-keto thioester is cleaved to acetyl CoA and the CoA thioester of a shorter-chain fatty acid. The sequence then repeats until the fatty acid is completely converted to acetyl CoA.*

The Tricarboxylic Acid Cycle

The metabolic paths of carbohydrates and fats come together in the *tricarboxylic acid cycle,* also known as the *Krebs cycle.* We have traced glucose metabolism as far as pyruvic acid, where we noted that in the absence of oxygen it is reduced to lactic acid or to ethanol. Only about 10% of the available energy from glucose metabolism is released in its conversion to pyruvic acid, and the remaining 90% is produced in the subsequent oxidation of pyruvic acid to CO_2. The first step in this process is the oxidation of pyruvic acid to acetyl CoA, and this step provides the link which joins carbohydrate and lipid metabolism (Fig. 15-25). This is a complicated reaction that involves no less

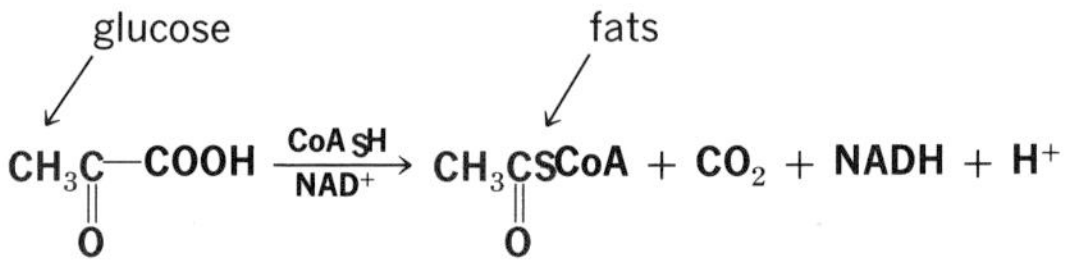

glucose ↘ fats ↘

$$CH_3\underset{\underset{O}{\|}}{C}{-}COOH \xrightarrow[NAD^+]{CoASH} CH_3\underset{\underset{O}{\|}}{C}SCoA + CO_2 + NADH + H^+$$

Figure 15-25. *Acetyl CoA is formed by oxidation of pyruvic acid, a step which links carbohydrate and fat metabolic pathways. Besides NAD^+, the vitamins lipoic acid and thiamin (vitamin B_1) are involved in this reaction.*

than three vitamins as cofactors, niacin, lipoic acid, and thiamin, and the details cannot be given here. Acetyl coenzyme A is, however, a key intermediate not only in metabolism, but also in biosynthesis; in Chapter 16 we shall see how it is used in biological synthesis.

In the tricarboxylic acid cycle, acetyl CoA is oxidized to CO_2. Nine reactions, some simple, some complex, are involved in this cycle, and at the end of each cycle two molecules of carbon dioxide have been formed. The cycle is given in Fig. 15-26. Although a detailed analysis of each reaction is beyond the scope of this book, we may examine some of them briefly. In step (1), for example, we can see an aldol condensation, in which a carbanion formed by removal of an acidic α-hydrogen of acetyl CoA adds to the carbonyl group of oxaloacetic acid to give citric acid. Step (2) is an isomerization, whose details are not yet clearly understood, and step (3) an oxidation by NAD^+ analogous to that we saw in Fig. 15-14 for glyceraldehyde-3-phosphate. The resultant oxalosuccinic acid loses carbon dioxide easily (Step (4)) because it is a β-ketoacid (see Fig. 15-8); in step (5) α-ketoglutaric acid is oxidized in the same way by which pyruvic acid is converted to acetyl CoA (Fig. 15-25). Hydrolysis, step (6), produces succinic acid, which is so symmetrical that the enzymes are unable to determine which carbons came from the last acetyl CoA to enter the cycle and which were there from a previous turn. The subsequent steps of oxidation, hydration, and further oxidation to oxaloacetic acid are analogous to those which occur in fatty acid metabolism.

Although ATP is not formed from ADP directly in any step in this cycle, the cycle is ultimately responsible for the production of numerous molecules of ATP. Four of the steps involve oxidations, including the conversion of

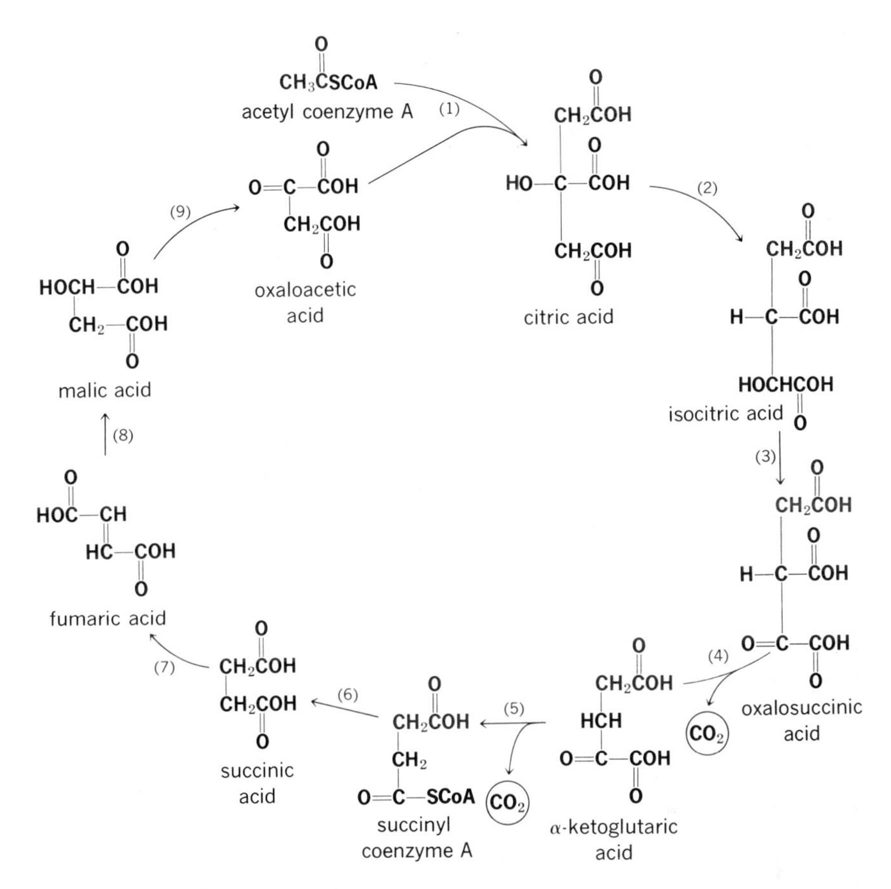

Figure 15-26. *The Krebs cycle for the conversion of acetyl coenzyme A to carbon dioxide. The two molecules of carbon dioxide produced are circled. Note that the first time through the cycle the carbon dioxide molecules arise from carbon atoms already in the cycle.*

pyruvic acid to acetyl CoA. The oxidizing agent is usually NAD^+, which is reduced to NADH. Since the ultimate oxidizing agent must be oxygen from the air, NAD^+ is regenerated in a series of reactions known as *oxidative phosphorylation*, which may be summarized as in Fig. 15-27. Note that three molecules of ATP are produced with each molecule of NAD^+ formed in the reaction. In all, 38 net molecules of ATP are formed in the complete metabolism of glucose to CO_2 and H_2O, and this amounts to the trapping of about 40% of the energy available in the form of ATP, with 60% being liberated as heat.

$$NADH + H^+ + \tfrac{1}{2}O_2 + 3ADP + 3H_3PO_4 \longrightarrow NAD^+ + 3ATP + 4H_2O$$

Figure 15-27. *Although ATP is not generated directly in the tricarboxylic acid cycle, it is formed by byproducts of the cycle, e.g., from NADH.*

Metabolism of Amino Acids

Because there are so many different amino acids, it is not possible to write a single metabolic pathway for them all. Ultimately, however, most of them are converted to carbon dioxide and water by way of the tricarboxylic acid cycle. To accomplish this, the amino acid must be *deaminated,* i.e., freed of its nitrogen. In many cases this is accomplished by oxidation and hydrolysis, as shown in Fig. 15-28. In some cases the product of this reaction is a compound in the tricarboxylic acid cycle. For example, oxidative deamination of glutamic acid gives α-ketoglutaric acid, and that of alanine gives pyruvic acid (Fig. 15-29). Other amino acids are metabolized through complicated reaction sequences in which many of the carbons are converted to compounds in the tricarboxylic acid cycle. In the next chapter we shall see how amino acids also serve as the source of carbon for many naturally occurring, physiologically active compounds.

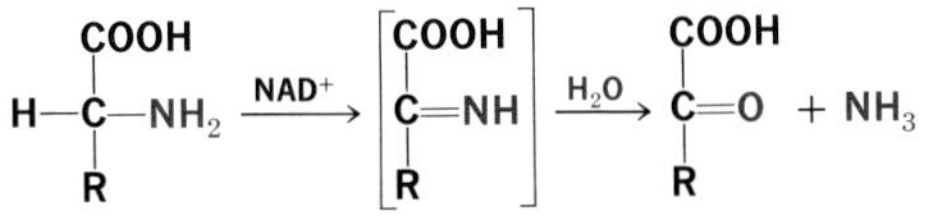

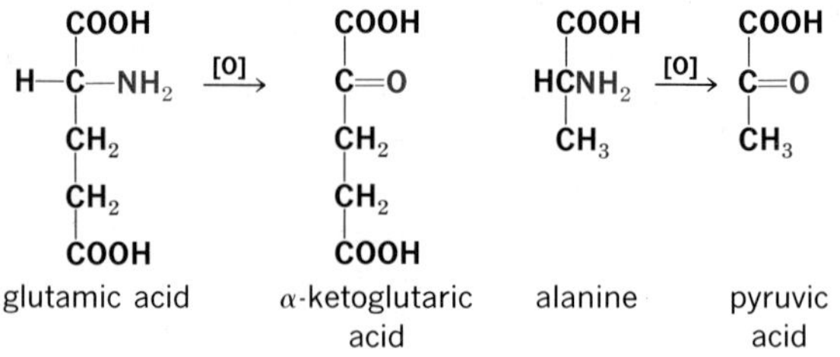

$$H{-}C(COOH)(R){-}NH_2 \xrightarrow{NAD^+} [\,(COOH)(R)C{=}NH\,] \xrightarrow{H_2O} (COOH)(R)C{=}O + NH_3$$

Figure 15-28. *In oxidative deamination an amino acid is converted to an α-keto acid and ammonia.*

$$HOOC{-}CH(NH_2){-}CH_2{-}CH_2{-}COOH \xrightarrow{[O]} HOOC{-}C(=O){-}CH_2{-}CH_2{-}COOH$$

glutamic acid → α-ketoglutaric acid

$$HOOC{-}CH(NH_2){-}CH_3 \xrightarrow{[O]} HOOC{-}C(=O){-}CH_3$$

alanine → pyruvic acid

Figure 15-29. *Oxidative deamination of glutamic acid and alanine gives compounds that are part of the tricarboxylic acid cycle.*

PROBLEMS

1. One way we can explore the acidity of hydrogens within an organic molecule is to dissolve the compound in D_2O in the presence of NaOD. If a hydrogen is removed by the base, it will be replaced by deuterium. For example, acetaldehyde, CH_3CHO, in D_2O/NaOD is quickly converted to CD_3CHO. What deuterio compounds would you get from each of the following compounds under those conditions?
   (a) acetone
   (b) methyl ethyl ketone
   (c) ethyl alcohol
   (d) benzaldehyde
   (e) cyclohexanone
   (f) ethyl acetate (using $C_2H_5OD/NaOC_2H_5$)
2. Explain by the use of resonance structures why the following exchange occurs. Why does exchange not occur in the other methyl group?

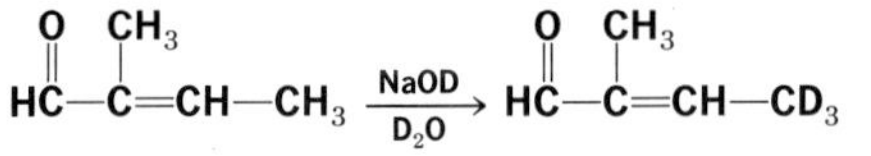

$$HC(=O)-C(CH_3)=CH-CH_3 \xrightarrow[D_2O]{NaOD} HC(=O)-C(CH_3)=CH-CD_3$$

3. Starting with ethyl acetoacetate and the halides of your choice, show how you could make each of the following:
   (a)  $C_6H_5CH_2CH_2C(=O)CH_3$
   (b) $C_6H_5CH_2CH_2COOH$
   (c) cyclopropyl methyl ketone: CH_2—CH_2 ring with CH—$C(=O)$—CH_3
   (d) 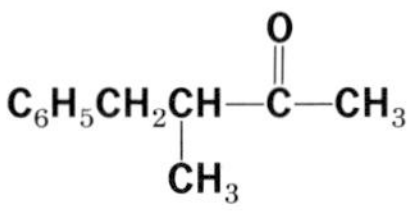 $C_6H_5CH_2CH(CH_3)-C(=O)-CH_3$
4. Write equations which show that mannose and glucose have the same enol form.
5. In the metabolic disease diabetes, acetone accumulates in the body and can be smelled on the breath of the patient. Acetoacetic acid is also found. These are thought to arise from acetyl CoA by Claisen condensation, hydrolysis, and loss of CO_2. Write equations for the production of acetoacetic acid and acetone from acetyl CoA.
6. Show clearly how protein, carbohydrate, and lipid metabolisms are interrelated.
7. Suppose that you prepared some glucose specifically labeled with radioactive carbon (^{14}C) in the aldehyde group. If you followed its metabolism as far as lactic acid, in which position would the ^{14}C be?

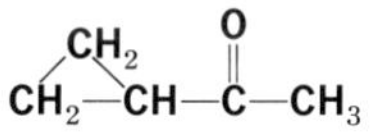

16
biosynthesis and the chemistry of natural products

Over the centuries, thousands of organic compounds have been isolated from natural sources in more or less pure form. Many of these compounds, usually called natural products, have been used as medicines, dating back to the salicylates extracted from willow leaves or to earlier herbal medicines. Such folk remedies have yielded many useful drugs. For example, doctors in India have known for many years that chewing the leaves of the *Rauwolfia serpentina* plant brings on a state of euphoria. Recent investigation has revealed that these leaves contain reserpine, a compound now marketed as a tranquilizer and widely used in the treatment of mental diseases.

Other natural products have been used for their pleasurable or hallucinatory effects, e.g., alcohol, opium, and mescaline. Still others are violent poisons, like strychnine, which is often used to kill rats, or the hemlock extract coniine used by the ancient Greeks to kill Socrates.

One thing about many of these natural products should be emphasized: To have commercial importance, they must normally be produced in reasonably large quantities by a plant or animal; they are not mere transient metabolic intermediates like many of the compounds mentioned in the last chapter, but end products of synthetic or metabolic pathways. A reasonable question would be: Why does an organism go to the trouble to synthesize a complex molecule like penicillin or morphine? Strangely enough, many of the products that we shall be discussing have no known physiological function within the plant or animal from which they are isolated. A few may play a protective role—penicillin, for example, may protect the fungus that produces it from bacterial infection. However, some other antibiotics actually poison the very organisms that produce them. Others appear simply to be offshoots of the main life processes, blind alleys that might once have been main metabolic highways. This is a subject about which there has been a great deal of speculation but the economic and medicinal importance of the compounds has made them the objects of an enormous amount of chemical and biochemical research.

Classification of Natural Products

Natural products have a fantastic range of organic structures, but they can be organized into a number of classes according to their ultimate uses or because of similarities in their chemical structures. *Alkaloids* are nitrogen-containing plant products; most of them are basic and can be extracted by acids. Strychnine, morphine, and quinine are representative members of this potent class. *Steroids* are a class of compounds containing a tetracyclic carbon skeleton. They are usually found in the ether extracts of plant and animal tissue, along with fats and other lipids. When this extract is saponified, the fats and oils are hydrolyzed, but the steroids are *nonsaponifiable* and remain soluble in ether. Cholesterol is the most abundant steroid in animal systems; many important hormones, including cortisone, and vitamin D are also steroids. *Terpenes* are plant products that bear a particular structural relationship to one another. Camphor and menthol belong to this class of compounds.

Antibiotics, on the other hand, are classified on the basis of their use. Various definitions are used, but the most commonly accepted one describes an antibiotic as a compound produced by one microorganism that is toxic to another microorganism. Since this is a wholly biological definition, it implies nothing about the chemical nature of the compounds. Indeed, antibiotics constitute no well-defined class of chemically related substances, but instead possess exceedingly varied structures. Many other natural products also do not fall into any typical class.

Rather than simply list formulas for different types of natural products in this chapter, we shall show correlations among groups of these products based either on the biosynthetic pathways by which they are formed (biogenesis) or on their relationships to the lipids, amino acids, and carbohydrates that make up the primary building blocks of living systems.

Biosynthesis of Fatty Acids from Acetate

We saw in Chapter 15 that fatty acids are degraded metabolically by the successive removal of two-carbon fragments after β-oxidation. Fatty acids are built up by an essentially reverse process (Fig. 16-1). Feeding sodium acetate,

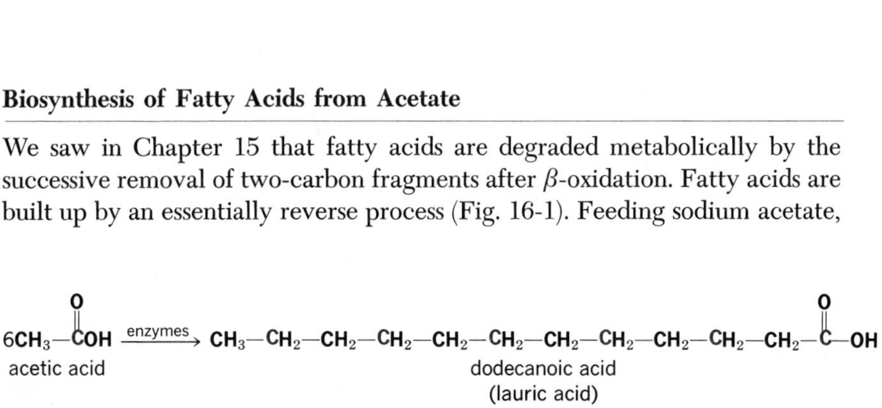

$$6CH_3\text{—}\overset{O}{\overset{\|}{C}}OH \xrightarrow{\text{enzymes}} CH_3\text{—}CH_2\text{—}CH_2\text{—}CH_2\text{—}CH_2\text{—}CH_2\text{—}CH_2\text{—}CH_2\text{—}CH_2\text{—}CH_2\text{—}CH_2\text{—}\overset{O}{\overset{\|}{C}}\text{—}OH$$

acetic acid — dodecanoic acid (lauric acid)

Figure 16-1. *Experiments with isotopically labeled acetate show that fatty acids can be synthesized directly from this substrate.*

isotopically labeled with ^{14}C in the carboxyl group, to a variety of living systems produces fatty acids with even numbers of carbon atoms. The isotope is incorporated with high efficiency, and is distributed in the carboxyl group and the other odd-numbered carbon atoms (C-3, C-5, etc).

The biosynthesis does not proceed by exactly the same steps as the metabolism, however. The crucial difference lies in the very first step of the biosynthetic pathway, which is the last step of metabolism. In the metabolic pathway, this step involves the cleavage of a β-ketoester derivative to acetyl coenzyme A. This step is theoretically reversible, but the equilibrium lies so far on the side of cleavage that the same reaction cannot be used to synthesize fatty acids. Energy must somehow be supplied to the system to overcome this unfavorable equilibrium.

One important source of energy in biological systems is ATP (adenosine triphosphate). In Chapter 15 we discussed how energy released in the metabolism of carbohydrates was stored in this molecule for eventual use. One such use is in the biosynthesis of fatty acids; in this process, acetyl coenzyme A reacts with carbon dioxide and ATP to form malonyl coenzyme A (Fig. 16-2).

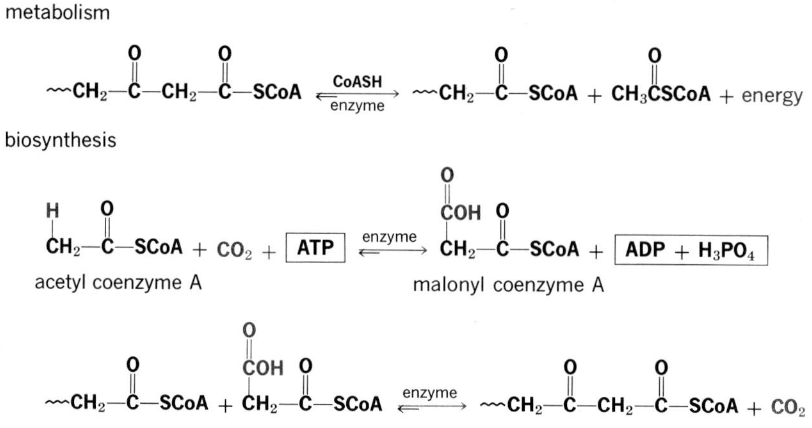

metabolism

$$\sim CH_2-\overset{O}{\overset{\|}{C}}-CH_2-\overset{O}{\overset{\|}{C}}-SCoA \xrightleftharpoons[\text{enzyme}]{\text{CoASH}} \sim CH_2-\overset{O}{\overset{\|}{C}}-SCoA + CH_3\overset{O}{\overset{\|}{C}}SCoA + \text{energy}$$

biosynthesis

$$\overset{H}{\overset{|}{C}}H_2-\overset{O}{\overset{\|}{C}}-SCoA + CO_2 + \boxed{ATP} \xrightleftharpoons{\text{enzyme}} \overset{\overset{O}{\|}}{\overset{COH}{\overset{|}{C}}}H_2-\overset{O}{\overset{\|}{C}}-SCoA + \boxed{ADP + H_3PO_4}$$

acetyl coenzyme A $\qquad\qquad$ malonyl coenzyme A

$$\sim CH_2-\overset{O}{\overset{\|}{C}}-SCoA + \overset{\overset{O}{\|}}{\overset{COH}{\overset{|}{C}}}H_2-\overset{O}{\overset{\|}{C}}-SCoA \xrightleftharpoons{\text{enzyme}} \sim CH_2-\overset{O}{\overset{\|}{C}}-CH_2-\overset{O}{\overset{\|}{C}}-SCoA + CO_2$$

Figure 16-2. *The final step in fatty acid metabolism releases energy. In order to accomplish the reverse, fatty acid biosynthesis, this energy must be supplied by ATP, which aids in the formation of malonyl coenzyme A.*

During the process, ATP is converted to ADP and phosphoric acid. Malonyl coenzyme A will then condense under the catalytic influence of an enzyme to give a β-ketoester *plus carbon dioxide,* whose elimination provides the driving force for the reaction.

Subsequent steps in the biosynthesis of a fatty acid are substantially the reverse of those of its metabolism. Thus the β-keto acid is reduced to a β-hydroxy acid, water is eliminated, and the resulting double bond is reduced. The steps are then repeated with another molecule of malonyl coenzyme A. The exact sequence of condensation, reduction, and elimination is not always the same, nor have many steps been completely defined; it is possible, for instance, to have two or more condensations take place before reduction. As we shall see, this is important in the synthesis of many other products originating from acetate.

Of the pathways available for formation of lipids containing unsaturated fatty acids in animals, the simplest pathway involves introducing a double bond by direct oxidation (dehydrogenation) of the chain of a saturated fatty acid; this reaction has no laboratory analogy. It is interesting that a double bond introduced in this way in animals is always at least six carbon atoms from the methyl end of the chain, reflecting, presumably, a structural demand of one of the enzymes involved. Unsaturated acids of the oleic and palmitoleic acid family are formed in this way (Fig. 16-3).

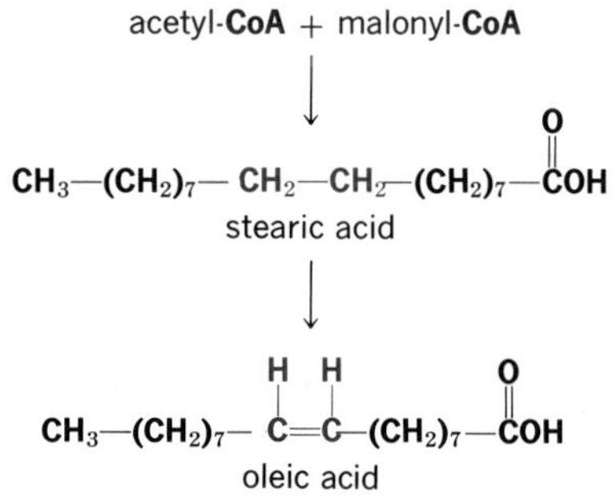

acetyl-**CoA** + malonyl-**CoA**

↓

$$CH_3{-}(CH_2)_7{-}CH_2{-}CH_2{-}(CH_2)_7{-}\overset{O}{\overset{\|}{C}}OH$$

stearic acid

↓

$$CH_3{-}(CH_2)_7{-}\overset{H}{\overset{|}{C}}{=}\overset{H}{\overset{|}{C}}{-}(CH_2)_7{-}\overset{O}{\overset{\|}{C}}OH$$

oleic acid

Figure 16-3. *Saturated fatty acids like stearic acid, derived from acetate and malonate units, can be dehydrogenated directly in animals to unsaturated acids like oleic acid. However, linoleic and linolenic acids must come from animals' diets.*

Unsaturated acids like linoleic or linolenic acids, in which the double bond is fewer than six carbons from the methyl end of the chain, are biologically synthesized only in plants; they must be taken in by animals in their diets, since they cannot be synthesized in animal tissues. However, acids of the oleic, linoleic, and linolenic types can all be further dehydrogenated by animals to polyunsaturated acids, in which additional double bonds are introduced nearer the carboxyl group. These latter acids play important roles in cell membranes.

Other Acetate-Derived Natural Products

If certain molds are allowed to grow in the presence of labeled acetate, a number of products in addition to fatty acids are found to contain radioactivity. For instance, 6-methylsalicyclic acid contains a high level of activity. Degradative studies have shown that the label is distributed in a way that is consistent with the biosynthetic scheme shown, in which four molecules of acetyl coenzyme A (or their equivalent, malonyl coenzyme A) condense into the triketo thioester shown. Such a molecule would cyclize easily; subsequent reduction, elimination of water, and enolization would lead to the observed product, although the exact sequence of these latter reactions is not known. Many other natural products arise by similar, though more involved, cyclizations. Juglone, a quinone, probably is produced from five molecules of acetate by the sequence shown in Fig. 16-4.

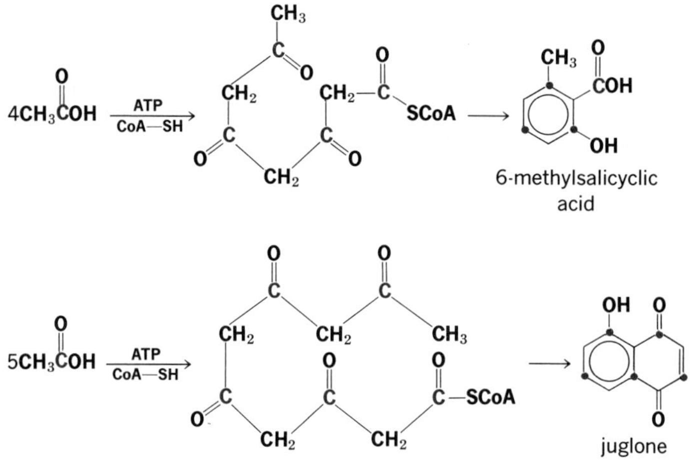

Figure 16-4. *Many phenolic natural products, including salicyclic acid derivatives and juglone, are derived from acetate units by cyclization.*

Acetate is also the precursor of more complicated aliphatic, heterocyclic, and aromatic-aliphatic natural products. The *tetracyclines* are examples of the latter class (Fig. 16-5). These are broad spectrum antibiotics that are active against a large number of bacteria. In addition to the parent compound, *aureomycin* and *terramycin* are tetracycline antibiotics. The basic skeleton of these antibiotics is formed from acetate units. *Erythromycin* is a representative of the important class of *macrolide antibiotics,* a group in which all members contain large lactone rings. Magnamycin and nystatin are other macrolide antibiotics. Erythromycin is especially interesting from the bio-

tetracyclines

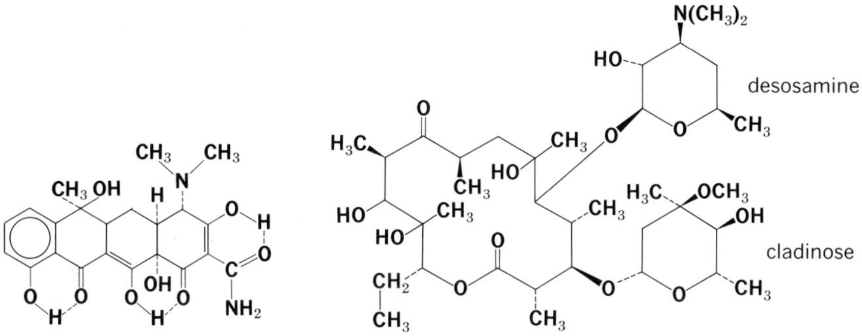

erythromycin

desosamine

cladinose

Figure 16-5. *Tetracycline (R_1=R_2=H), aureomycin (R_1=Cl, R_2=H), and terramycin (R_1=H, R_2=OH) are examples of widely used broad-spectrum antibiotics formed mainly from acetate. Erythromycin is a macrolide antibiotic in which the lactone portion is formed from propionate.*

synthetic standpoint, since it appears to be formed by the combination of propionate units rather than acetate units. This accounts for the large number of C—CH_3 units present in the molecule.

Terpenes

Three acetate units can be combined biosynthetically in a nonlinear way to form the important biosynthetic intermediate *mevalonic acid* (Fig. 16-6). To form this compound, the third acetate undergoes what is essentially an aldol or Claisen condensation and condenses at the ketone group of acetoacetate;

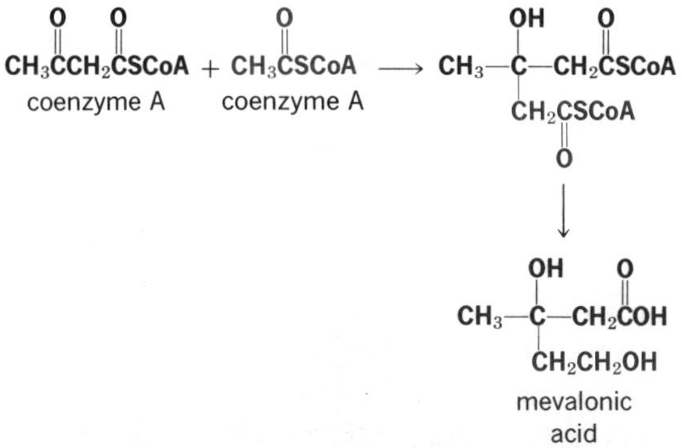

$$\underset{\text{coenzyme A}}{CH_3\overset{O}{\overset{\|}{C}}CH_2\overset{O}{\overset{\|}{C}}SCoA} + \underset{\text{coenzyme A}}{CH_3\overset{O}{\overset{\|}{C}}SCoA} \longrightarrow CH_3-\underset{CH_2\underset{\underset{O}{\|}}{C}SCoA}{\overset{OH}{C}}-CH_2\overset{O}{\overset{\|}{C}}SCoA$$

$$\downarrow$$

$$\underset{\text{mevalonic acid}}{CH_3-\underset{CH_2CH_2OH}{\overset{OH}{C}}-CH_2\overset{O}{\overset{\|}{C}}OH}$$

Figure 16-6. *Mevalonic acid, a biological precursor of terpenes and steroids, is formed from three acetate units.*

subsequent reduction leads to mevalonic acid. We shall see presently how mevalonic acid is related to biosynthesis of the class of compounds known as *terpenes*, but first we must define that class.

For a long time organic chemists had realized that many naturally occurring substances could be considered to be formed by appropriate combination of the five-carbon unit isoprene (the relationship of natural rubber to isoprene has already been discussed in Fig. 4-8). Compounds with the most obvious relationship to isoprene are known as terpenes and occur widely in plants. Structures of other typical terpenes are shown in Fig. 16-7, together with their relationship to isoprene.

isoprene

geraniol

camphor

β-selinene

vitamin A

abietic acid

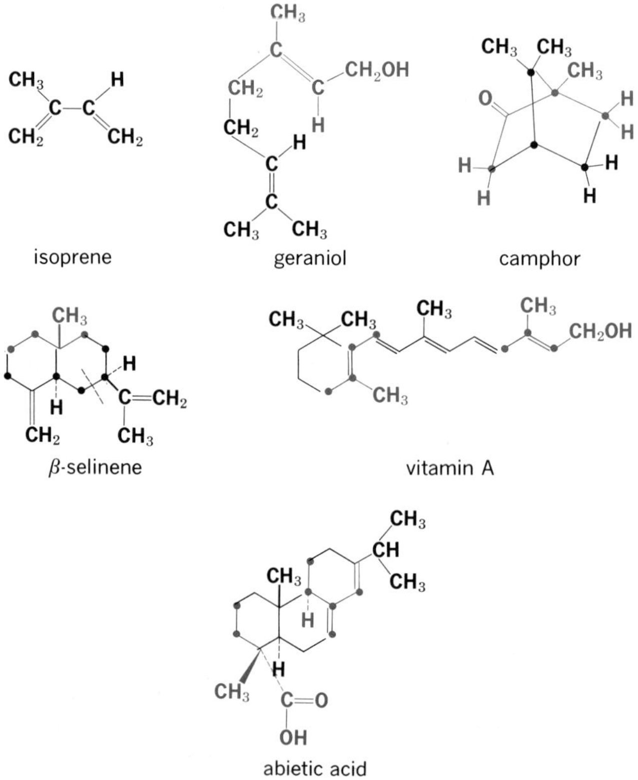

Figure 16-7. *Terpenes are formed biosynthetically from five-carbon units with the skeleton of isoprene. Alternate isoprene units are shown in red.*

Terpenes are classified as monoterpenes if they can be formally derived from two isoprene units, diterpenes if they can be derived from four isoprene units, etc. Geraniol, from geraniums, and the well-known camphor are monoterpenes (C_{10}); β-selinene, from oil of celery, is a sesquiterpene (C_{15}, sesqui $= 1\frac{1}{2}$); vitamin A and abietic acid are diterpenes (C_{20}). Lanosterol, which we shall encounter in Fig. 16-10, is a triterpene (C_{30}).

In biological systems, terpenes are now known not to be constructed from isoprene itself, but from the closely related five-carbon fragments shown in Fig. 16-8. These in turn are derived biosynthetically from mevalonic acid,

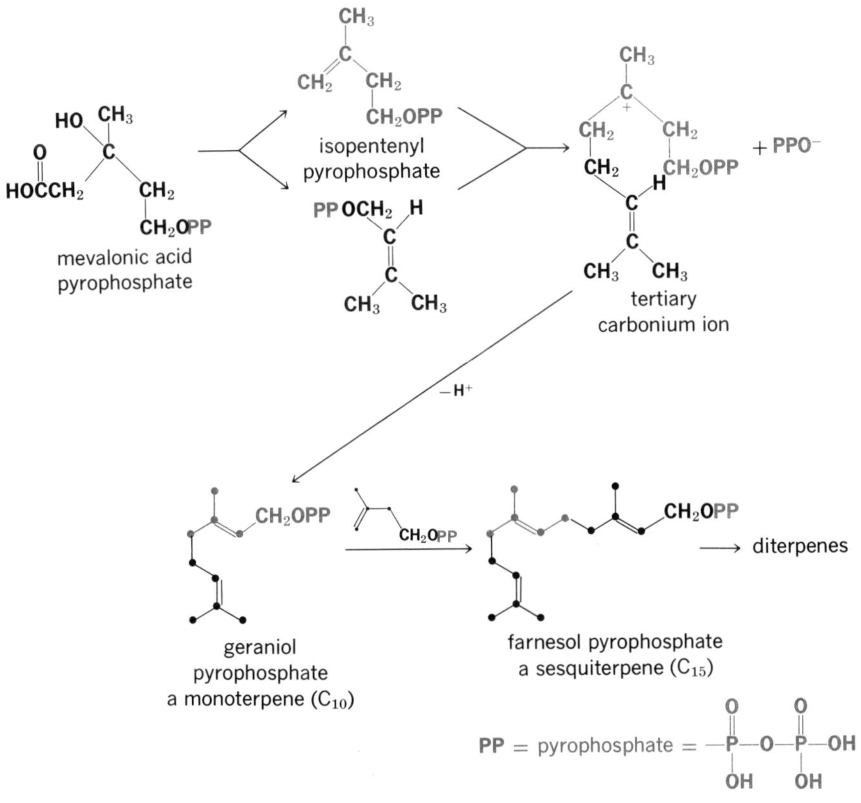

Figure 16-8 *The stepwise addition of C_5 isopentenyl fragments leads to the construction of terpene skeletons. In the compounds shown, the double bonds are* trans. *Similar steps with enzymes that give* cis *isomers lead to natural rubber.*

by dehydration, decarboxylation, and esterification with phosphoric acid. It is this conversion of mevalonic acid (from acetate) to isopentenyl pyrophosphate that interrelates the acetate and terpene paths of biosynthesis.

The processes involved in the conversion of the C_5 pyrophosphates into terpenes are thought to resemble mechanistically those used industrially for the polymerization of isobutylene to make butyl rubber (Chapter 4).

Thus addition of the carbonium ion, resulting from loss of the pyrophosphate group adjacent to the double bond, would, according to Markovnikov's rule, lead to the stable tertiary carbonium ion shown; this would yield geraniol pyrophosphate after the loss of a proton. Repetition leads to a sesquiterpene, farnesol pyrophosphate, and a third addition leads to diterpenes. These reactions are catalyzed by enzymes, of course, but similar reactions occur in nonenzymatic systems. Natural rubber is formed by the continuation of this process, although the enzymes involved lead to the exclusive formation of *cis* double bonds.

Two farnesyl pyrophosphates may be joined together reductively head-to-head through their pyrophosphate ends to form the important triterpene squalene, found in shark's liver oil and in the human liver and scalp. For purposes that will become apparent shortly, the farnesyl molecules are written in a slightly different way in Fig. 16-9. Triterpenes, of which many are known,

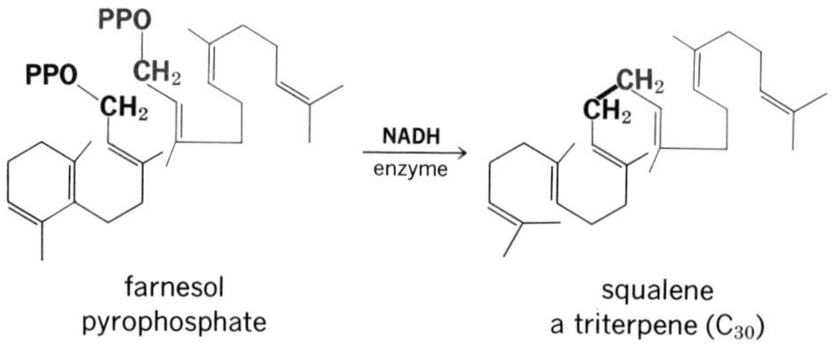

PPO PPO CH_2 CH_2 → NADH enzyme → CH_2 CH_2

farnesol pyrophosphate

squalene a triterpene (C_{30})

Figure 16-9. *The triterpene squalene is formed by the reductive dimerization of two farnesol pyrophosphate molecules.*

seem to be formed in this way rather than by linear polymerization, as might have been expected. Squalene is further converted by cyclization reactions into tetracyclic triterpenes like lanosterol, and eventually into steroids, as shown in Fig. 16-10.

Steroids

Many steroids play important physiological roles in the human body, and cholesterol, formed from acetate by way of squalene as shown in Fig. 16-10, serves as a precursor in their synthesis. Cholesterol is readily absorbed through the intestines so that any consumed in food is added to the body's supply. However, this is not the sole source of cholesterol in the body: Since we know the metabolic relationship between fats and squalene, it is easy to consider

squalene $\xrightarrow{\text{enzymes}}$ lanosterol $\xrightarrow{\text{enzymes}}$ cholesterol

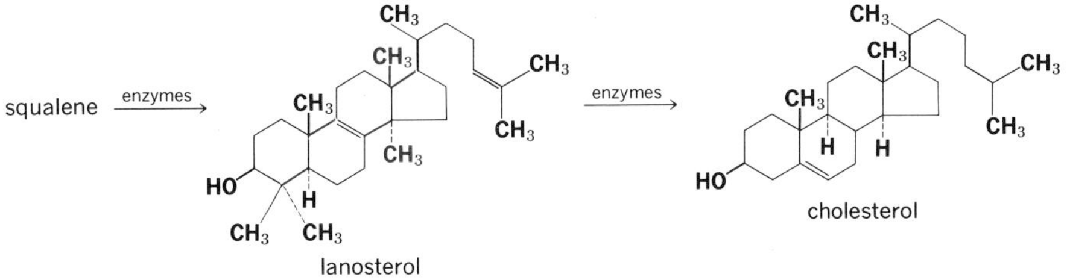

Figure 16-10. *Squalene is a precursor of lanosterol (a triterpene) and cholesterol. In the formation of lanosterol from squalene, the tetracyclic ring system has resulted from the formation of the bonds in red, and the two methyl groups in red have changed their positions by one carbon each.*

a relationship between fat intake and cholesterol levels in the blood. When too much cholesterol is present in the system, it begins to precipitate from solution. Crystalline cholesterol precipitates in the gall bladder as gall stones, while cholesterol precipitated in the veins and arteries constricts the flow of blood and leads to high blood pressure and related medical symptoms.

The body converts cholesterol, in turn, into a great number of other steroids. The bile acids, e.g., cholic acid, are secreted into the intestines and, acting as emulsifying agents, aid in the absorption of fats into the bloodstream by assisting in their passage through the intestinal wall (Fig. 16-11). Other steroids act as hormones. Progesterone is a progestin, a female sex hormone that represses ovulation and is important in the maintenance of pregnancies. Related synthetic compounds like norethynodrel, marketed as Enovid, are effective in preventing conception. Estradiol is an example of a second type of female hormone, called an estrogen.

Male sex hormones, androgens like testosterone, promote the development of secondary male sexual characteristics like beards and deep voices as well as virility. A further series of hormones, the corticosteroids, are produced in the adrenal cortex and have important roles in the regulation of a great many body functions. One of these hormones, cortisol, the principal hormone of the adrenal glands, is closely related to cortisone; the latter compound is known for its medical uses in treating skin diseases and arthritis. Cortisone differs from cortisol (hydrocortisone) only in having a keto group instead of a hydroxyl group at C-11.

Natural Products from Carbohydrates

A second large group of naturally occurring compounds is that formed from carbohydrate precursors. Some of these are relatively simple compounds, and

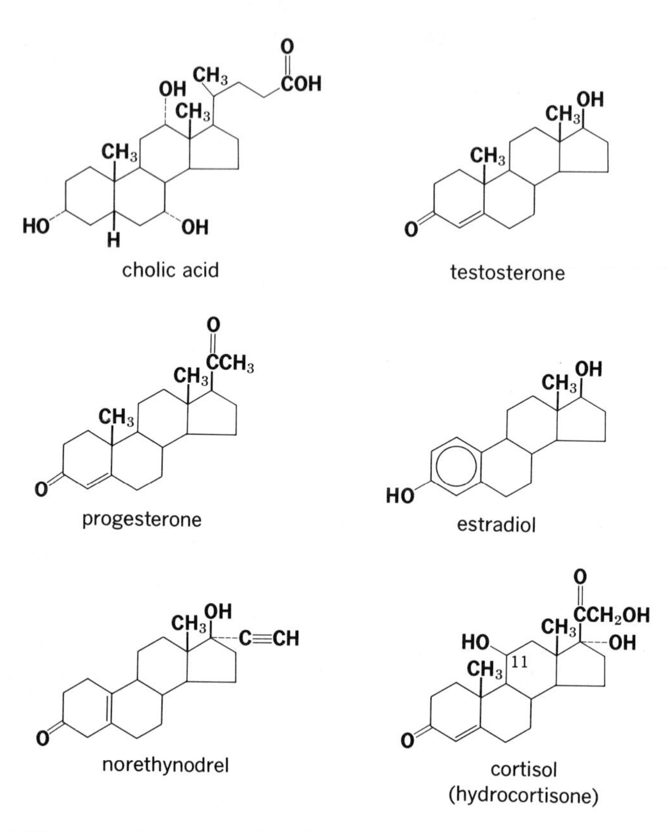

Figure 16-11. *Typical examples of the bile acid, male and female sex hormone, and corticosteroid families.*

their relationship to sugars is obvious; ascorbic acid (vitamin C) belongs to this group (Fig. 16-12) as do the sugars in erythromycin (Fig. 16-5). Others are quite complex and their carbohydrate origins are more obscure.

Ascorbic acid has been shown to be formed from D-glucose in the livers of a number of animals and also in some plants. The first step in this conversion is the oxidation of the C-6 primary hydroxyl group of D-glucose to a carboxyl group and its cyclization to a lactone, D-glucurono-γ-lactone. The aldehyde group is next reduced to an alcohol; the synthesis is completed by oxidation, involving the loss of two hydrogen atoms. Note that C-1 of glucose (red carbon atom) has become C-6 of ascorbic acid. When the formula of the compound is written with the principal functional group at the top (C=O), the bottom hydroxyl group of ascorbic acid is on the left so that it is designated L-ascorbic acid (see Fig. 16-12). The enzymes required for the individual steps of these transformations have not yet been completely isolated and characterized.

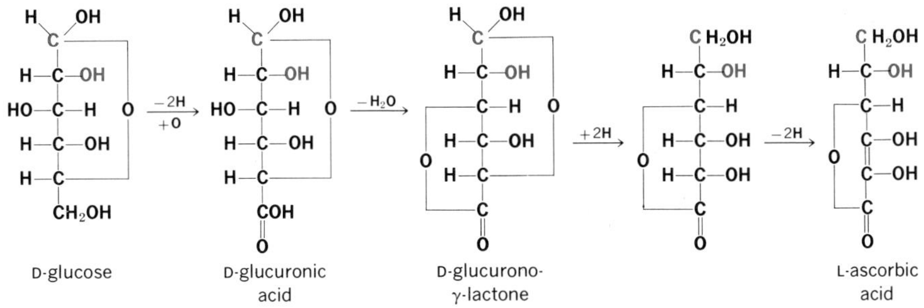

Figure 16-12. *The biological conversion of* D-*glucose to* L-*ascorbic acid.*

Another naturally occurring group of compounds obviously related to the hexoses are the inositols, 1,2,3,4,5,6-hexahydroxycyclohexanes. Of the several isomers of this structure, *myo*-inositol is the most widely distributed in nature. It has been shown to be formed in some plants from D-glucose by a series of steps in which the six-membered chain remains intact and C-1 of glucose becomes the carbon shown in red in *myo*-inositol (Fig. 16-13). Intermediates are known to be glucose 6-phosphate and a *myo*-inositol phosphate.

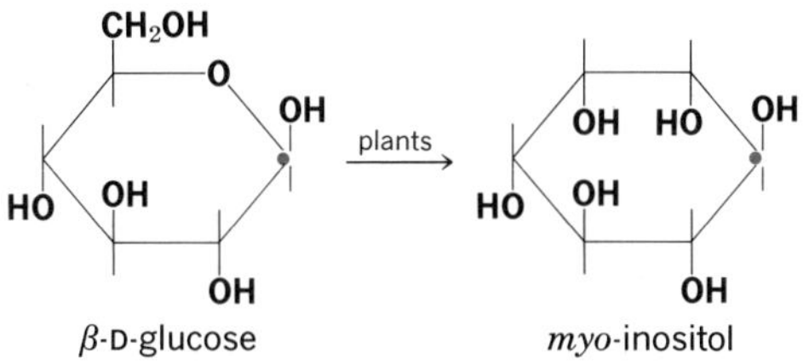

Figure 16-13. *The compound* D-*glucose can be converted to* myo-*inositol without breakdown of the carbon chain.*

An ever-increasing number of medically important antibiotics of carbohydrate origin are being used. The best known of these antibiotics are *streptomycin, neomycin, gentamicin,* and *kanamycin*. The structure of streptomycin was shown in Fig. 9-16. All three portions of streptomycin are biosynthesized from D-glucose. Both the aminohexose and the branched-chain sugar come directly from glucose, while the cyclohexane derivative streptomycin is derived from glucose via *myo*-inositol.

Shikimic Acid

Surprisingly, carbohydrates are also an important source of many naturally occurring aromatic compounds, including the aromatic amino acids. The crucial intermediate in this conversion is the cyclic compound shikimic acid (Fig. 16-14). This acid contains seven carbon atoms and is formed by an aldol

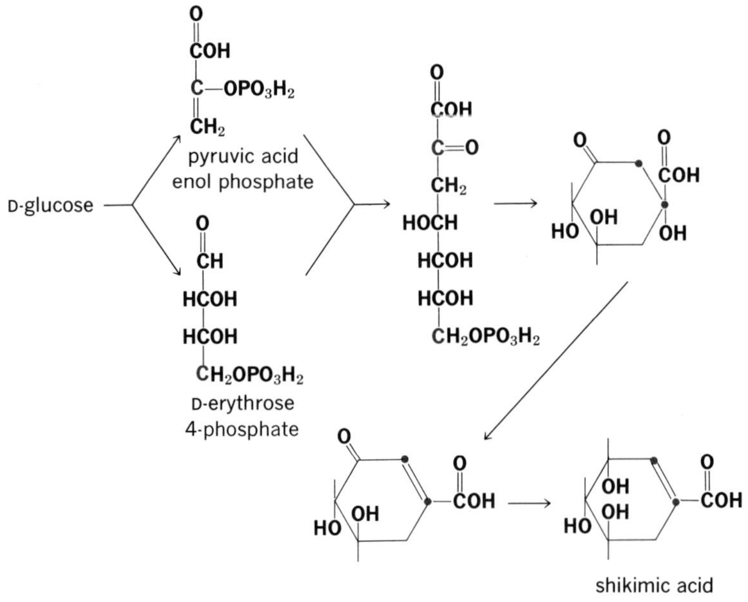

Figure 16-14. *In the biological synthesis of aromatic amino acids from carbohydrates, shikimic acid plays a central role. Its three- and four-carbon precursors are formed from glucose by separate metabolic pathways.*

condensation reaction and cyclization from two separate metabolic products of glucose, pyruvic acid enol phosphate (phosphoenolpyruvic acid) and the 4-phosphate of the four-carbon sugar erythrose. Elimination of water and reduction of the ketone group complete the conversion.

Some of the many natural products whose aromatic rings are derived from shikimic acid are shown in Fig. 16-15. These include the aromatic amino acids and the antibiotic chloramphenicol. Substituted benzoic acids, which often occur attached to other natural products, may also have their origin in shikimic acid.

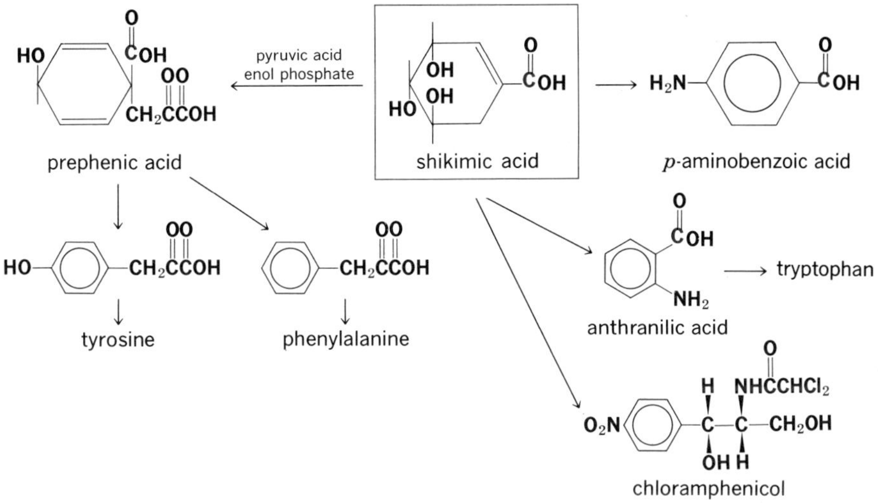

pyruvic acid enol phosphate

prephenic acid

shikimic acid

p-aminobenzoic acid

tryptophan

anthranilic acid

tyrosine

phenylalanine

chloramphenicol

Figure 16-15. *Some of the many aromatic natural products that arise from glucose by way of shikimic acid.*

Natural Products from Amino Acids

Just as many natural products arise from fatty acid and carbohydrate metabolic pathways, other large classes are closely related biosynthetically to amino acids and proteins. We have already noted that penicillin G and actinomycin D (Figs. 14-14, 14-16) bear obvious relationships to their amino acid precursors. Many other naturally occurring substances, however, especially alkaloids, hardly reveal their amino acid precursors at first glance. Nicotine is a typical example. This alkaloid contains both pyridine and pyrrolidine rings, and is found in tobacco. It is synthesized in the plant by a complex series of reactions beginning with two amino acids, tryptophan, which gives rise to the pyridine ring, and glutamic acid, which is converted to the pyrrolidine ring by the steps indicated (Fig. 16-16).

Many other alkaloids are derived, at least in part, from amino acids. Although the exact details by which these compounds are synthesized in plants are far from completely understood, the schemes outlined for the examples in Figs. 16-17 and 16-18 are probably close to those which actually take place; in any event, the scheme has been extremely useful in predicting the structures of new alkaloids. It makes use of two known reactions of amino acids: their conversion into amines by loss of carbon dioxide, and their degradation into aldehydes by oxidation and loss of ammonia. The resulting aldehyde and amine combine, often spontaneously even in the absence of enzymes, to form alkaloids or alkaloid precursors. For instance, the loss of water between trypta-

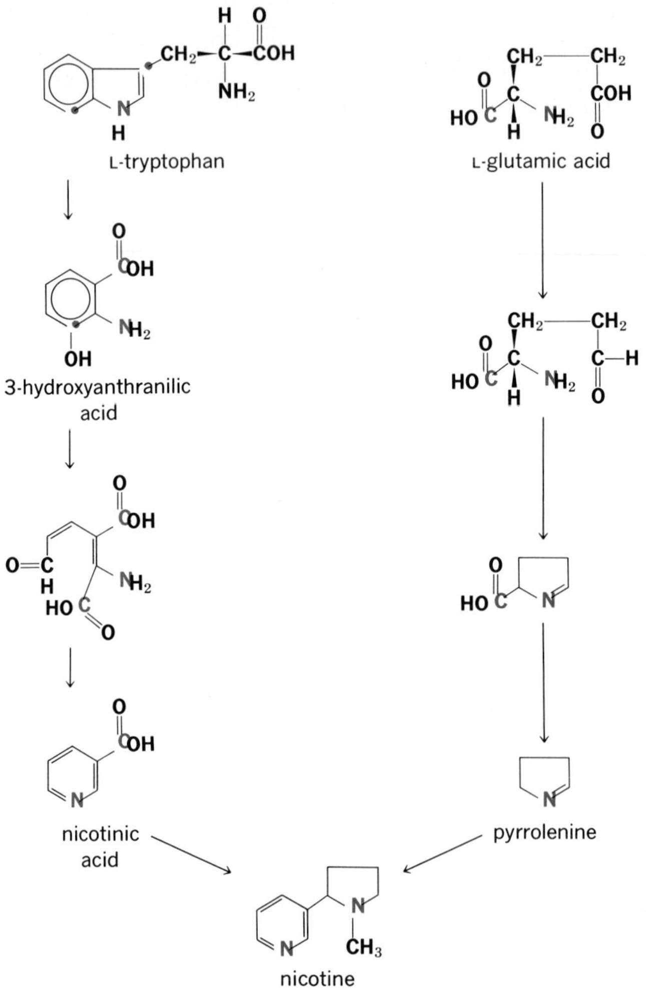

Figure 16-16. *Nicotine can be synthesized in plants from tryptophan and glutamic acid.*

mine, derived from tryptophan, and acetaldehyde, from alanine, could lead to the alkaloid ring system of harmane (Fig. 16-17).

Most of the enormous diversity of structure among alkaloids arises by subsequent reactions after this basically simple type of condensation. If we start with the aldehyde and amine derived from dihydroxyphenylalanine, cyclization involving the loss of a mole of water leads directly to the alkaloid *norlaudanosoline.* Ring closure without incorporation of additional carbon atoms can occur in several different ways: One of the ways leads to *morphine*

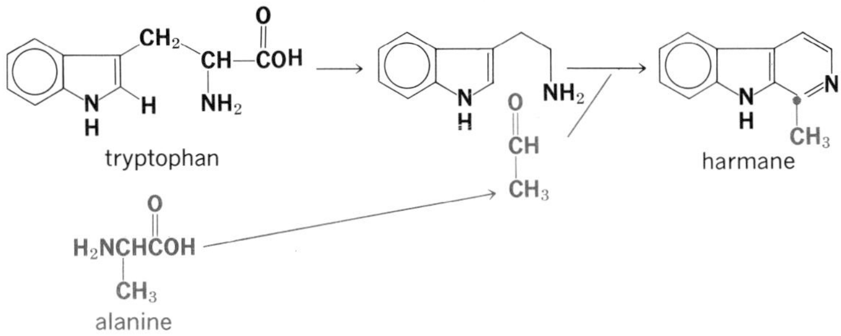

Figure 16-17. *Many alkaloids, like harmane, appear to be formed by condensations between amines and aldehydes related to amino acids.*

(Fig. 16-18); others lead to less well-known alkaloids. Despite the vast superficial differences in structure among alkaloids, many seem to arise by analogous biosynthetic pathways.

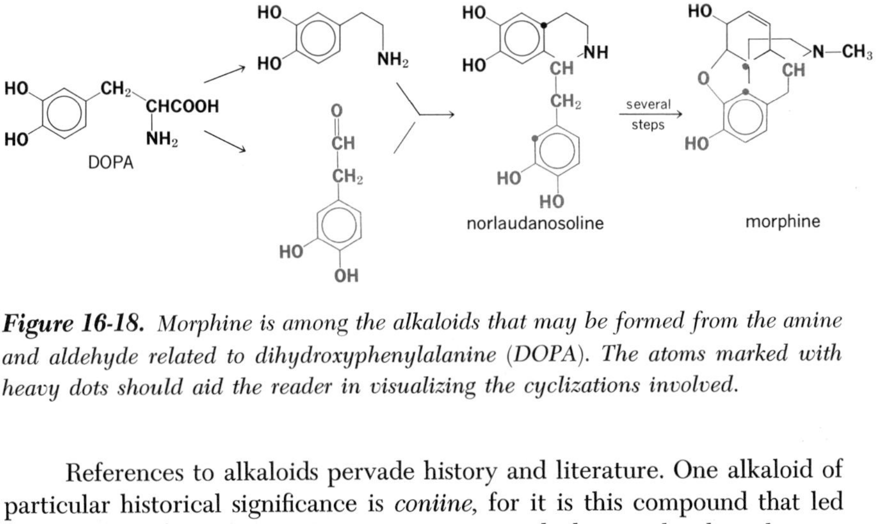

Figure 16-18. *Morphine is among the alkaloids that may be formed from the amine and aldehyde related to dihydroxyphenylalanine (DOPA). The atoms marked with heavy dots should aid the reader in visualizing the cyclizations involved.*

References to alkaloids pervade history and literature. One alkaloid of particular historical significance is *coniine,* for it is this compound that led to the death of Socrates in 399 B.C. *Cocaine,* with the peculiar bicyclic ring structure shown in Fig. 16-19, is a narcotic that Sherlock Holmes found relaxing ("Quick, Watson, the needle!"). *Quinine,* extracted from the bark of an East Indian tree, is used in the treatment of malaria.

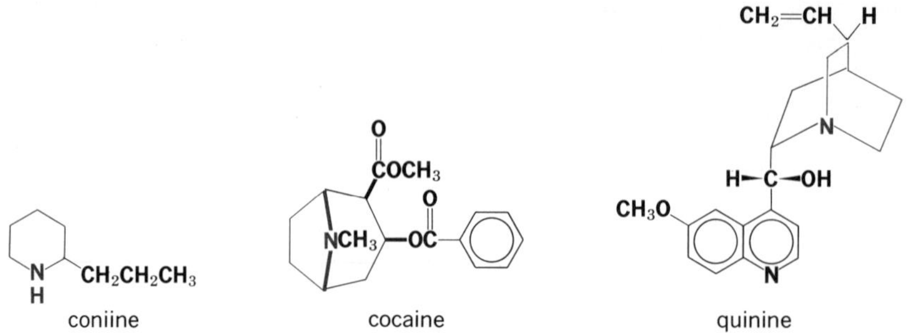

coniine cocaine quinine

Figure 16-19. *Three alkaloids of historical, literary, or medicinal importance.*

Nucleic Acid Bases

We shall complete our discussion of naturally occurring organic molecules with a few words about the biosynthesis of the purine and pyrimidine bases in DNA and RNA.

Extensive studies have identified the sources of the various carbon and nitrogen atoms in these compounds to be as indicated in the formulas in Fig. 16-20.

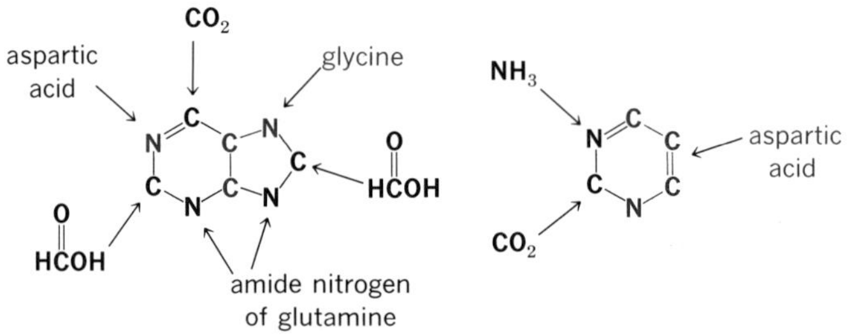

Figure 16-20. *The sources of the various atoms in the purine (adenine, guanine) and pyrimidine (cytosine, uracil, thymine) bases of nucleic acids.*

PROBLEMS

1. If oleic acid were formed in the presence of acetic and malonic acids containing ^{14}C-labeled carboxyl groups, should C-10 of oleic acid be labeled?
2. Suggest which carbon atoms of the fungal product curvulinic acid would be expected to be derived from the carboxyl group of acetate.

O OH
CH_3—C
O
HOC—CH_2 OH

curvulinic acid

3. How many stereoisomers are possible for tetracycline, aureomycin, terramycin, and erythromycin? *Hint:* How many asymmetric carbon atoms are in each molecule?
4. Identify the isoprene units in limonene and pinene (Fig. 3-29).
5. Give a possible mechanism by which limonene (Fig. 3-29) could be formed biologically from geraniol pyrophosphate.
6. If a shark were fed acetic acid containing ^{14}C in the methyl group, which carbons in squalene would become radioactive?
7. Draw the structure of cortisone and show which carbons would be radioactive if acetic acid with ^{14}C in the methyl group were converted biologically into cortisone.
8. The compound L-ascorbic acid (vitamin C) can also be formed by a series of reactions similar to those of Fig. 16-12, but starting from D-galactose. Write this reaction sequence.

answers to selected problems

Chapter 1

1-3. $CH_3CH_2CH_2Cl$; $CH_3CHClCH_3$; $CH_3CH_2CHCl_2$; $CH_3CHClCH_2Cl$; $CH_2ClCH_2CH_2Cl$; $CH_3CCl_2CH_3$

1-5. $CH_3CH_2CHCH_3$ (*sec*-butyl); $(CH_3)_2CHCH_2$— (isobutyl)

p. 28

4. **(a)** pentane **(b)** heptane **(c)** pentane

6. **(a)** 2-methylbutane **(b)** 3,4,5-trimethyloctane
(c) 2-methyl-4-ethylhexane

7. **(a)** 2,3,4,6-tetramethylheptane
(b) 2,2,5-trimethyl-4-ethylheptane
(c) 3-ethyl-6-isopropyl-4-*t*-butylnonane

9. **(e)** cycloheptane **(f)** cyclooctane

11. **(a)** 1,1-dichlorocyclopropane
(b) 1,4-dimethylcycloheptane
(c) 2,4-dimethyl-1-isopropylcyclohexane
(d) *t*-butylcyclopentane

12. **(a)** *trans*-1,2-dimethylcyclobutane

14. **(f)** *trans*-1,4-dimethylcyclohexane
(h) 1,1,2,2,3,3-hexafluorocyclopropane (common name: perfluorocyclopropane)

20. About 240 lbs of CO_2

21. About 16 lbs of CO

Chapter 2

p. 40

3. The *n*-pentane peak would be twice as large as before, thus identifying it.

4. $CH_4 + O\cdot \longrightarrow CH_3\cdot + HO\cdot \longrightarrow CH_3OH$

6. In the *trans*-1,2 isomer, both methyl groups can be equatorial in one chair form. In the *cis* isomer, one methyl group is always axial. So, the *trans* isomer is more stable.

Chapter 3

3-2. A triple bond is linear, and two more carbons are not enough to complete a six-membered ring (C—C≡C—C). Cyclooctyne is known.
3-4. The addition is begun by a Lewis acid. The only one present is Br_2. Chlorine is present only as chloride ion (Cl^-), not as a Lewis acid.
3-5. 1-bromo-2-chloropropane.
3-7. CH_3—CHCl—CH=CH—CH=CH_2; CH_3CH=CH—CH(Cl)—CH=CH_2
CH_3CH=CH—CH=CH—CH_2Cl

p. 67
1. **(j)** cyclobutene **(l)** 3-methyl-1-butyne
(m) 1,2-dimethylcyclohexene **(n)** 3-methylcyclopentene
2. **(e)** 3-ethyl-1-heptyne
4. **(e)** 1,3-pentadiene **(i)** 1,4-cyclohexadiene
(f) 2,3-diethyl-1,4-hexadiene **(j)** 1,5-dimethyl-1,3-cyclopentadiene
5. **(d)** 2-hexen-5-yne
9. Conjugated: (a); (c).
13. (c)

Chapter 4

p. 79
7. F:B̈:F (with F above B) Boron is missing an electron pair that would make it a complete octet.
8. ·N̈::Ö: Nitric oxide is a free radical that combines with other free radicals to terminate polymerization before it gets started.

10. RO· + $HCCl_3$ ⟶ ROH + ·CCl_3 $\xrightarrow{CH_2=CHR}$ Cl_3C—CH_2—ĊHR
$\downarrow HCCl_3$
·CCl_3 + Cl_3C—CH_2CH_2R (·CCl_3 returns to the chain)

Chapter 5

5-1. 9,10 bond $1\frac{1}{3}$ (double in one structure, single in two); 1,9 bond $1\frac{1}{3}$
5-3. 61.0 kcal (Note that this is less than twice that of benzene.)
5-6. CH_2Cl—CH_2NO_2, $CH_3CHBrCl$
5-7. **(a)** nonaromatic **(b)** aromatic

p. 106
2. **(a)** *p*-bromochlorobenzene or *p*-chlorobromobenzene
(b) *o*-dinitrobenzene
(c) *p*-iodotoluene
(d) anthracene

(e) ethylbenzene (or phenylethane)
(f) 2-phenyl-1-chloroethane
(g) 1,3-dichloro-4-nitrobenzene
(h) *m*-nitrotoluene
(i) naphthalene

3. (b) $CH_3CHClCH_2NO_2$

(g) HOOC—⟨benzene ring⟩—COOH

(j) ⟨benzene ring⟩—$CHBrCH_3$

8. Only in the meta isomer can attack occur so that the positive charge is on both carbons bearing methyl groups.

9. 33 kcal

Chapter 6

6-3. Dehydrate to 1-butene; hydrate to 2-butanol.
6-4. Dehydrate to propene; add BH_3; oxidize with H_2O_2.
6-5. Disulfonate; fuse with potassium hydroxide.

p. 129

2. (e) *cis*-2-methylcyclopentanol
(f) 2-(*m*-nitrophenyl)-1-propanol

4. (e) [naphthoquinone structure: O, O] (g) [cyclopentane-1,2-diol structure: H, OH, OH, H]

(j) [1-methylcyclopentene, CH_3] (major) + [3-methylcyclopentene, CH_3] (minor)

12. IV < I < II < III

Chapter 7

7-4. $C_2H_5O^-$ + $(CH_3)_3CBr$ would give only isobutylene.
7-5. Magnesium reacts with CCl_4 to form a Grignard reagent. The heat generated in this reaction would make the fire burn more vigorously.

p. 151

2. (g) ⟨cyclopentane⟩—OH + ⟨cyclopentene⟩

(h) [cyclopentanol structure: CH_3, H, H, H, C_2H_5, OH] + CH_3—⟨cyclopentene⟩ [H, C_2H_5]

Chapter 8

8-2. 3-methyl-3-hexene
8-4. *n*-propyl Grignard + formaldehyde; ethyl Grignard + propionaldehyde

p. 169
2. **(e)** 2-methyl-2-cyclohexenone
(h) *p*-chloroacetophenone

5. **(e)** 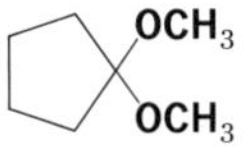 OCH_3 OCH_3
(f) $C_6H_5CH_2CH_2CHO$
(g) $C_6H_5CH{=}CH{-}CH_2OH$
9. **(a)** $LiAlH_4$; H_2SO_4, heat
(b) $LiAlH_4$; HBr
(c) $LiAlH_4$; H_2SO_4, heat; H_2O,H^+
(d) $LiAlH_4$; H_2SO_4, heat; O_3; Zn
(e) CH_3MgBr; $K_2Cr_2O_7$; CH_3MgBr
(f) H_2O,H^+; $K_3Cr_2O_7$; NH_2OH

Chapter 9

9-3. 32; 16, both in open-chain form
9-5. D- and L-galactose

p. 191
6. Two different osazones
7. Any two aldoses that differ in configuration only at the 2-position will give the same osazone (allose and altrose, glucose and mannose, etc.).
10. Mannose, talose

Chapter 10

10-1. Less rapidly because in acid most of the aniline would be present as the ammonium salt

p. 203
2. **(b)** diphenylamine
(e) *p*-bromo-*N*-ethylaniline

3. **(e)** $CH_2{=}CH_2$ + $N(CH_3)_2$

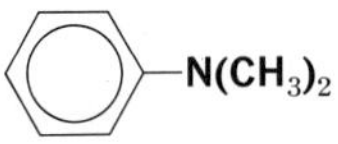

(f) $CH_3CH_2CH(CH_3){-}CH{=}CH_2$ (major)
8. The enzyme acetylcholinesterase apparently mistakes muscarine for acetylcholine, and so nerve transmission is prevented.

Chapter 11

11-4. $CH_3{}^{18}OH$
11-6. A tertiary amine has no hydrogen on its nitrogen atom to be replaced in amide formation.

11-7. (a) Treat with CH_3NH_2
(b) Hydrolyze to propionic acid, convert to propionyl chloride, and treat with sodium propionate

p. 222
2. (a) isopropyl acetate; methyl pentanoate; ethyl formate; and *n*-propyl benzoate
(b) benzoic anhydride
(c) benzamide
(d) urea
(e) *p*-chlorobenzonitrile
(f) sodium acetate

3. (f) $HOOC-C_6H_4-COOH$ (g) 6,6-nylon

(h) *t*-butyl alcohol (i) acetone

6. The Grignard reagent reacted with the acidic hydrogen on the carboxyl group.

Chapter 12

p. 236
3. Models will show that a long hydrocarbon chain with a *trans* double bond can stretch very straight and so pack into a crystal easily.

Chapter 15

p. 289
1. (a) CD_3COCD_3 (b) $CD_3COCD_2CH_3$
(c) C_2H_5OD (d) none

(e) cyclohexanone with D, D on C-2 and D, D on C-6 (=O) (f) $CD_3COOC_2H_5$

2. $H-C(=O)-C(CH_3)=CH-\underline{C}H_2 \longleftrightarrow HC(=O)-\underline{C}(CH_3)-CH=CH_2 \longleftrightarrow H-C(O^-)=C(CH_3)-CH-CH_2$

If a proton is removed from the other methyl group the negative charge cannot be shared with the oxygen atom by resonance.

8. In the methyl group.

index